MERCEDES - BENZ

VITO & VIANO

Petrol & Diesel Models

WORKSHOP MANUAL

2004 -2010

COVERING

MERCEDES-BENZ VITO & VIANO SERIES 639

109 CDI, 111 CDI & 115 CDI WITH 2.2 LITRE DIESELS - ENGINE TYPE 646

3.2 & 3.7 LITRE V6 PETROL MODELS - ENGINE TYPE 112

Vito models covered: 639.601, 639.604, 639.605, 639.701, 639.703 & 639.705.
Viano models covered: 639.711, 639.713, 639.811, 639.813 & 639.815.

Second edition - February 2015

Brooklands Books

BROOKLANDS BOOKS LTD.
P.O. BOX 904, AMERSHAM,
BUCKS, HP6 9JA. UK

MBV4WH www.brooklandsbooks.com 7W5/2227

About this Manual

This Owners Edition - Workshop Manual covers the following Mercedes-Benz Vito & Viano petrol and diesel models from the start of model year 2004 to 2010.

Covered are the Vito petrol models 119, 122 and 123 with 140, 160 and 170 kW (190, 218 and 231 b.h.p.) engines (type 646 with 3.2 & 3.7 litre) and the three diesel models 109 CDI, 111 CDI and 115 CDI, all fitted with the latest common rail 2.2 litre diesel engine (type 112). Depending on the version the engine has a performance of 65 kW (88 b.h.p.), 80 kW (109 b.h.p.) or 110 kW (150 b.h.p.). In Sections 0, 1 & 10 can be found further details of the various models and engines dealt with here.

This manual has been written for the practical owner who wants to maintain their vehicle in first-class condition and carry out the bulk of their own servicing and repairs. Comprehensive step-by-step instructions are provided for service and overhaul operations to guide the reader through what might otherwise be unfamiliar and complicated tasks and hundreds of illustrations are included to amplify the text.

With the aid of this manual, many aspects of service, overhaul and repair are within the scope of an owner with a reasonable degree of mechanical aptitude. Some operations however demand more skill. Other jobs require the use of special tools and in some cases, testing facilities and techniques that are not generally available. Only you can judge whether a job is within your capabilities. We do however try to assist the reader to come to an informed decision. Whilst every effort has been made to ensure that the information provided is correct, it is obviously not possible to guarantee complete freedom from errors or omissions.

Information to be found in the driver's handbook is not necessarily duplicated here and it is not possible within this volume to cover every aspect to be found in the manufacturer's own workshop manual which is of much greater size and complexity. However, it should be consulted if more detailed information is needed.

Always remember that you are responsible for your own safety and that of others when working with you on a vehicle. Take particular care with safety-related systems like the brakes and steering and seek professional advice if in any doubt. Never work under a vehicle unless it is properly supported (a single jack is not enough). Take care with power tools, also regard as potentially harmful fuel, lubricants, solvents and sealers which should always be stored in labelled, sealed containers. Always obtain your spare parts from an officially appointed Mercedes-Benz dealer.

With care and common sense, the practical owner can make an excellent job of maintenance and overhaul. The benefits include money saved and the satisfaction of work well done. You will be adding to your knowledge too, knowing more about the vehicle you own will help you to make logical decisions about what needs to be done, even if it does in some instances have to go into a professional repair shop.

The Mercedes-Benz Vito and Viano vehicles are built with care and precision. With regular care and maintenance they will provide long, reliable and faithful service.

CONTENTS

Brooklands Books Ltd., PO Box 904, Amersham, Bucks, HP6 9JA, England
E-mail: sales@brooklands-books.com www.brooklandsbooks.com

ISBN 9781783180103 Ref: MBV4WH

www.brooklandsbooks.com

0. INTRODUCTION

Our 'Owners Edition' Workshop Manuals are based on easy-to-follow step-by-step instructions and advice, which enables you to carry out many jobs yourself. Moreover, now you have the means to avoid these frustrating delays and inconveniences which so often result from not knowing the right approach to carry out repairs which are often of a comparatively simple nature.

Whilst special tools are required to carry out certain operations, we show you in this manual the essential design and construction of such equipment, whenever possible, to enable you in many cases to improvise or use alternative tools. Experience shows that it is advantageous to use only genuine parts since these give you the assurance of a first class job. You will find that many parts are identical in the range covered in this manual, but our advice is to find out before purchasing new parts.

Always buy your spare parts from an officially appointed dealer.

0.0. General Information

The manual covers the listed Mercedes-Benz commercial vehicles with the following engines, fitted to Vito and Viano models. Vito models are known as type 119, 122 and 123 (petrol), 109 CDI, 111 CDI and 115 CDI (diesel). Viano models are fitted with the CDI or petrol engine. All Vito and Viano models belong to the model series "639", with three numbers behind, indicating the precise model. Vito models covered are 639.601, 604, 605, 701, 703 and 705. Viano models covered are 639.711, 713, 811, 813 and 815.

With petrol engine. Also fitted to other MB vehicles. The engine is of the type "112" but has various features which will be briefly referred to:

- Both V6 engines operates with 18 valves in a light-alloy cylinder head. The valve seat inserts are made of hardened steel and are pressed into the cylinder head. The overhead valves operate in brass valve guides.
- Two overhead camshafts ensure the opening and closing of the valves. The camshaft on the R.H. side, seen in the direction of drive is the exhaust valve camshaft, the L.H. shaft is the inlet valve camshaft. The camshaft bearings are not machined directly into the cylinder head.
- Hydraulic tappets are inserted between the ends of the valves and the cams. The various cams of the camshaft push down the bucket tappet in question and open the valve.
- Adjustment of the valves is no longer necessary on these engines. The hydraulic valve elements ensuring the correct valve clearance under all conditions, i.e. the valves operate without any play. The use of the hydraulic tappets also has the advantage to ensure a quieter engine operation. The function of the hydraulic valve clearance compensating elements is to eliminate valve clearance, i.e. the dimensional changes in the valve train (valve lash) due to heat expansion and wear are compensated by the elements. The compensating elements cannot be repaired, but can be checked for correct functioning as described later on in the manual.
- The camshafts are driven by means of a long-life Duplex timing chain. The chain is fitted between the crankshaft timing sprocket, the exhaust valve camshaft and the inlet valve camshaft. To prevent vibration of the long chain run, a hydraulically operated tensioning rail and glide rails are used.
- The engine performance for the engines in question are given in section 10.0.

With CDI Diesel engines: The following models are covered:

109 CDI model are fitted with the 2.2 litre four-cylinder diesel engine with a performance of 65 kW (88 BHP) at 3800 rpm (engine type 646).

111 CDI model are fitted with the 2.2 litre four-cylinder diesel engine with a performance of 80 kW (109 BHP) at 3800 rpm (engine type 646).

115 CDI model are fitted with the 2.2 litre four-cylinder diesel engine with a performance of 110 kW (150 BHP) at 3800 rpm (engine type 646).

The front suspension consists of spring struts with integral shock absorbers, wishbones and a stabiliser (anti-roll) bar. A connecting linkage is fitted between each end of the stabiliser bar and the spring strut on the side in question. The connecting linkages have a ball joint on each end and are secured by means of a nut to the stabiliser bar and the spring strut. The two wishbones, i.e. the suspension arms are attached at their inner ends to the subframe and on the outside with a removable ball joint to the steering knuckle.

The connection between the spring struts and the steering knuckle (wheel bearing housing) is by means of two clamp bolts with nuts on the other side. The upper bolt is an eccentric bolt, used to adjust the front wheel geometry.

Vito and Viano models have no actual rear axle. The rear suspension consists of swing arms, attached to the rear crossmember. The up and down movement of the swing arms are controlled by means of two-way acting, vertically arranged hydraulic shock absorbers. Coil springs between the upper ends of the swing arms and the frame floor ensure the suspension. V models have air suspension as standard fitment. Vito models are also available with air suspension.

Rear wheel bearings and wheel hubs are fitted to the inside of the swing arms. The wheel bearings are pressed into the inside of the swing arms and onto the wheel hubs. Wheel bearings and wheel hubs are held together by means of a long bolt, inserted from the inside into the swing arm and secured on the outside by means of a nut. A compression piece at the rear of the suspension arm is fitted with a toothed wheel if ASS is fitted.

Note: Note that models without air suspension and models with air suspension are covered in separate chapters.

Disc brakes at the front and rear make up the brake system, assisted by a brake servo unit. The vacuum for the servo unit is taken from a special exhauster pump (vacuum pump), in the case of a diesel engine or from the inlet manifold, in the case of a petrol model.

The rack and pinion steering is servo-assisted.

0.1. Identification

When you purchased the vehicle you will have received various data cards which have all important vehicle details, as for example chassis number and component numbers as well as identification of optional equipment, etc. Your Owner's Manual will give you further instructions on the use of these cards.

You will know the location of the type identification plate, giving you the max. permissible weight, max. permissible weight on front and rear axles and other important data. The type identification plate is located in the door frame of the front door on the L.H. side in the direction of travel. The chassis number can be found under the seat trim on the R.H. front seat, again viewed in the direction of travel. The engine number is engraved into the cylinder block.

Always quote the complete chassis number and/or engine number and the model year, when ordering parts. Important is also the actual model number, which will indicate the construction of the vehicle.

0.2. Filling Capacities

Engine:
- Oil change and filter change – Petrol engine: .... 10.0 litres
- Engine dry: .... 9.5 litres
- Oil change and filter change – Diesel engine: .... 8.5 litres
- Engine dry: .... 8.3 litres

Difference between max./min.: .... approx. 2.0 litres

Cooling System:
- With CDI engine: ....8.6 or 10.5 litres, refer to owners manual, depending on models
- With petrol engine: .... 9.9 litres, refer to owners manual

Manual Transmission:
- Type 716.637: .... 1.2 litres
- Type 716.652: .... 1.5 litres

Automatic Transmission:
- Except type 722.682 (complete fill): .... 7.8 litres
- Type 722.682 (complete fill) .... 8.7 litres

Power-assisted steering: .... 0.8 – 1.0 litres

Brake system:
- Without ESP: .... 0.8 litre
- With ESP: .... 1.0 litre

0.3. General Servicing Notes

The servicing and overhaul instructions in this Workshop Manual are laid out in an easy-to-follow step-by-step fashion and no difficulty should be encountered, if the text and diagrams are followed carefully and methodically. The "Technical Data" sections form an important part of the repair procedures and should always be referred to during work on the vehicle.

In order that we can include as much data as possible, you will find that we do not generally repeat in the text the values already given under the technical data headings. Again, to make the best use of the space available, we do not repeat at each operation the more obvious steps necessary - we feel it to be far more helpful to concentrate on the difficult or awkward procedures in greater detail. However, we summarise below a few of the more important procedures and draw your attention to various points of general interest that apply to all operations.
Always use the torque settings given in the various main sections of the manual. These are grouped together in separate sub-sections for convenient reference.

Bolts and nuts should be assembled in a clean and very lightly oiled condition and faces and threads should always be inspected to make sure that they are free from damage burrs or scoring. DO NOT degrease bolts or nuts.

All joint washers, gaskets, tabs and lock washers, split pins and "O" rings must be replaced on assembly. Seals will, in the majority of cases, also need to be replaced, if the shaft and seal have been separated. Always lubricate the lip of the seal before assembly and take care that the seal lip is facing the correct direction.

References to the left-hand and right-hand sides are always to be taken as if the observer is at the rear of the vehicle, facing forwards, unless otherwise stated.

Always make sure that the vehicle is adequately supported and on firm ground, before commencing any work on the underside of the car. A small jack or a make shift prop can be highly dangerous and proper axle stands are an essential requirement for your own safety.

Dirt, grease and mineral oil will rapidly destroy the seals of the hydraulic system and even the smallest amounts must be prevented from entering the system or coming into contact with the components. Use clean brake fluid or one of the proprietary cleaners to wash the hydraulic system parts. An acceptable alternative cleaner is methylated spirit, but if this is used, it should not be allowed to remain in contact with the rubber parts for longer than necessary. It is also important that all traces of the fluid should be removed from the system before final assembly.

Always use genuine manufacturer's spares and replacements for the best results.

Since the manufacturer uses metric units when building the cars it is recommended that these are used for all precise units. Inch conversions are given in most cases but these are not necessarily precise conversions, being rounded off for the unimportant values.

Removal and installation instructions, in this Workshop Manual, cover the steps to take away or put back the unit or part in question. Other instructions, usually headed "Servicing", will cover the dismantling and repair of the unit once it has been stripped from the vehicle. It is pointed out that the major instructions cover a complete overhaul of all parts but, obviously, this will not always be either necessary and should not be carried out needlessly.

There are a number of variations in unit parts on the range of vehicles covered in this Workshop Manual. We strongly recommend that you take care to identify the precise model, and the year of manufacture, before obtaining any spares or replacement parts.

Std.: To indicate sizes and limits of components as supplied by the manufacturer. Also to indicate the production tolerances of new unused parts.

O/S: Parts supplied as Oversize or Undersize or recommended limits for such parts, to enable them to be used with worn or re-machined mating parts.

U/S: O/S indicates a part that is larger than Std. size U/S may indicate a bore of a bushing or female part that is smaller than Std.

Max.: Where given against a clearance or dimension indicates the maximum allowable. If in excess of the value given it is recommended that the appropriate part is fitted.

TIR: This indicates the Total Indicator Reading as shown by an dial indicator (dial gauge).

TDC: Top Dead Centre (No. 1 piston on firing stroke).

MP: Multi-Purpose grease.

0.4. Recommended Tools

To carry out some of the operations described in the manual you will need some of the tools listed below:

As basic equipment in your tool box you will need a set of open-ended spanners (wrenches) to reach most of the nuts and bolts. A set of ring spanners is also of advantage. To keep the costs as low as possible we recommend a set of combined spanners, open-ended on one side and a ring spanner on the other side. Fig. 0.1 shows a view of the spanners in question. Sockets are also a useful addition to your tool set.

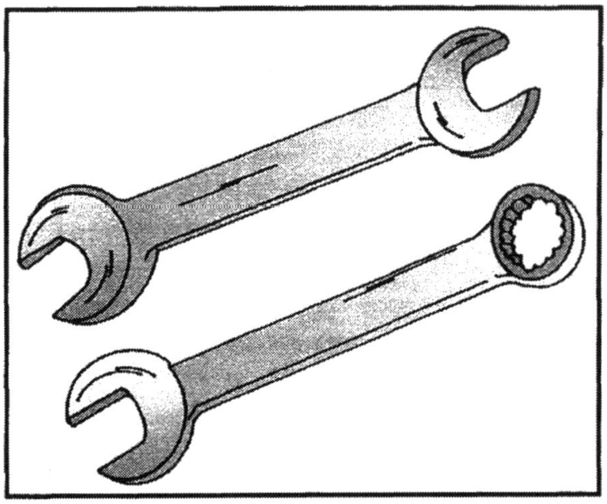

Fig. 0.1 – A double open-ended spanner in the upper view and an open-ended/ring spanner in the lower view. Always make sure that the spanner size is suitable for the nut or bolt to be removed and tightened.

A set of cross-head screwdrivers, pliers and hammers or mallets may also be essential. You will find that many bolts now have a "Torx" head. In case you have never seen a "Torx" head bolt, refer to Fig. 0.2. A socket set with special t "Torx" head inserts is used to slacken and tighten these screws. The size of the bolts are specified by the letter "T" before the across-flat size.

Fig. 0.2 – A graduated disc is used to "angle-tighten" nuts and bolts. "Torx" head bolts are shown on the R.H. side.

Circlip pliers may also be needed for certain operations. Two types of circlip pliers are available, one type for external circlips, one type for internal circlips. The ends of the pliers can either be straight or angled. Fig. 0.3 shows a view of the circlip pliers. Apart from the circlip pliers you may also need the pliers shown in Fig. 0.4, i.e. side cutters, combination pliers and water pump pliers.

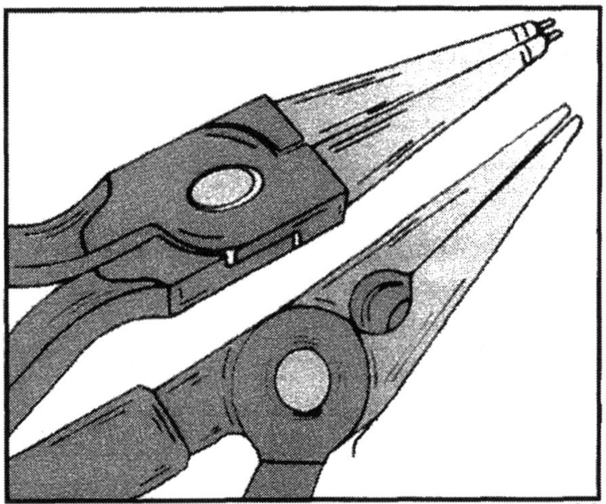

Fig. 0.3 – Circlip pliers are shown in the upper view. The type shown in suitable for outside circlips. The lower view shows a pair of pointed pliers.

Every part of the vehicle is tightened to a certain torque value and you will therefore need a torque wrench which can be adjusted to a certain torque setting. In this connection we will also mention a graduated disc, shown in Fig. 0.2, as many parts of the vehicle must be angle-tightened after having been tightened to a specific torque. As some of the angles are not straight-forward (for example 30 or 60 degrees), you will either have to estimate the angle or use the disc.

Finally you may consider the tool equipment shown in Fig. 0.5 which will be necessary from time to time, mainly if you intend to carry out most maintenance and repair jobs yourself.

0.5. Before you start

Before you carry out any operations on your vehicle it may be of advantage to read the following notes to prevent injuries and damage to the vehicle:

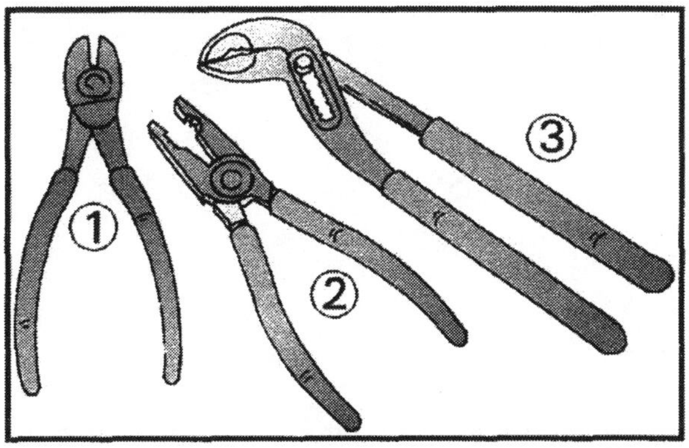

Fig. 0.4 – Assortment of pliers suitable for many operations.

1 Side cutter
2 Combination pliers
3 Water pump pliers

- Never carry out operations underneath the vehicle when the front or rear is only supported on the jack. Always place chassis stands in position (refer to next section). If no chassis stands are available and if the wheels are removed place on wheel on top of the other one and place them under the side of the vehicle where you work. If the jack fails the vehicle will drop onto the two wheels, preventing injury.

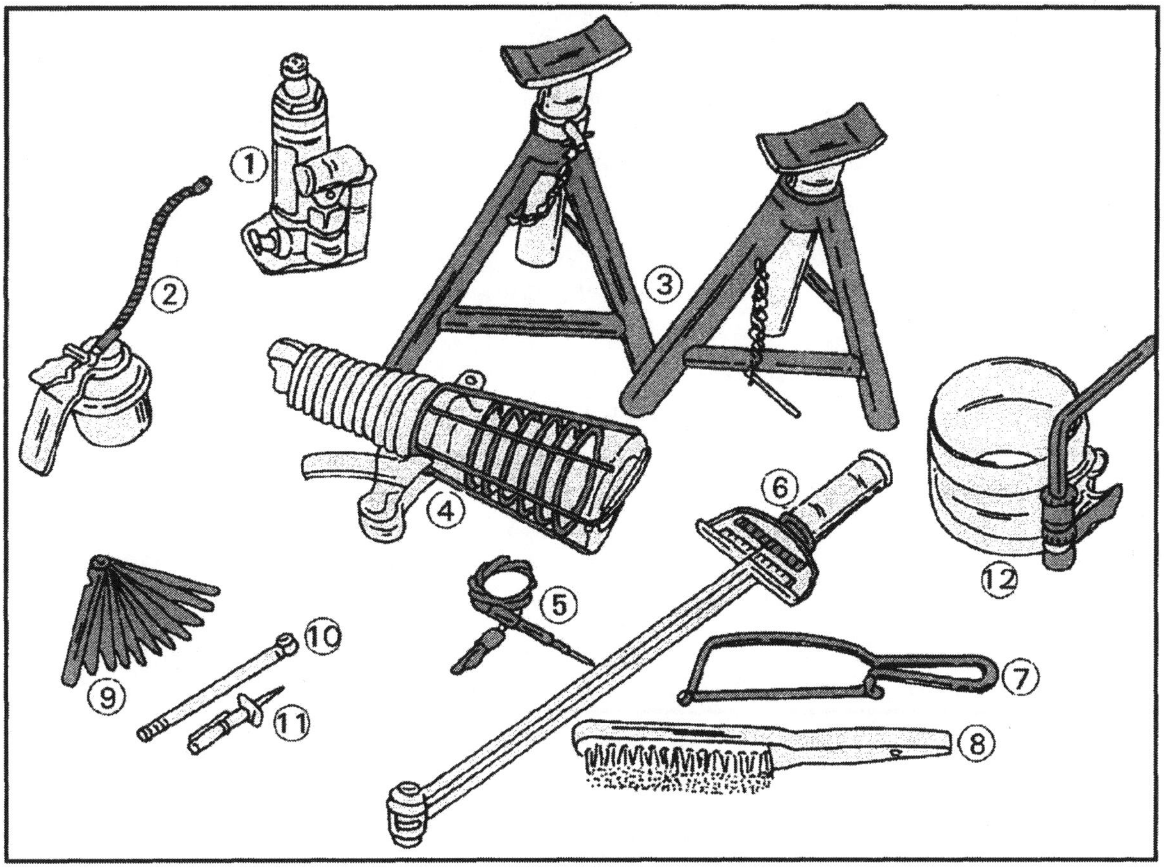

Fig. 0.5 – Recommended tools to service and repair your vehicle.

1 Hydraulic jack
2 Oil can
3 Chassis stands
4 Electric hand lamp
5 Test lamp (12 volts)
6 Torque wrench
7 Small hand saw
8 Wire brush
9 Feeler gauges
10 Tyre pressure gauge
11 Tyre profile depth checker
12 Piston ring clamp band

- Never slacken or tighten the axle shaft nuts or wheel bolts when the vehicle in resting on chassis stands.
- Never open the cooling system when the engine is hot. Sometimes it may, however, be necessary. In this case place a thick rag around the cap and open it very slowly until all steam has been released.

- Never drain the engine oil when the engine is hot. Drained engine oil must be disposed of in accordance with local regulations.
- Never allow brake fluid or anti-freeze to come in contact with painted areas.
- Never inhale brake shoe or brake pad dust. If compressed air is available, blow off the dust whilst turning the head away. A mask should be worn for reasons of safety.
- Remove oil or grease patches from the floor before you or other people slip on it.
- Do not work on the vehicle wearing a shirt with long sleeves. Rings and watches should be removed before carrying out any work.
- If possible never work by yourself. If unavoidable ask a friend or a member of the family to have a quick look to check that's everything is OK.
- Never hurry your work. Many wheel bolts have been left untightened to get the vehicle quickly back on the road.
- Never smoke near the vehicle or allow persons with a cigarette near you. A fire extinguisher should be handy, just in case.
- Never place a hand lamp directly onto the engine to obtain a better view. Even though the metal cage will avoid direct heat it is far better if you attach such a lamp to the open engine bonnet.

0.6. Jacking up of the Vehicle

Always remember the height of the vehicle when you jack it up inside a normal garage. Make sure that sufficient head room is available.

To prevent damage to the underside of the vehicle, apply the jack only in accordance with the instructions in your owner's manual. Never place a jack underneath the front or rear suspension arms.

For small jobs, jack up the vehicle as described in the Owner's Manual, with the jack supplied with the vehicle. If a mobile jack is used, make sure it can take the weight of the vehicle. Additional weights should always be removed from the cargo area of the vehicle.

To jack up the vehicle completely, first jack up the front end. Chock the rear wheels to prevent the vehicle from rolling off the jack or apply the handbrake. It is also advisable to engage the first gear (manual transmission) or to change into the "Parking" position (automatic transmission).

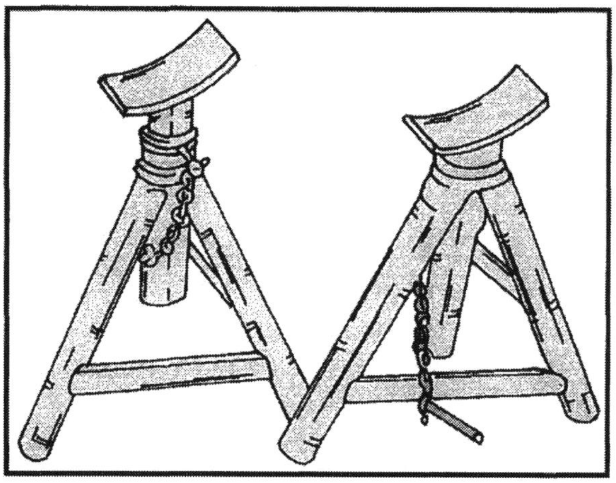

Fig. 0.6 – Three-legged chassis stands are the safest method to support the vehicle when work has to be carried out on the underside of the vehicle.

Chassis stands should only be placed on the L.H. and R.H. sides under the side of the body so as not to damage to the paint work. Chassis stands of the construction shown in Fig. 0.6, should be used, but again make sure that they are strong enough to carry the weight of the vehicle. Make sure the vehicle cannot slip off the stands.

Note: It is always difficult to raise a vehicle first on one side and then on the other. Take care that the vehicle cannot tip-over when the first side is lifted. Ask a helper to support the vehicle from the other side. Never work underneath the vehicle without adequate support.

1. DIESEL ENGINE

1.0. Technical Data

Fitted Engines:	642.990, 642.982
Number of Cylinders:	Four
Injection Order:	1 – 3 – 4 – 2
Arrangement of cylinders:	In-line
Camshafts:	Two overhead camshafts, No. marked in end face
Arrangement of valves:	Overhead
Cylinder bore:	88.00 mm (3.4962 in.)
- Piston stroke:	88.4 mm (3.5121 in.)
Capacity:	2148 c.c.
Compression Ratio:	19.0 : 1
Max. kW/B.H.P. (DIN):	
- 109 CDI:	65 kW (88 BHP) at 3800 rpm
- 111 CDI, Viano CDI:	80 kW (109 BHP) at 3800 rpm
- 115 CDI, Viano CDI:	110 kW (150 BHP) at 3800 rpm
Max. Torque (variations possible):	
- 109 CDI:	240 Nm (173 ft.lb.) at 1400 – 2400 rpm
- 111 CDI, Viano CDI:	290 Nm (208 ft.lb.) at 1600 – 2400 rpm
- 115 CDI, Viano CDI:	330 Nm (238 ft.lb.) at 1800 – 2400 rpm
-	
Crankshaft bearings:	5 friction bearings
Cooling system:	Thermo system with water pump, thermostat, cooling fan without fluid clutch, tube-type radiator
Lubrication:	Pressure-feed lubrication with gear-type oil pump, driven with chain from crankshaft. With full-flow and by-pass oil filter
Air cleaner:	Dry paper element air cleaner

1.1. Engine – Removal and Installation

1.1.0. REMOVAL OF THE POWER UNIT

Note: The following instructions are only valid for models without automatic air conditioning system. The system must be discharged during the removal of the front part (front module) of the vehicle. Discharging and charging of the air conditioning system can only be carried out in a specialist workshop.

The engine and transmission is removed from the vehicle as a complete unit, as shown in Fig. 1.2. Engine and transmission are a heavy unit and a suitable hoist or crane is required to lift out the assembly, remembering that the weight is more than 200 lbs.

The following description is a general guide line, as we cannot refer to every possible variation and/or equipment that may be fitted to your vehicle. The engine bonnet must be in an upright position and tied to the vehicle. Note that various operations are only referred to in brief, i.e. remove the propeller shaft, signifies for example that you will have to refer to Section "Propeller Shaft" for further instructions.

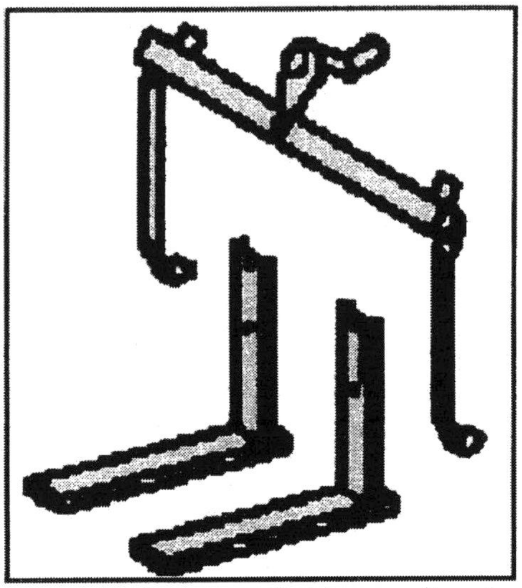

Fig. 1.1 – Bracket assembly for the removal and installation of the front module. A similar equipment can perhaps be made up.

The front end (front module) of the vehicle must be removed to gain access to various parts. This is a rather complicated operation as many parts must be removed and/or disconnected which you will have to locate. The workshop uses a lifting device, consisting of the parts shown in Fig. 1.1, which is bolted to the inside of the module to lower it to the ground after removal. Attempt the removal as follows with the bonnet open:

- Remove the air cleaner housing.
- Draw the fluid out of the steering fluid container, using a suitable suction tool or hand pump and disconnect two fluid lines from the steering unit. Some fluid will run out. Soak it up with a piece of cloth.
- Disconnect the coolant hose from the thermostat housing.
- Locate the wiring harness at the front of the module and unclip it. Place the wiring harness to one side.
- If a headlamp cleaning system is fitted unclip a quick-type lock, located below the coolant reservoir.
- Unhook the engine bonnet cable from the brackets and remove the engine bonnet catch.

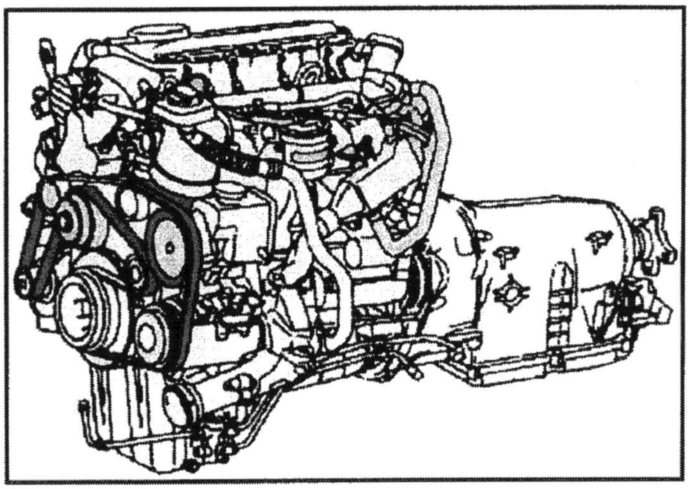

Fig. 1.2 – Engine and transmission are lifted out together.

- Disconnect the electrical lines for the removal of the front module from the top and from the bottom.
- From underneath the vehicle remove two expansion clips on both sides of the vehicle.
- Remove the charge air hose.
- On a vehicle with automatic transmission disconnect the fluid feed line (see from below on the L.H. side) and the fluid return line (see from below on the R.H. side). Some fluid may run out and must be collected. Tighten the union nuts during installation of the pipes to 1.8 kgm (13 ft.lb.).
- Remove four bolts securing the crossmember to the longitudinal member. Two of the bolts are next to each other. One is on the L.H. side and the other one on the R.H. side. During installation tighten the M8 bolts to 2.5 kgm (18 ft.lb.) and the M10 bolts to 4.0 kgm (29 ft.lb.).
- Inside the wheel house remove the bolts and at the upper end of the front module remove the bolts
- If available, fit the lifting device shown in Fig. 1.1 to the front of the module, hooking the legs into the openings of the bumper. Otherwise make the necessary

arrangements to attach the front end with ropes or chains to a lifting device. After the bolts securing the front module to the frame lift the models up and out. The two brackets in the lower view of Fig. 1.1 can be bolted to the inside of the module so that the module can be placed on the ground without tipping over.

The installation is a reversal of the removal procedure. Note the following points:

- Fill the power steering system and bleed it.
- Check the fluid level of an automatic transmission and if necessary correct.
- Adjust the headlamps or have them adjusted and check the headlamp washer system (if fitted) for correct operation.

Removal of the engine and transmission can now be attempted as follows, after the front module has been removed as described above:

- On some models (639.7.. and 639... after VIN number 055347) the lower engine compartment panelling must be removed. Only the smaller of the two cover panels must be removed.
- Disconnect all hoses and lines from the engine if not already removed.
- Disconnect the exhaust system from the primary catalytic converter. Note that new sealing cones should be used during installation.
- Remove the exhaust pipe bracket held in position by bolts. Tighten the bolts to 2.3 kgm (16.5 ft.lb.) during installation.
- Support the transmission from below on a suitable mobile jack and remove the rear engine crossmember. The crossmember is secured by two bolts below the propeller shaft on the L.H. and R.H. side and two further bolts in the centre. Mark the exact position of the crossmember before you take it out. During installation tighten the bolt securing the rear engine mount to the transmission carrier to 2.0 kgm (14.5 ft.lb.) and the bolt securing the transmission carrier to the longitudinal member to 4.0 kgm (29 ft.lb.).
- Remove the propeller shafts as described in the relevant section. After removal push the shaft to one side and tie it up with a piece of wire. The propeller shaft bolts are tightened to 1.5 kgm (11 ft.lb.) and from the final position a further quarter of a turn.
- Remove the two gear selector cables on a model with manual transmission by levering the cables out of the ball socket and remove from the protection. Remove the single selector cable on a model with automatic transmission in a similar manner. The ball joints must be lubricated with long-term grease during installation.
- Disconnect the hydraulic clutch fluid line from the transmission.
- Attach suitable chains to the engine lifting points and the lifting device and lift the engine/transmission assembly until just under tension. Remove the engine mounting bolts on the L.H. and R.H. side of the engine and lift the assembly. Before removing any of the engine mountings ensure that the weight of the assembly is well supported. The mounting bolts are tightened to 5.8 kgm (42 ft.lb.) during installation.
- Lift the engine with the lifting tackle from the subframe and out towards the front. Continuously check that none of the cables, hoses, pipes, etc. have been forgotten and that none of the parts can get caught.

1.1.1. INSTALLATION OF THE POWER UNIT

The installation of the power unit is a reversal of the removal procedure. Here are some of the points, which should be observed:

- Do not connect any of the parts to the engine or transmission until the engine and transmission mountings have been refitted and the engine is fitted free of any load or tension.
- If work has been carried out on the engine or transmission, make sure that the oil drain plug of the respective unit has been tightened and the oil level is correct.
- Keep oil or grease away from all rubber parts.
- All bolts and nuts must be tightened to the correct tightening torque. Some important torque values: Engine carrier to engine mounting 5.8 kgm (42 ft.lb.).
- If the original coolant is refilled to the engine, check the anti-freeze strength. If the coolant has been in the engine for a considerable time, replace it. If possible check the cooling system for leaks before starting the engine.
- Check the air cleaner element for contamination before it is refitted.

Note: As the vehicle is fitted with a self-diagnosis and fault memory system it is possible that certain faults have been logged into the memory. The fault memory system can only be checked and cleared in a workshop.

1.2 Engine - Dismantling

Before commencing dismantling of the engine, all exterior surfaces should be cleaned, as far as possible, to remove dirt or grease. Plug the engine openings with clean cloths to prevent any foreign matter entering the cavities and openings. Detailed information on engine dismantling and assembly is given in the sections dealing with servicing and overhaul (sections commencing at 1.3.) and these should be followed for each of the sub-assemblies or units to be dealt with.
Follow the general dismantling instructions given below:

- Dismantling must be carried out in an orderly fashion to ensure that parts, such as valves, pistons, bearing caps, shells, tappets and so on, are replaced in the same positions as they occupied originally. Mark them clearly, but take care not to scratch or stamp on any rotating or bearing surfaces. A good way to keep the valves in order is by piercing them through an upside-down cardboard box and writing the number against each valve. Segregate together the tappets, the springs and retainers with collets for each valve, if possible in small plastic bags for each individual valve.
- If a proper engine dismantling stand is not available, it will be useful to make up wooden support blocks to allow access to both the top and bottom faces of the engine. The cylinder head, once removed from the block, should be supported by a metal strap, screwed to the manifold face and secured by two nuts onto the manifold studs.

1.3. Engine - Overhaul

1.3.0. Cylinder Head and Valves – Technical Data

Cylinder Head:

Cylinder head height:	126.85 - 127.15 mm
Max. Distortion of Cylinder Head Faces:	
- Longitudinal direction:	0.08 mm
- Across the face:	0.00 mm
Max. deviation of faces between upper and lower sealing faces (parallel to each other):	0.10 mm

Depth of valve head faces and cylinder head sealing face:

- Inlet valves	+0.17 to 0.23 mm

- Exhaust valves:	+0.12 to 0.28 mm
- With re-cut valve seats:	1.0 mm – all valves
Valves	
Valve Head Diameter:	
- Inlet valves:	28.60 – 28.80 mm
- Exhaust valves:	26.10 – 28.80 mm
Valve seat angle:	45° + 15'
Valve Stem Diameter:	
- Inlet valves:	6.960 – 6.975 mm
- Exhaust valves:	6.955 – 6.990 mm
Valve Length:	
- Inlet valves:	103.90 – 104.30 mm
- Exhaust valves:	104.30 – 104.70 mm
Valve Seat Width:	
- Inlet valves:	1.1 – 1.4 mm
- Exhaust valves:	1.1 – 1.4 mm
Valve Identification:	
- Inlet valves:	E 628 00 or E 646 00
- Exhaust valves:	A 626 00 05

Valve Seats

Valve seat width:	
- Inlet valves:	1.1 – 1.4 mm
- Exhaust valves:	1.1 – 1.4 mm
Valve seat angles:	45° + 30'
Upper correction angle:	15°
Lower correction angle:	60°

Valve Timing

Inlet valves open:	19° after TDC
Inlet valves close:	7° after BTDC
Exhaust valves open:	9° before BTDC
Exhaust valves close:	22° before TDC

Valve seat basic bore in cylinder head (for reference only):

- Inlet valves – Std.:	40.000 – 40.016 mm
- Exhaust valves – Std.:	37.000 – 37.016 mm
Valve seat height, all valves:	6.97 – 7.00 mm

Upper edge of valve seat rings to cylinder head face:

- Inlet valves:	2.37 – 2.25 mm
- Exhaust valves:	2.44 – 2.25 mm

Valve Springs

Colour code:	yellow-green or purple-green
Outer diameter:	33.20 mm
Wire diameter:	4.25 mm
Free length:	50.80 mm
Length under load of 72 to 77 kg:	27.00 mm
- Wear limit:	27.00 mm at 65 kg

Valve Guides

Inlet valve Guides:

- Outer diameter – Std.:	14.044 – 14.051 mm
- Outer diameter – Repair size:	14.214 – 14.222 mm
- Inner diameter:	7.000 – 7.015 mm

Basic bore in cylinder head:

- Std.:	14.030 – 14.035 mm
- Repair size:	14.198 – 14.203 mm

Exhaust Valve Guides:

- Outer diameter – Std.:	14.044 – 14.051 mm
- Outer diameter – Repair size:	14.214 – 14.222 mm
- Inner diameter:	7.000 – 7.015 mm

Basic bore in cylinder head:

- Std.:	14.030 – 14.035 mm
- Repair size:	14.198 – 14.203 mm

Interference fit of valve guides – All guides:

- Std.:	0.009 – 0.021 mm
- Repair size:	0.011 – 0.024 mm

Camshaft

Camshaft Bearings:	4
Journal diameter:	Not available
Camshaft Bearing Clearance:	
- New condition:	0.050 – 0.091 mm
- Wear limit:	0.11 mm
Camshaft End Float:	
- New condition:	0.07 – 0.15 mm
- Wear limit:	0.18 mm

1.3.0.1. Cylinder Head – Working on the Cylinder Head

The following information should be noted when work is carried out on a cylinder head:

- The cylinder head is made of light-alloy. Engine coolant, engine oil, the air required to ignite the fuel and the exhaust gases are directed through the cylinder head. Glow plugs, injectors and valve tappets are fitted to the cylinder head. Also in the cylinder heads you will find the camshafts.
- The exhaust manifold and the inlet manifold are bolted to the outside of the head. The fuel enters the head on one side and exits on the other side.
- The cylinder head is fitted with various sender units, sensors and switching valves, responsible for certain functions of the temperature control.
- As the cylinder head is made of light alloy, it is prone to distortion if, for example, the order of slackening or tightening of the cylinder head bolts is not observed. For the same reason never remove the cylinder head from a hot engine.
- A cylinder head cannot be checked in fitted position. Sometimes the cylinder head gasket will "blow", allowing air into the cooling system. A quick check is possible after opening the coolant reservoir cap (engine fairly cold). Allow the engine to warm-up and observe the coolant. Visible air bubbles point in most cases to a "blown" gasket. Further evidence is white exhaust smoke, oil in the coolant or coolant in the engine oil. The latter can be checked at the oil dipstick. A white,

grey emulsion on the dipstick is more or less a confirmation of a damaged cylinder gasket.

- If you are convinced that water has entered the engine and you want to get home or to the nearest garage, unscrew the injectors and crank the engine with the starter motor for a while to eject the water. Refit the injectors, start the engine and drive to your destination without switching off the engine. This is the only method to avoid serious engine damage (bent connecting rods for example).

The cylinder head must only be removed when the engine is cold. The head is removed together with the exhaust manifold, but the inlet manifold must be separated from the cylinder head before the head can be lifted off. New cylinder head gaskets are wrapped in plastic and must only be unwrapped just before the gasket is fitted. The cylinder head can be removed with the engine fitted and these operations are described below, but note that operations may vary, depending on the equipment fitted. Many secondary operations are necessary before the actual cylinder head can be removed. The following description refers therefore in detail to the various jobs.

Cylinder Head Cover – Removal and Installation

Fig. 1.3 shows the parts to be removed fitted to the top of the cylinder head cover.

- Disconnect the battery earth cable.
- Remove the injectors (15) as described in section "Diesel Injection".
- Remove the bolts (13) securing the bracket (7) for the EGR valve from the cylinder head cover (5). The bolts are tightened to 1.6 kgm (11.5 ft.lb.) during installation.

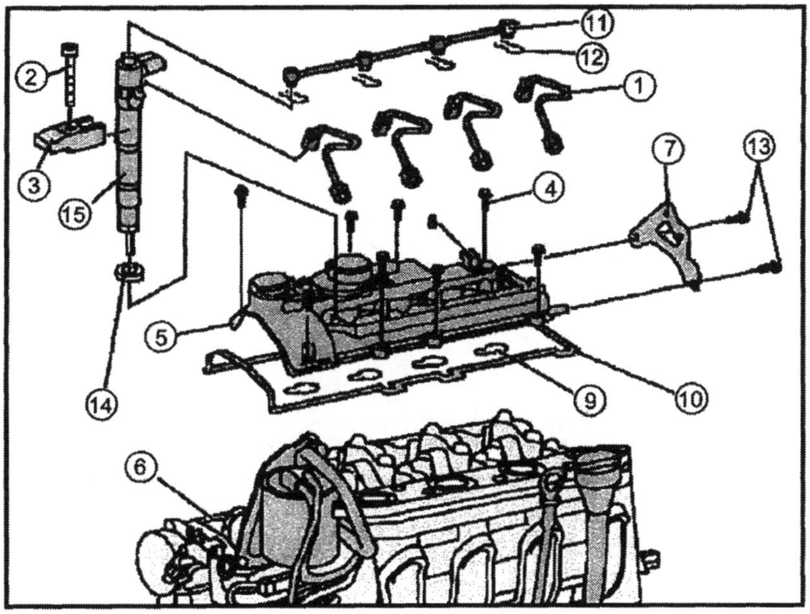

Fig. 1.3 – Details for the removal and installation of the cylinder head cover. The numbers are referred to in the text.

- Unplug the plug connector from the camshaft Hall sensor (8).
- Unscrew the bolts, unclip the wiring at the cylinder head cover and push the wiring to one side.
- Detach the crankcase ventilation hose (7 in Fig. 1.4) from the side of the cylinder head cover.
- Disconnect the fuel lines shown in the illustration above the cylinder head cover and place the lines to one side. Again this operation is described in section "Diesel Injection".
- Remove the bolts (4) securing the cylinder head cover (5). During installation do not tighten the bolts (4) fully immediately, as the injectors (15) are aligned by means of the cylinder head cover. Seized injectors and injection lines under stress may be the result.
- Remove the cylinder head cover (5). The camshaft Hall sensor (8) must be removed if a new cylinder head is to be fitted.

The installation is a reversal of the removal procedure. Gaskets (9) and (10) for the cylinder head cover must always be replaced. Clean the sealing faces thoroughly before installation.

Trim Panel of Cylinder Head Cover – Removal and Installation

The items to be removed are shown in Fig. 1.4.

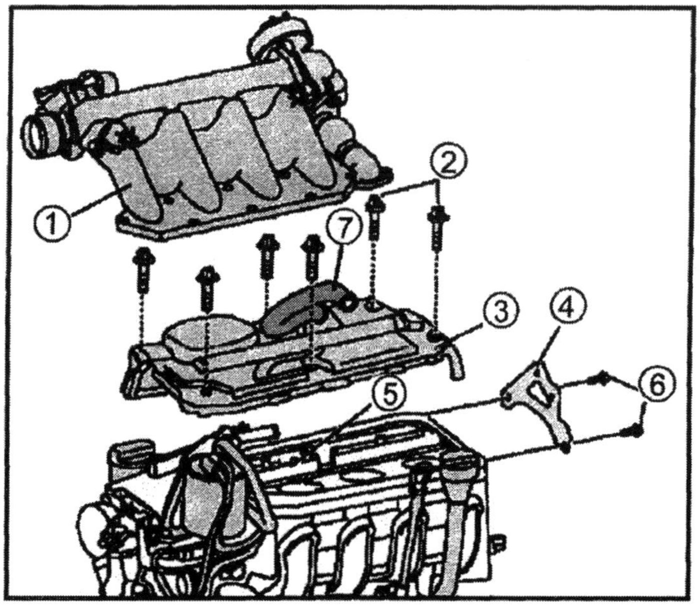

Fig. 1.4 – Details for the removal and installation of the cylinder head cover trim panel. The numbers are referred to in the text.

1 Mixing chamber
2 Bolts, 1.0 kgm
3 Trim panel of cover
4 Bracket for EGR valve
5 Cylinder head cover
6 Bolts, 1.6 kgm
7 Crankcase ventilation hose

- Remove the mixing chamber together with the exhaust gas re-circulation (1).
- Remove the bolts (6) securing the bracket for the EGR valve (4) to the cylinder head cover (5). The bolts (6) are tightened to 1.6 kgm (11.5 ft.lb.) during installation.
- Disconnect the crankcase ventilation hose (7) from the trim panel (3). During installation ensure that the hose is inserted correctly.
- Remove the trim panel (3) after removal of the securing bolts. Tighten the bolts to 1.0 kgm (7 ft.lb.) during installation.

The installation is a reversal of the removal procedure, noting the points already given above.

Front Cover of Cylinder head – Removal and Installation

The following operations must be carried out before the front cover can be removed.

- Remove the cylinder head cover as already described.
- Remove the chain tensioner as described later on.

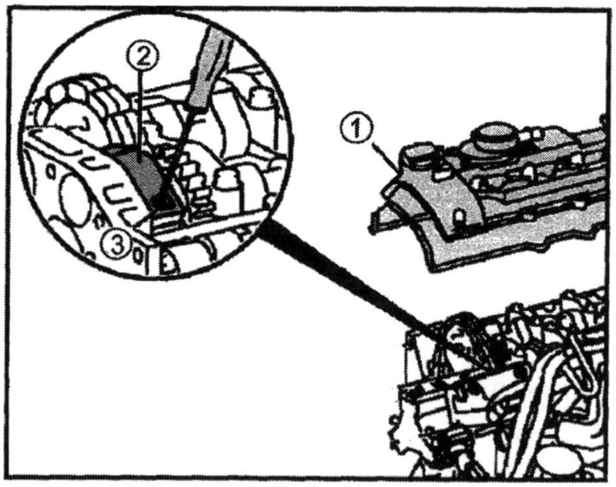

Fig. 1.5 – The locking pawl (3) for the upper guide rail (2) is raised as shown. The cylinder head cover (1) must be removed.

- Remove the vacuum pump as described near the end of section "Brakes".
- Remove the pre-delivery pump as described below.
- Remove the bolts securing the cover to the front of the cylinder head. The bolts are tightened to 1.4 kgm (10 ft.lb.).
- The next operation is more complicated, but can be carried out by referring to Fig. 1.5. Insert a screwdriver as shown and raise the locking pawl (2) for the upper guide rail at the position shown until free. The rail must not be removed.
- The cover can now be removed. Note the dowel pins in the cover/cylinder head during installation.

During installation refer to Fig. 1.6. Clean the faces of cover (1) and cylinder head and coat the sealing face of the cover with "Loctite 5970" sealing compound (width of 1.5 mm). The cover must be fitted within 10 minutes after the application of the sealing compound. Place the cover over the dowels (3), (4) and (5) and push it against the cylinder head face and tighten the bolts (2) to the torque given above.

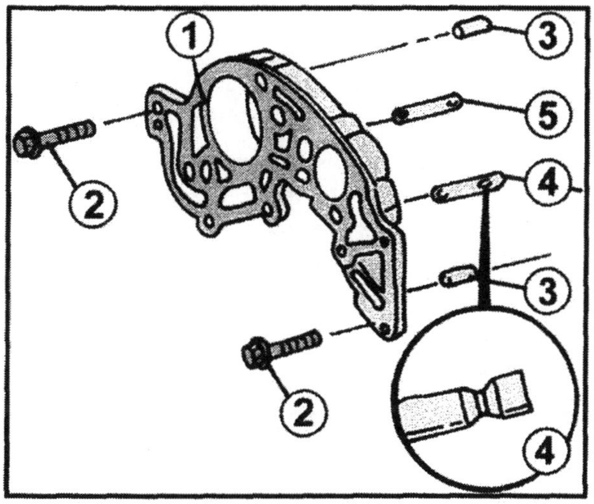

Fig. 1.6 – The front cover of the cylinder head. Note the position of the dowel pins.

The guide rail in the cylinder head must be replaced if a new front cover is fitted.

Pre-Delivery Pump – Removal and installation

The fitted position of the pump is shown with (1) in Fig. 1.7 and can be removed after the vacuum pump has been removed as described above. Then disconnect the two fuel lines (2) and (3) from the pump (1) and unscrew the bolts (5) in Fig. 1.7.

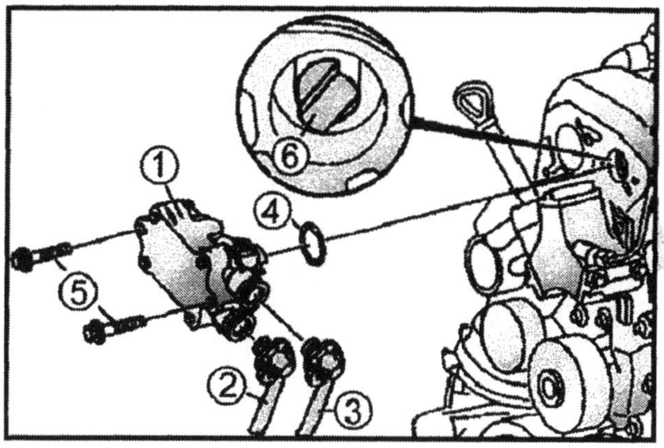

Fig. 1.7 – Removal and installation of the pre-delivery pump. Its fitting position is shown in Fig. 1.8 with (9).

Remove the pump. The poly V-belt should be protected with a thick rag as fuel will run out. During installation replace the sealing ring (4). Make sure that the sealing surfaces are clean. Ensure that the driver of the pump enters the driver (6) at the end of the inlet camshaft. Tighten the bolts (5) to 0.9 kgm (7 ft.lb.).

Chain Tensioner – Removal and Installation

The chain tensioner is located on the outside of the cylinder head at the position shown in Fig. 1.8 and also shown later on in the section dealing with the timing mechanism. From the outside only the head of a hexagonal bolt is visible, with a sealing washer below.

Fig. 1.8 – The chain tensioner can be found at the position shown by the arrow. The items mentioned below are also located on this side.

To remove the tensioner you will have to know the engine number of a 646.982 or 646.983 engine. Remove the tensioner as follows:

- Remove the air cleaner housing.
- Remove the air hose between the turbo charger and the air cleaner housing.
- Remove the oil dipstick guide tube from the side of the cylinder head on the engines mentioned above before engine end number 007337. Remove the securing bolt and press the guide tube carefully to one side. On these engines also remove a hose clamp immediately below the bolt and turn it to one side if necessary to create more space.
- Unscrew the chain tensioner plug.

The installation is a reversal of the removal procedure. Replace the sealing ring during installation. To facilitate the installation turn the crankshaft slightly in the direction of rotation. This will slacken the tension of the timing chain near the tensioning rail. The tensioner is tightened to 8.0 kgm (58 ft.lb.).

1.3.0.2. Cylinder Head – Removal and Installation

The camshaft sprockets must be locked against rotation when the head is removed. We therefore recommend to read the instructions before you attempt with the removal. Figs. 1.9 and 1.10 show the parts to be disconnected and/or removed before the final removal can take place. Again we describe all necessary operations under separate headings when applicable. Proceed as follows:

Fig. 1.9 – Details for the removal and installation of the cylinder head. Also refer to Fig. 1.10.

1 Trim cover
2 Injector
3 Cylinder head cover
4 Chain tensioner
5 Sealing ring
6 Top guide rail
7 Vacuum pump
8 Top R.H. engine bracket
9 Pre-delivery pump
10 Front cylinder head cover
11 High pressure pump
12 Thermostat housing

- Disconnect the battery earth cable.
- Remove the air cleaner housing.
- Drain the cooling system as described in section “Cooling System”.

Fig. 1.10 – Details for the removal and installation of the cylinder head and the camshafts.

1 Fuel filter
2 Camshaft bearing cap
3 Exhaust camshaft
4 Fuel filter bracket
5 Inlet camshaft
6 High pressure pump gear
7 Sprocket bolts,
8 Mixing chamber
9 Camshaft housing
10 Exhaust manifold
11 Bolts, 2.0 kgm
12 Inlet manifold

- Remove the injectors (section "Diesel Injection").
- Unscrew the turbo charger to exhaust manifold bolt at the bottom.
- Remove the charge air distribution pipe.
- Remove the trim panel of the cylinder head cover as described earlier on.
- Disconnect the coolant hoses from the thermostat housing (1) in Fig. 1.9. Check the condition of hose clamps and hoses before re-using them. Also detach the vent line from the coolant reservoir.
- Detach the large hose leading from the engine to the intake duct (top left, standing in front of the engine compartment) and also disconnect the vent hose next to the oil dipstick handle.
- Detach the fuel lines.
- Take off the bracket (4) in Fig. 1.10 for the fuel filter from the cylinder head.
- Separate the plug connector from the coolant temperature sensor, rail pressure sensor and the pressure control valve.
- Remove the exhaust manifold (10) in Fig. 1.10.
- Remove the top R.H. engine mounting bracket (8) in Fig. 1.9.

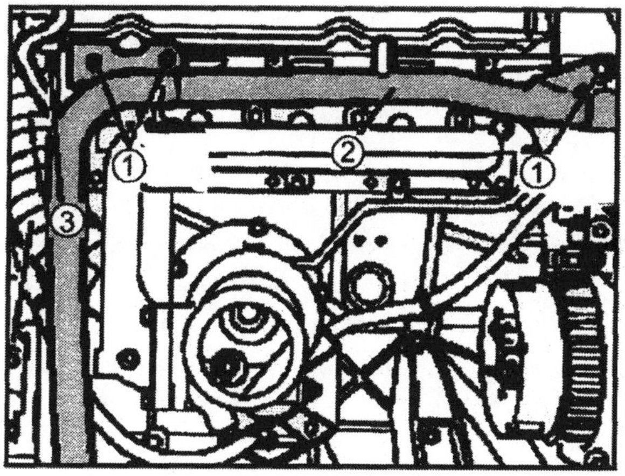

Fig. 1.11 – Items (1) to (3) must be removed as explained below.

- Refer to Fig. 1.11 and remove the bolts (1) at the positions shown and remove the coolant pipe (2) from the cylinder head.
- Remove the inlet manifold (12) in Fig. 1.10.
- Unscrew the nut (3) in Fig. 1.11 securing the charge air hose to the cylinder head.
- Remove the cylinder head cover as described above.

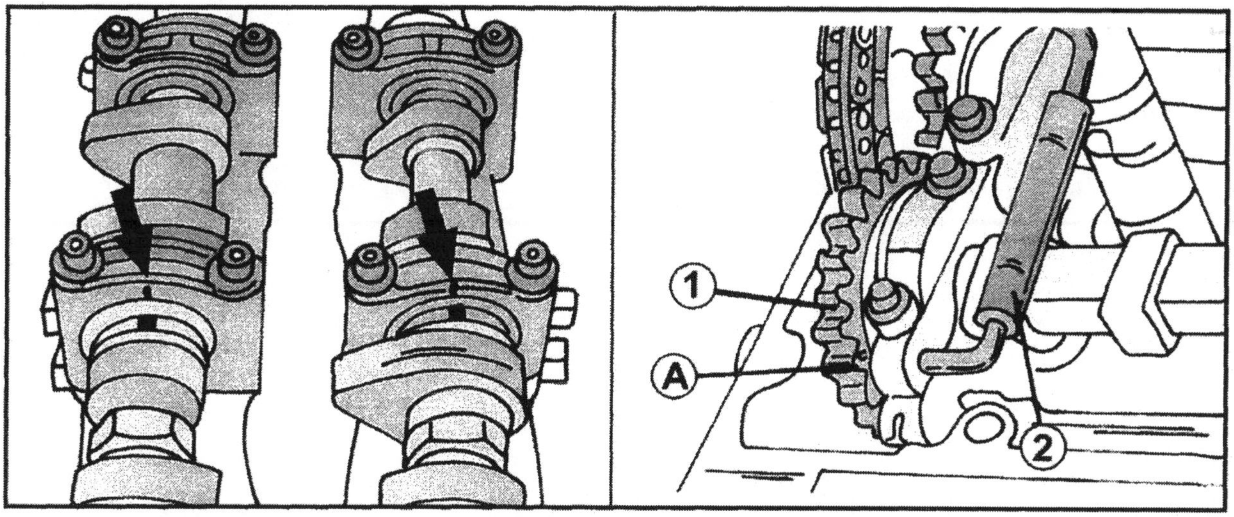

Fig. 1.12 – With No. 1 piston at top dead centre the marks in camshafts and camshaft bearing caps must be aligned as shown. The R.H. view shows where the retaining pin (2) is inserted into the bore (A) of the inlet camshaft sprocket (1). See also Fig. 1.14.

- Position the No. 1 piston to the top dead centre position. To do this rotate the crankshaft in a clockwise direction - never backwards - until the markings in the camshaft and camshaft bearing cap are aligned as shown in L.H. view of Fig. 1 .12. Do not rotate the camshaft(s) to obtain the setting. The "OT" (TDC) mark in the vibration damper must be in line with the mark in the damper pulley.

- Remove the chain tensioner (4) in Fig. 1.9 as described earlier on.
- Remove the front cover (10) of the head cover as already described.
- Remove the top slide rail (6) in Fig. 1.9 as described later on.
- Remove the inlet camshaft (5) and the exhaust camshaft (3) as described below.

Removal and Installation of Camshafts

The details of the camshafts are shown in Fig. 1.13. Note the identification of the camshafts, which will be explained later on.

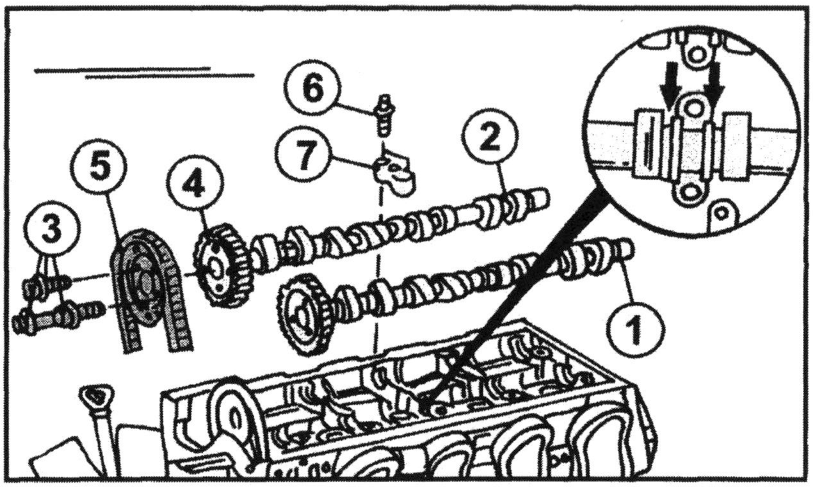

Fig. 1.13 – The camshafts. The arrows in the circle show where the alignment of the camshaft on the bearing cap is checked. The numbers are referred to in the text.

- Lock the inlet camshaft (1) in Fig. 1.13 against rotation. The workshop uses a retaining tool, shown on the R.H. side of Fig. 1.12 which is inserted from the rear of the No. 1 bearing cap into the bore "A" in the inlet camshaft sprocket, as shown in Fig. 1.14. The markings (B) in the two sprockets must be opposite each other as shown in the same illustration.

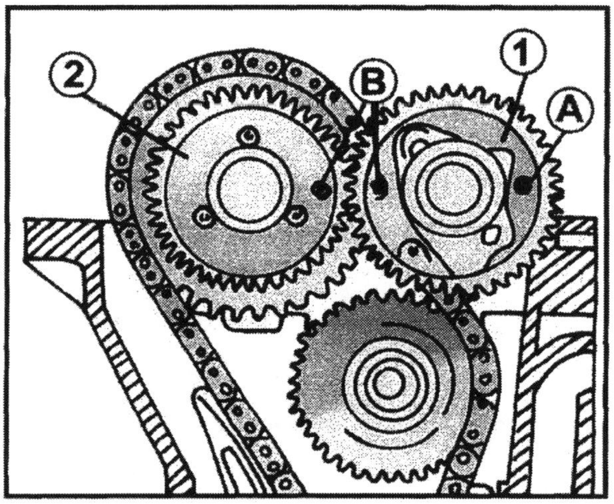

Fig. 1.14 – View of the inlet camshaft sprocket (1) and the exhaust camshaft sprocket (2). The retaining pin in Fig. 1.12 is inserted into bore (A). The marks (B) must be opposite each other.

- Remove chain tensioner and the front cylinder head cover as already described (see also Fig. 1.6). Then unscrew the driver out of the inlet camshaft and take off the slide rail. Details of the operation are also given in the section dealing with the timing mechanism.

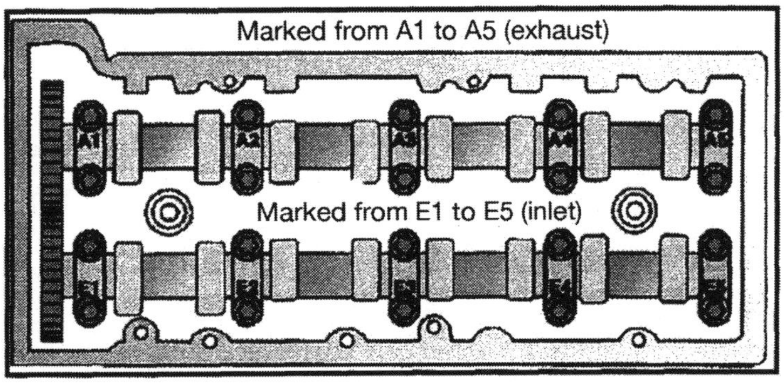

Fig. 1.15 – The camshaft bearing caps are marked from 1 to 5. Caps must not be inter-changed during installation.

- Remove the sprocket (5) in Fig. 1.13 from the exhaust camshaft (2) together with the timing chain after removal of bolts (3). If necessary use an open-ended spanner and prevent the camshaft from rotating by applying the spanner to a hexagon on the camshaft.
- Remove the camshaft bearing caps (7). Slacken the "Torx" head bolts (6) securing the bearing caps (7) evenly in steps of one turn until the pressure onto

the bearings is released. The bearing caps are marked as shown in Figs. 1.15. Remove the caps and lift out the camshafts (1) and (2).

- Remove the hydraulic valve clearance and compensating elements and lift off the camshaft bearing housing (9) in Fig. 11.7. The dowel pins must engage when the housing is fitted.

The installation of the camshafts is carried out as follows:

- Place the camshaft housing (9) in Fig. 1.10 over the dowel pins and tap the housing in position, using a rubber mallet.
- Lubricate the hydraulic valve clearance and compensating elements with engine oil, place the camshafts into their bearing bores and rotate the shafts until the alignment shown in Fig. 1.13 is obtained (arrows). The shafts must now be rotated until the two bores (B) in the sprockets in Fig. 1.14 are opposite each other and the marks shown by the arrows in the L.H. view of Fig. 1.12 are in line. *Remember, if new camshafts are fitted quote the engine type and the model year.*
- Place the bearing caps in Fig. 1.13 in accordance to their numbers in Fig. 1.15 and tap them in position with a mallet. Fit the camshaft bearing bolts and tighten them in several stages by one turn each, using a "Torx" head socket and a torque wrench until all bolts are tightened to 0.9 kgm (7 ft.lb.). Finally check once more that all bearing caps are fitted in accordance with their numbers in Fig. 1.15.

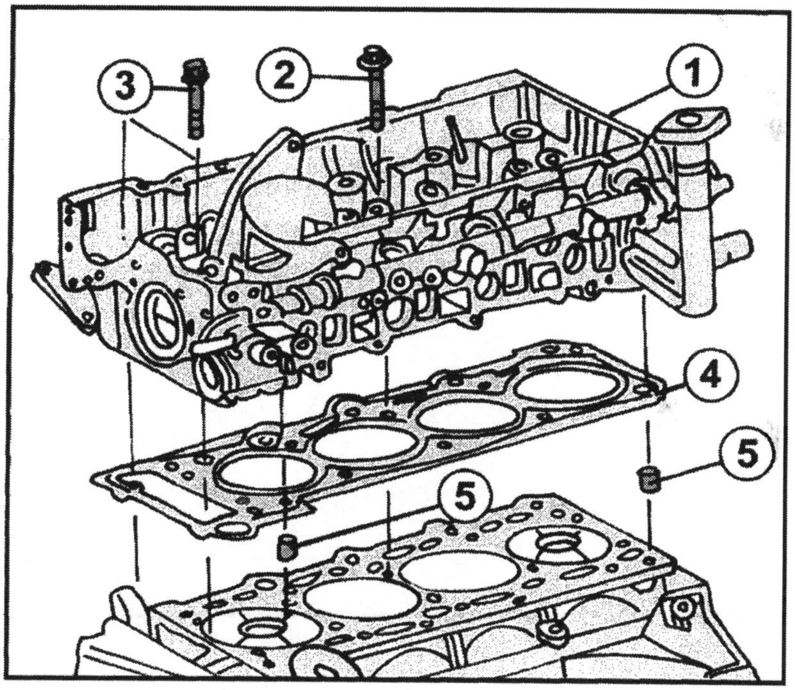

Fig. 1.16 – Removal and installation of the cylinder head. The numbers are referred to below.

- Fit the exhaust camshaft sprocket (5) to the shaft (2) with the timing chain around the sprocket, making sure the dowel pin (4) engages. Insert the bolts (3) – always new bolts must be used – and tighten them evenly to 1.8 kgm (13 ft.lb.).
- Fit the glide rail and the driver to the inlet camshaft, the front cover to the cylinder head and the cylinder head cover as described earlier on.

Final Removal of Cylinder Head (and installation)

The following operations are now carried out by referring to Fig. 1.10.

- Remove the bolt, withdraw the bush and take off the high pressure pump drive gear (6).

The following operations are now carried out by referring to Fig. 1.16:

- Remove the two bolts (11) in Fig. 1.10 securing the cylinder head to the timing case cover.
- Slacken the cylinder head bolts (3) from the timing cover and the bolts (2) in stages from the outside towards the inside, i.e. in reverse order to the order shown in Fig. 1.17 in several stages, until they all slack and can be removed without force.

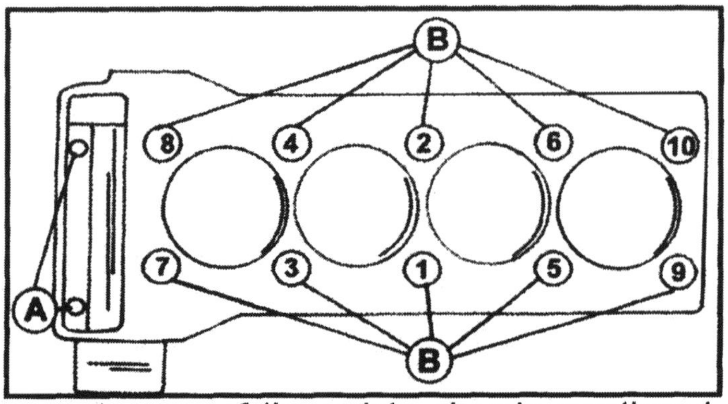

Fig. 1.17 – Tightening sequence for the cylinder head.

- Lift the cylinder head (1) from the cylinder block and take off the cylinder head gasket (4). The head is located on the dowel sleeves (5), i.e. must be removed in an upward direction.

The installation of the cylinder head is carried out as follows, taking into account some of the points already mentioned during removal.

- Measure the length of each cylinder head bolt between the end of the thread to the underside of the bolt head. The standard length is 102 mm. Any bolt longer than 104 mm must be replaced.
- Coat the bolt thread and the underside of the bolt heads with engine oil. The bores in the cylinder block must be free of oil.
- Place a new cylinder head gasket (4) in Fig. 1.16 in position over dowels (5).
- Place the cylinder head (1) in position and tap it down with a mallet. Insert the cylinder head bolts and tighten them hand-tight.

The bolts are now tightened in the following order. Fig. 1.17 shows the tightening sequence. Bolts (A) are Allen-head bolts (M8 thread), bolts (B) have a "Torx" head and a suitable socket is required. Bolts (A) secure the cylinder head to the timing cover. Tighten the bolts as follows:

- Tighten the bolts (B) to 1.5 kgm (11 ft.lb.) in the order shown.
- Tighten the bolts (B) to 6.0 kgm (43 ft.lb.) in the order shown.
- Tighten bolts (A) to 2.0 kgm (14.5 ft.lb.).
- Tighten bolts (B) by a quarter of a turn in the order shown and re-check the torque of bolts (A) once more.
- Tighten the bolts (B) again by a quarter of a turn in the order shown.
- A re-tightening of the cylinder head bolts is not necessary.
- All other operations are carried out in reverse order.

1.3.0.3. Cylinder Head - Dismantling

The following description assumes that the cylinder head is to be overhauled in an engine shop. If only a top overhaul is asked for, you can remove the valves as described below and can then decide if a complete overhaul of the head is required. The cylinder head must be removed.

The valve stem oil seals can be replaced with the cylinder head fitted. Signs of worn oil seals are blue exhaust smoke when the vehicle is coasting (gear engaged), when the engine is accelerated after idling for a while or blue smoke when starting the swap round. If the oil consumption has reached 1 litre per 600 miles, replace the valve stem oil seals as described under a separate heading. The valve stem oil seals can also be replaced with the engine fitted to the vehicle, but various special tools are required. It is therefore a job for the Mercedes-Benz dealer.

The valves can be removed as follows. After removal you will have the parts shown in Fig. 1.18.

- Remove all auxiliary parts from the cylinder head, including the exhaust manifold.
- A valve spring compressor is required to remove the valves. Fig. 1.19 shows such a compressor. Different types are available, but it must be possible to

compress the springs. Valves are held in position by means of valve cotter halves. Compress the springs and remove the valve cotter halves (2) with a pair of pointed pliers or a small magnet (1).

Fig. 1.18 – Valves removed from the cylinder head.
1 Valve
2 Lower spring seat
3 Valve stem oil seal
4 Valve spring
5 Upper valve spring cup
6 Valve cotter halves

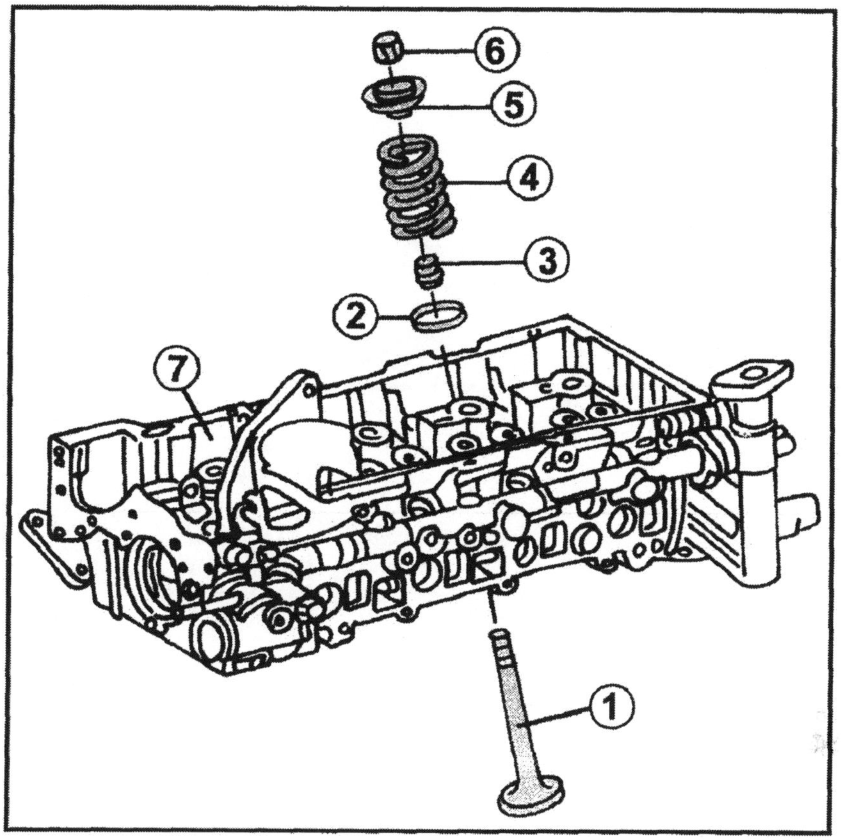

- If a valve spring compressor is not available, it is possible to use a short piece of tube to remove the valve cotter halves. To do this, place the tube over the upper valve spring collar and hit the tube with a blow of a hammer. The valve cotter halves will collect in the inside of the tube and the components can be removed. The valve head must be supported from the other side of the cylinder head. Keep the hammer in close contact with the tube to prevent the cotter halves from flying out.

Fig. 1.19 – Removal of valves with a valve lifter (3). The valve cotter halves (2) can be removed with a small magnet (1).

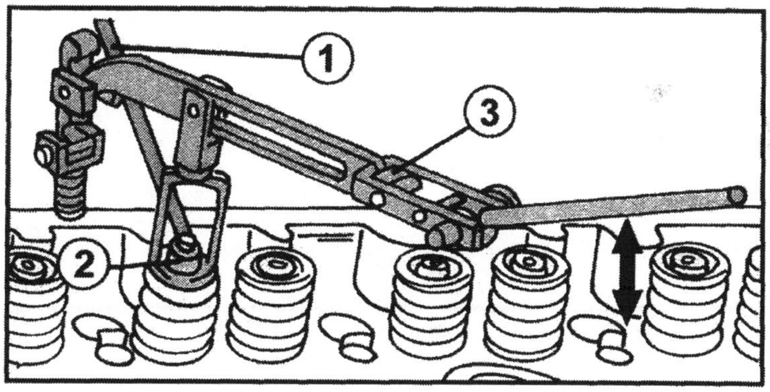

- Remove the valve spring collar (5) and the valve spring (4) in Fig. 1.18. The valve springs (one spring per valve) are identified with a paint spot and only a spring with a paint spot of the same colour must be fitted.
- Remove valve stem oil seals (1) in the L.H. view of Fig. 1.20 carefully with a screwdriver or a pair of pliers (2).
- Remove the valves one after the other out of the valve guides and pierce them in their fitted order through a piece of cardboard. Write the cylinder number against each valve if they are to be re-used.

A few words should be said about the camshafts. The camshafts have five bearings. The lower part of the bearing location is machined directly into the cylinder head. The Removal and installation of the camshafts has already been described in connection with the cylinder head.

The camshaft sprockets are marked with timing marks, which indicate the top dead centre position of piston No. 1 when they are aligned in then position shown in Fig. 1.14. Camshafts are marked with a number, referred to in the section dealing with the timing mechanism. If a shaft is replaced, only fit a shaft with the same number. The parts lists will have the correct camshaft(s) for the engine in question.

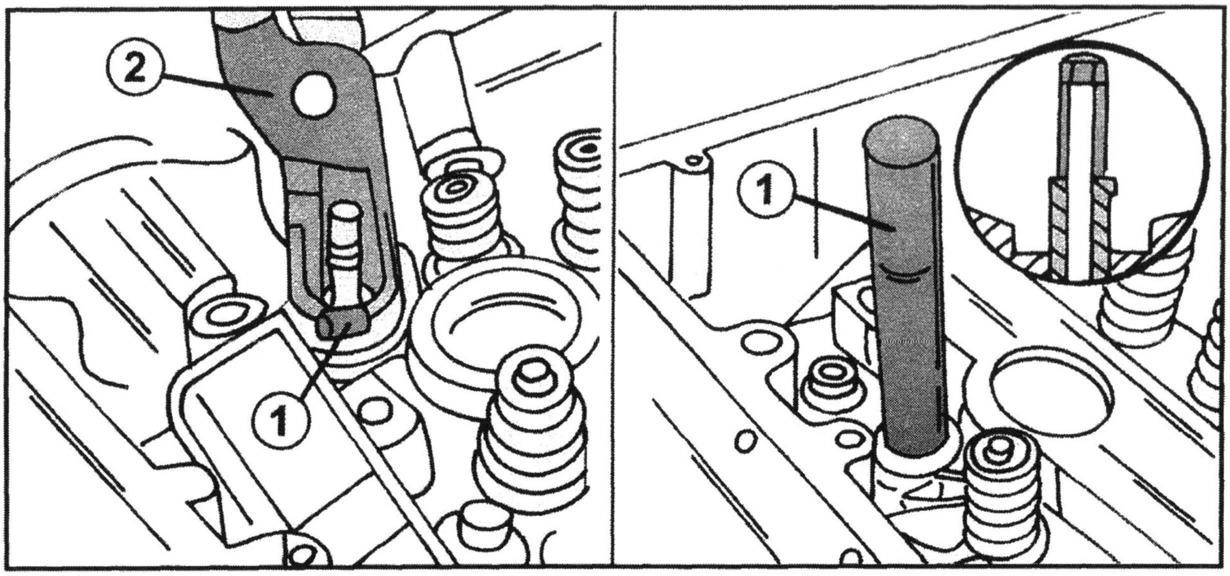

Fig. 1.20 – Removal of valve stem oil seals on the left. Use a pair of pliers (2) to remove the seal (1). The R.H. view shows the installation of a seal with the piece of tube.

1.3.0.4. Cylinder Head - Overhaul

The cylinder head must be thoroughly cleaned and remains of old gasket material removed. The checks and inspections are to be carried out as required.

Valve Springs: If the engine has a high mileage, always replace the valve springs as a set. To check a valve spring, place the old spring and a new spring end to end over a long bolt (with washer under bolt head) and fit a nut (again with a washer). Tighten the nut until the springs are under tension and measure the length of the two springs. If the old spring is shorter by more than 10%, replace the complete spring set.
The springs must not be distorted. A spring placed with its flat coil on a surface must not deviate at the top by more than 2 mm (0.08 in.).

Valve Guides: Valve guides for inlet and exhaust valves are made of cast iron and have different diameters, Guides for exhaust valves are shorter and have a larger inner diameter.
Clean the inside of the guides by pulling a petrol-soaked cloth through the guides. Valve stems can be cleaned best by means of a rotating wire brush. Measure the inside diameter of the guides. As an inside micrometer is necessary for this operation, which is not always available, you can insert the valve into its guide and withdraw it until the valve head is approx. level with the cylinder head face. Rock the valve to and fro and check for play. Although no exact values are available, it can be assumed that the play should not exceed 1.0 - 1.2 mm (0.04 -0.047 in.). Mercedes workshops use gauges to check the guides for wear.
Guides are removed with a shouldered mandrel from the combustion chamber side of the cylinder head. If guides with nominal dimension 1 can be used, drive them in position, until they rest against the cylinder head. If repair guides are fitted, the locating bores in the cylinder head must be reamed out to take the new guides. As dry ice is required to fit the new valve guides, we recommend to have the work carried out in a workshop.
Before a valve guide is replaced, check the general condition of the cylinder head. The guides must be reamed after installation, and after the cylinder head has cooled down,

if applicable, to their correct internal diameter, given in "Technical Data", noting the different diameters for inlet and exhaust valves.

Valves must always be replaced if new valve guides are fitted. The valve seats must be re-cut when a guide has been replaced. If it is obvious that seats cannot be re-ground in the present condition, new valve seat inserts must be fitted. Again this is an operation for a specialist and the work should be carried out in a workshop.

Valve Seats: If the camshaft bearings are excessively worn, fit a new or exchange cylinder head. In this case there is no need to renovate the valve seats.

Check all valve seats for signs of pitting or wear. Slight indentations can be removed with a 45° cutter. If this operation is carried out properly, there should be no need to grind-in the valves. Valve seat inserts can be fitted to the cylinder head. Replacement of valve seat inserts will require that the old seat insert is removed by machining. The machining must not damage the bottom face of the head recess. As this is a critical operation, we advise you to bring the cylinder head to your Mercedes Dealer who has the necessary equipment and experience to do the job. It may be possible to obtain a reconditioned cylinder head in exchange for the old one to avoid time delay. In this case remove all ancillary parts from the old head and refit them to the new head.

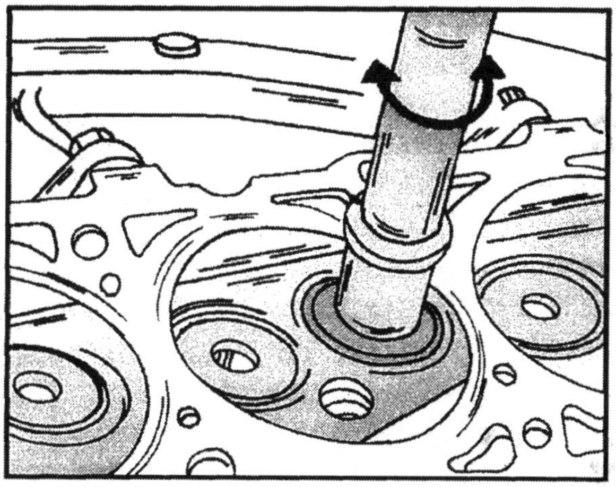

Fig. 1.21 – Grinding-in of valves, using a suitable sucker.

Valves can be ground into their seats in the conventional manner. Fig. 1.21 shows details. Attach the suction tool to the valve and move it towards the left and right. Here and then lift the valve, rotate it by a quarter of a turn and then re-grind as described and shown. If necessary apply additional grinding paste.

After completing the operation clean all affected parts from grinding paste and check the seat on valves and valve seats. An uninterrupted seat must be visible on valves and seats. Otherwise have the seats re-machined in an engine shop or in severe cases replace the cylinder head.

Fig. 1.22 – Checking a camshaft for run-out at the centre bearing journal.

Valves: The stems of inlet valves and exhaust valves have been specially treated, i.e. the ends of the valve stems must not be ground off.

Valves can be cleaned best with a rotating wire brush. Check the valve faces for wear or grooving. If the wear is only slight, valves can be re-ground to their original angle in a valve grinding machine, but make sure that there is enough material left to have an edge on the valve head. The valve head thickness must be 0.5 - 0.7 mm in the case of the inlet valves and 0.5 - 0.6 mm in the case of the exhaust valves. Again we recommend an engine shop.

Check the valve stem diameters and in this connection the inside diameters of the valve guides. If there is a deviation from the nominal values, it may be necessary to replace the valve guides (see above). Also check the end of the valve stems. There should be no visible wear in this area.

Always quote the model year and the engine number when ordering new valves, as different valves are used. These are marked by means of a number in the end of the valve stem.

Sometimes it is only required to replace the exhaust valves, if these for example are burnt out at their valve head edges.

Cylinder Head: Thoroughly clean the cylinder head and cylinder block surfaces of old gasket material and check the faces for distortion. To do this, place a steel ruler with a sharp edge over the cylinder head face and measure the gap between ruler and face with feeler gauges. Checks must be carried out in longitudinal and diagonal direction and across the face. If a feeler gauge of more than 0.10 mm (0.004 in.) can be inserted, when the ruler is placed along or across the cylinder head, have the cylinder head face re-ground. The dimension between the face of the valve heads and the cylinder head face will change after regrinding, but the workshop will correct it accordingly.

Camshaft: Place the camshaft with both end journals into "V" blocks or clamp the shaft between the centres of a lathe and apply a dial gauge to the centre journal, as shown in Fig. 1.22 on the previous page. Slowly rotate the shaft to check for run-out. If the dial gauge reading exceeds 0.01 mm, fit a new shaft. Make sure to fit the correct shaft if the shaft is to be replaced. Check the identification number when ordering a new shaft.

Replacing Valve Stem Oil Seals (Cylinder Head fitted): Valve stem oil seals are available in repair kits. Included in the repair kits are protective sleeves which must be pushed over the valves during installation of the seals.

Normally a special tool is used to fit the seals, as shown in Fig. 1.20 on the R.H. side. but a well fitting piece of tube of suitable diameter can be used. Take care not to damage the sealing lip and the spring.

The valve cotter halves and the valve springs must be removed to replace the seals. To prevent the valves from dropping into the combustion chambers, the pistons must be at the top dead centre position. Start by setting the piston of the No. 1 cylinder to the top dead centre position. Valve stems seals of cylinders Nos. 1 and 4 can be replaced in this position.

- Remove the camshafts (see description earlier on).
- Remove the valve cotter halves of the first cylinder as described in Section 1.3.0.3. Remove the valve springs.
- Use a pair of pliers (Fig. 1.20, L.H. view) or a screwdriver to remove the valve stem seals, without damaging the valve stems.
- Coat the new seals with engine oil and carefully push them over the valve stems. The protective sleeve must be used for the inlet valves. Push the seals over the valve guides until properly in position.
- Fit the valve springs with the paint spot towards the bottom, fit the upper spring cup and compress the springs until the valve cotter halves can be inserted. Make absolutely sure that the cotter halves have properly engaged before the crankshaft is rotated.
- Replace the valve stem seals of cylinder No. 4 in the same manner.
- Lift the camshaft sprocket slightly to prevent disengagement of the timing chain and rotate the crankshaft by half a turn. The valve stem seals of cylinders Nos. 2 and 3 can be replaced in the same manner.

- Refit the camshaft and associated parts.

Note: Operate the valve spring compressor very slowly, as valve cotter halves sometimes stick to the valve stems. Observe the valve spring during compression. Only the spring should move, not the valve. Prevent by all means the valve pressing against the piston.

1.3.0.5. Cylinder Head - Assembly

The assembly of the cylinder head is a reversal of the dismantling procedure. Note the following points:

- Lubricate the valve stems with engine oil and insert the valves into the correct valve guides.

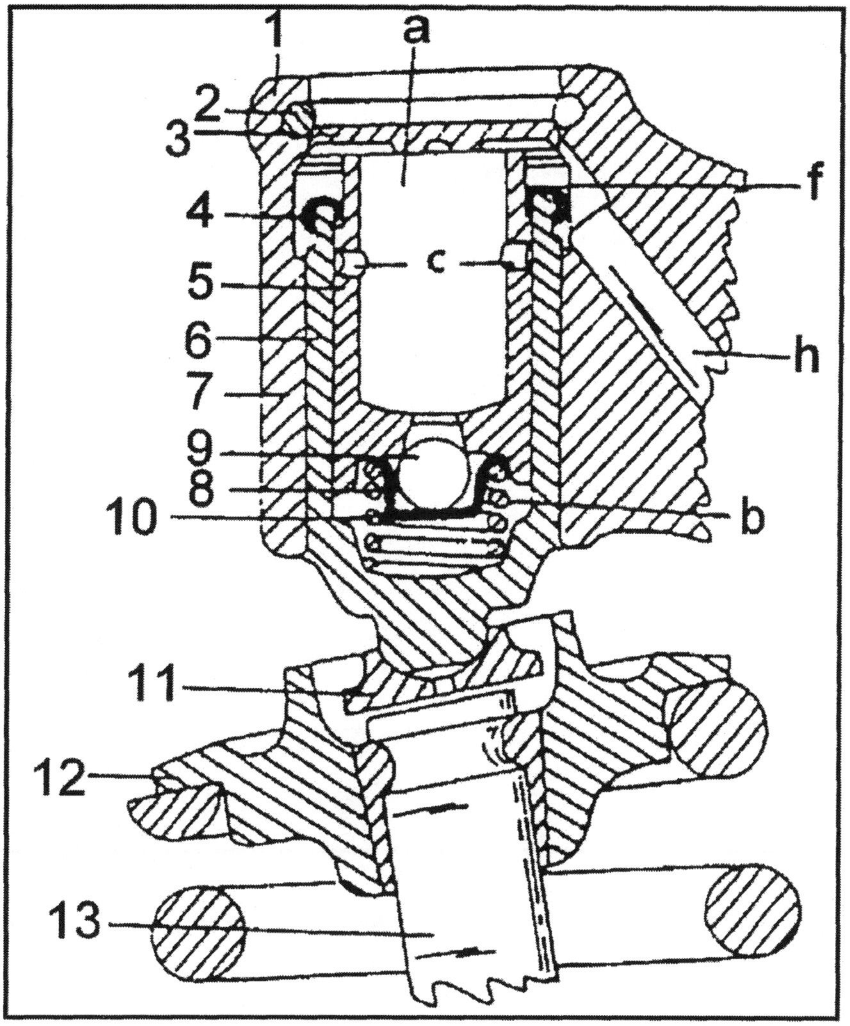

Fig. 1.23 - Sectional view of a valve clearance compensating element.

1 Rocker lever	7 Ball guide	13 Valve
2 Lock ring	8 Thrust spring	a Oil chamber
3 Washer	9 Ball, 4 mm	b Work chamber
4 Closing cap	10 Thrust spring	c Return bores
5 Thrust pin	11 Ball socket	f Ring groove
6 Guide sleeve	12 Valve spring retainer	h Oil bores

- Valve stem seals are different for inlet and exhaust valves. Make sure to order the correct seals. The repair kit contains fitting sleeves and these must be used to fit the seals (see last Section).
- The sleeves are fitted over the valve stem before the seal is pushed in position.

- Fit the valve spring and valve spring collar over the valve and use the valve lifter to compress the spring. Insert the valve cotter halves and release the valve spring lifter. Make sure that the cotter halves are in position by tapping the end of the valve stem with a plastic mallet. Place a rag over the valve end - just in case.
- Fit the camshafts as described and carry out all other operations in reverse order to the dismantling procedure.

1.3.0.6. Hydraulic Valve Clearance Compensation

The function of the hydraulic valve clearance compensating elements is to eliminate valve clearance, i.e. the dimensional changes in the valve train (valve lash) due to heat expansion and wear are compensated by the elements. The rocker arm is in constant contact with the cam.

The compensating elements cannot be repaired, but can be checked for correct functioning as described below. Figs. 1.23 and 1.24 show sectional views of a valve with clearance compensation. We will give a short description of the operation. All references refer to Fig. 1.23.

The hydraulic valve compensating element are fitted into the rocker levers and operate the valves directly via a ball socket (11). Each element consists of the following components:

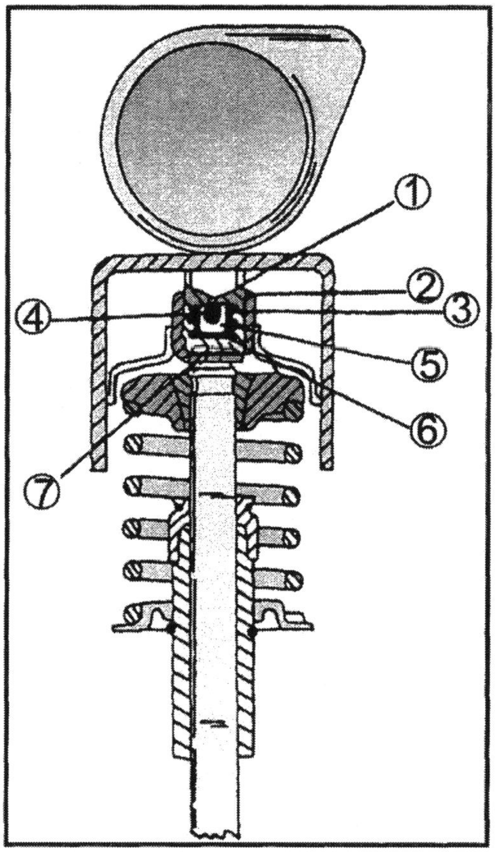

Fig. 1.24 – Sectional view of a valve and a compensating element.

1 Thrust pin
2 Lock Ring
3 Thrust spring
4 Ball guide
5 Ball
6 Thrust spring
7 Guide bush

- The thrust pin (5) with oil supply chamber and the return bores and the ball valve (check valve), i.e. items (9), (7) and (10). The ball valve separates the supply chamber from the work chamber.
- The guide sleeve (6) with the work chamber (b), the thrust spring (8) and the closing cap (4).

When the engine is stopped and the tappet is held under load from the cam, the element can completely retract. The oil displaced from the work chamber (b) flows through an annular gap, i.e. the clearance between the guide sleeve and the thrust pin to the oil supply chamber (a).

When the cam lobe has moved past the valve tappet, the thrust pin (5) will be without load. The thrust spring (8) forces the thrust pin upwards until the valve tappet rests against the cam.

The vacuum resulting from the upward movement of the thrust pin in the work chamber (b) opens the ball valve and the oil can flow from the supply chamber (h) into the work chamber (b). The ball valve closes when the valve tappet presses against the cam and puts the thrust pin under load. The oil in the work chamber acts as a "hydraulic rigid connection" and opens the valve in question.

When the engine is running and depending on the engine speed and the cam position, the thrust pin is only pushed down slightly.

The oil contained in the oil supply chamber (a) is sufficient to fill the work chamber (b) under all operating conditions of the engine. Oil or leak oil which is not required, as well as air are able to escape via the annular gap between the washer (3) and the rocker lever. The oil ejected from the work chamber flows via the annular gap between the guide sleeve and the thrust pin and the two return bores (c) into the oil supply chamber.

If the tappets are removed, note the following points:

- Always keep the tappets in an upright position, i.e. the open side towards the top.
- After removal of a tappet (see below), mark the cylinder number and the compensating element in suitable manner. Always fit original parts in their same locations.

Checking a hydraulic compensating element: As the elements are in continuous contact with the camshaft, you will rarely hear noises from the area of the hydraulic elements. If noises can be heard, check the elements as follows:

- Start the engine and run it approx. 5 minutes at 3000 rpm.
- Remove the cylinder head cover.
- Rotate the crankshaft until the cam for the tappet to be checked is pointing vertically towards the top.

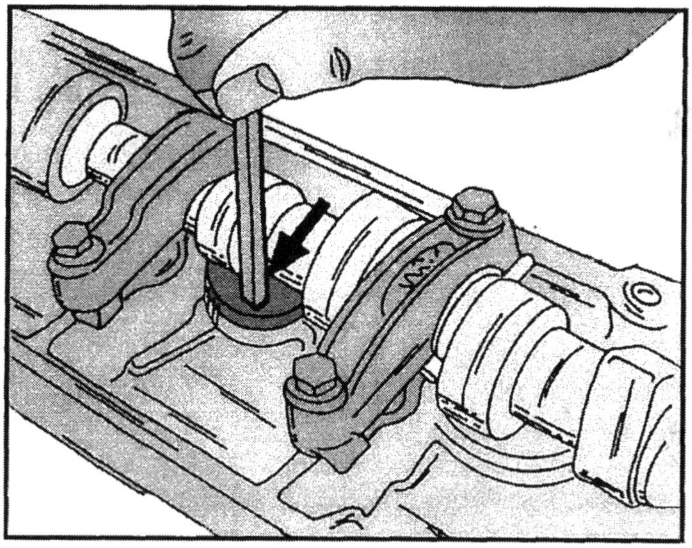

Fig. 1.25 – Checking of a hydraulic compensating element.

- Use a drift, as shown in Fig. 1.25 and push the tappet towards the inside, or try to move the tappet with the fingers.
- If the tappet cannot be depressed or excessive clearance can be felt between the tappet and the back of the cam, replace the tappet. The tappet is supplied together with the hydraulic compensating element. Mercedes workshops can reset the tappet to its original position, but this operation is beyond the scope of the home mechanic.

Tappet Removal and Installation:

- Remove the camshaft as described later on.
- Use a suction tool to remove the tappets. Mark them, if they are to be refitted.
- Fit the tappets into their original bores, if re-used. Refit the camshafts as described and carry out all other operations in reverse order to the removal procedure.

1.3.1. PISTON AND CONNECTING RODS

1.3.1.0. Technical Data

All dimensions are given in metric units.

Pistons

Piston Diameter - Std.:

Class A: . . . . 87.970 - 87.976 mm

Class X: .... More than 87.975 - 87.983 mm
Class B: .... More than 87.982 - 87.988 mm

Piston Running Clearance:
- New:.. 0.017 – 0.043 mm
- Wear limit:.. 0.12 mm

Max. weight difference within engine:
- New..4 grams
- Wear limit:.. 10 grams

Piston Pins:
Pin diameter:..26.995 – 27.000 mm
Piston pin running clearance:
- In small end bush:..0.018 – 0.029 mm
- In piston:..0.004 – 0.015 mm

Piston Rings:
Piston Ring Gaps:
- Upper rings:..0.22 – 0.42 mm
  - Wear limit:..1.5 mm
- Centre rings:..0.20 – 0.40 mm
  - Wear limit:..1.0 mm
- Lower rings:..0.20 - 0.40 mm
  - Wear limit:..1.0 mm

Side Clearance in Grooves:
- Upper rings:..0.12 - 0.16 mm
  - Wear limit:.. 0.20 mm
- Centre rings:..0.050 - 0.090 mm
  - Wear limit:.. 0.15 mm
- Lower rings:..0.030 - 0.07 mm
  - Wear limit:.. 0.10 mm

Connecting Rods
Distance from centre small end bore to
centre big end bore:..148.970 – 149.030 mm
Width of con rod at big end bore:.. 24.000 mm
Basic bore diameter of big end bore:.. 47.95 mm
Basic bore diameter of small end bore:.. 29.500 mm

Small End Bush:
- Outer diameter: ....29.50 mm
- Inner diameter: ....27.000 mm
Max. twist of connecting rods: ....0.10 mm per 100 mm
Max. bend of connecting rods: ....0.045 mm per 100 mm
Max. weight difference in same engine:..5 gram (per set)
Connecting Rod Bolts:
- Thread: .... M8 x 1
- Length of shank:.. 47.0 mm
- Max. permissible length of shank: .... 48.0 mm

Connecting rod bearing details: .... See under "Crankshaft"

1.3.1.1. Piston and Connecting Rods – Removal

The pistons are made of light-alloy.

Three piston rings are fitted to each piston. The two upper rings are the compression rings, i.e. they prevent the pressure above the piston crown to return to the crankcase. The lower ring is the oil scraper ring. Its function is to remove excessive oil from the cylinder bore, thereby preventing the entry of oil into the combustion chamber. The three rings are not the same in shape. The upper ring has a rectangular section, the centre ring has a chamfer on the inside and the lower ring is chrome-plated on its outside and also has a coil spring. Only the correct fitting of the piston rings will assure a proper piston seal.

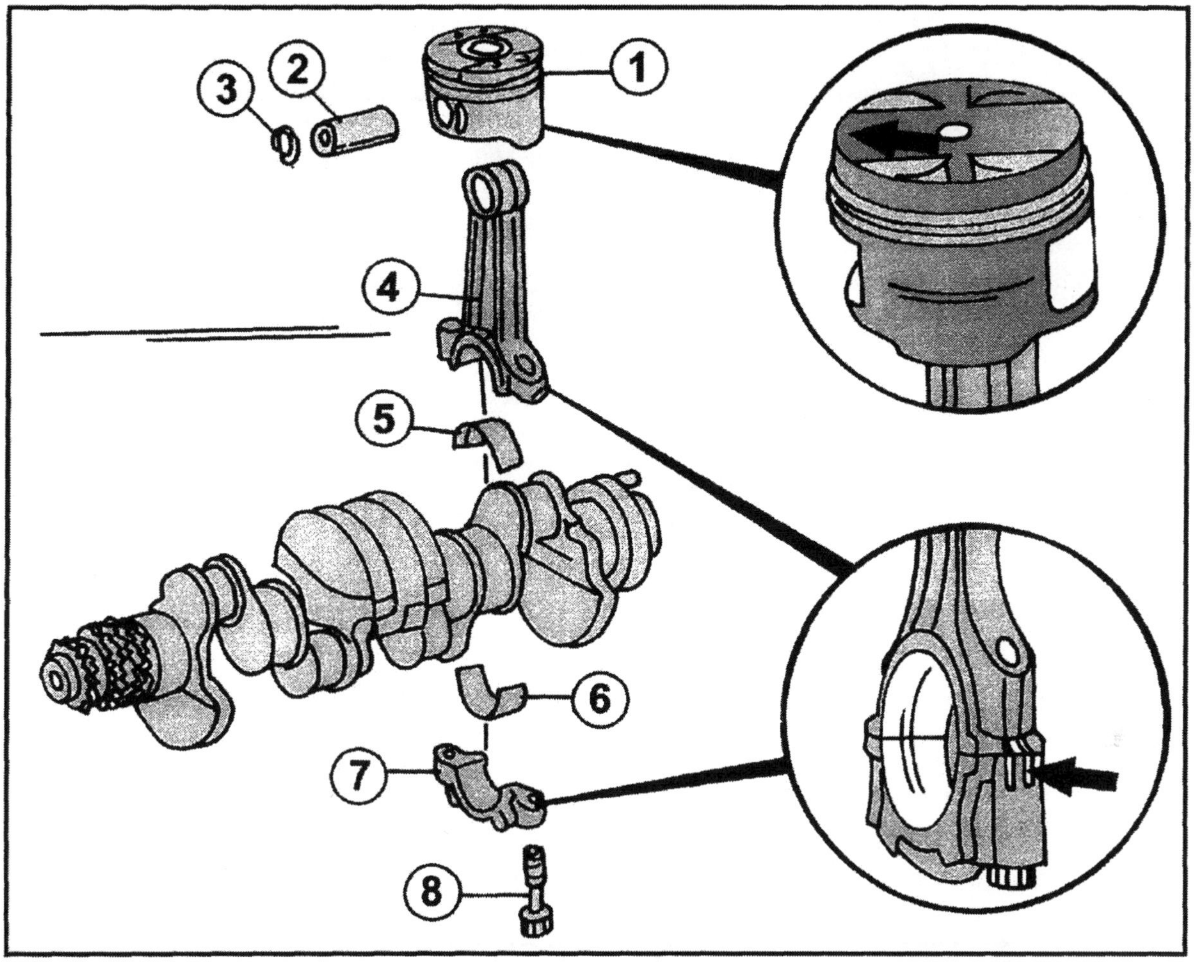

Fig. 1.26 – Details for the removal and installation of the pistons.

1 Piston
2 Piston pin
3 Circlip
4 Connecting rod
5 Upper bearing shell
6 Lower bearing shell
7 Big end bearing cap
8 Bearing cap bolt

The connecting rods connect the pistons with the crankshaft, a piston pin connects the piston with the connecting rod. Fig. 1.26 shows details of the various parts.

Pistons and connecting rods are pushed out towards the top of the cylinder bores, using a hammer handle after connecting rod bearing caps and shells have been removed. Before removal of the assemblies note the following points:

- The engine must be removed, oil sump, oil pump and cylinder head must be removed.
- Pistons and cylinder bores are graded in three diameter classes within specified tolerance. The class number is stamped into the upper face of the cylinder block, next to the particular cylinder bore, as shown in Fig. 1.27.

Fig. 1.27 – Identification (marking) of piston crown and cylinder block face with the group code letters.

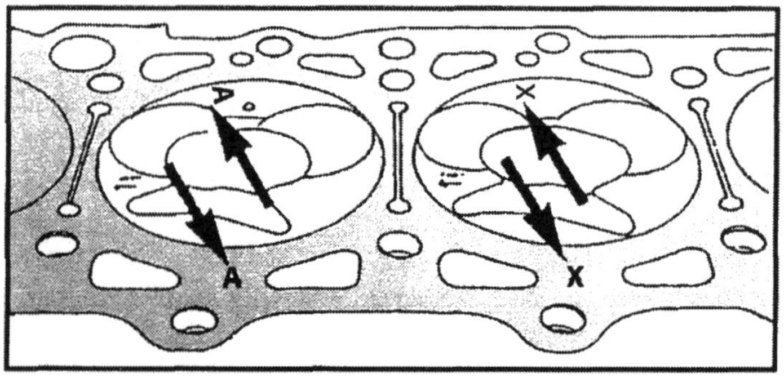

- Each piston is marked with an arrow to indicate the fitting direction, as shown in the upper circle in Fig. 1.27. The arrow points in the direction of drive.
- Mark each piston and the connecting rod before removal with the cylinder number. This can be carried out by writing the cylinder number with paint onto the piston crown. Also mark an arrow, facing towards the front of the engine (the arrow in the piston crown will be covered by the carbon deposits). When removing the connecting rod, note the correct installation of the big end bearing cap. immediately after removal mark the connecting rod and the big end bearing cap on the same side (exhaust side). This is best done with a centre punch (cylinder No. 1 one punch mark, etc., see Fig. 1.26, lower circle).
- Mark the big end bearing shells with the cylinder number. The upper shells have an oil drilling (to lubricate the piston pin).

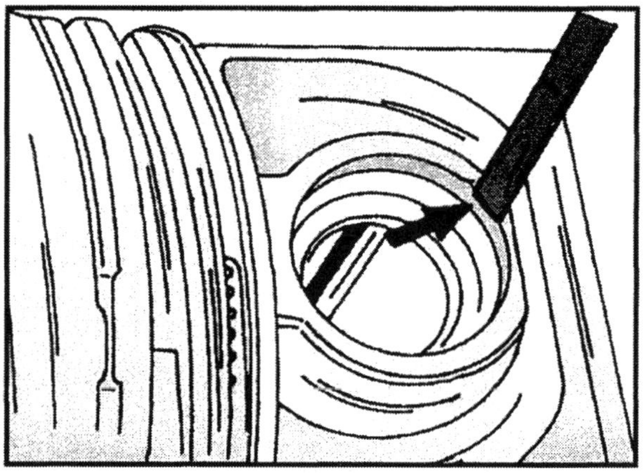

Fig. 1.28 – Removal of the securing clip for the piston pins. Apply the screwdriver blade at the ring gap.

- Big end bearing journals can be reground to four undersizes (in steps of 0.25 mm between sizes). Corresponding bearing shells are available.
- Remove the bearing caps and the shells and push the assemblies out of the cylinder bore. Any carbon deposits on the upper edge of the bores can be carefully removed with a scraper.

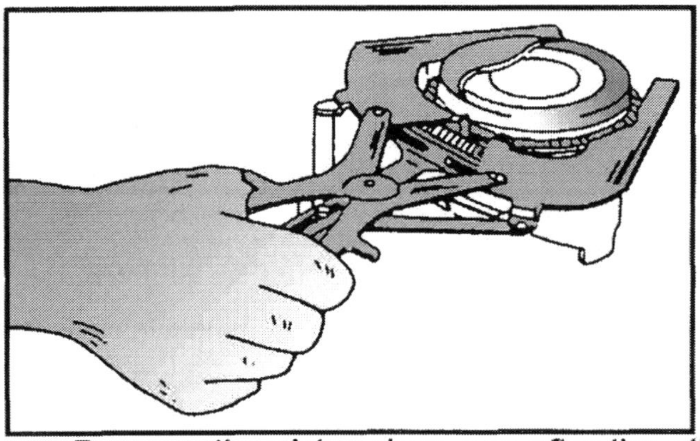

Fig. 1.29 – Removal or installation of piston rings with a pair of piston ring pliers. Never expand rings more than necessary to prevent breakage.

- Remove the piston pin snap rings. A notch in the piston pin bore enables a pointed drift to be inserted, as shown in Fig. 1.28, to remove the rings. Press the piston pins out of the pistons. If necessary heat the piston in boiling water.
- Remove the piston rings one after the other from the pistons, using a piston ring pliers if possible (Fig. 1.29). if the rings are to be re-used, mark them in accordance with their pistons and position.

1.3.1.2. Measuring the Cylinder Bores

An inside caliper is necessary to measure the diameter of the cylinder bores. The following operations are not possible if none is available or cannot be hired.

Cylinder bores must be measured in longitudinal and transverse direction and at three positions down the bore, i.e. 10 mm (0.4 in.) from the upper bore edge, 10 mm from the lower bore edge and once in the centre, totalling 6 measurements. The worst measurement must be taken when deciding on the size for the pistons to be fitted (Fig. 1.30). Fig. 1.31 shows how an internal micrometer is used.

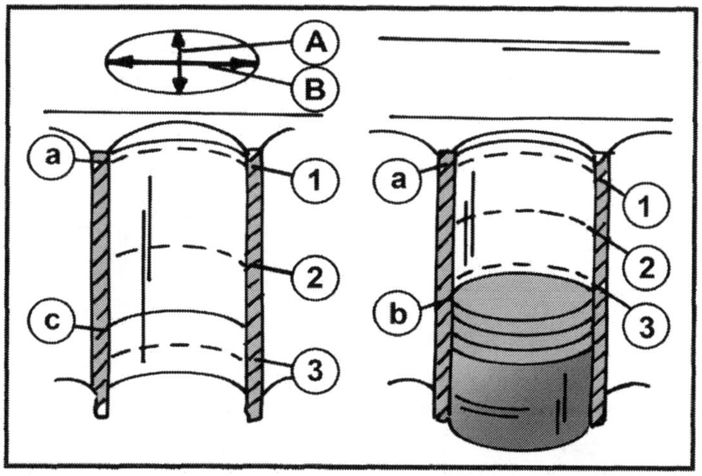

Fig. 1.30 – Measuring diagram for the cylinder bores. Numbers 1, 2 and 3 indicate the three levels where measurements should take place.

A = Measurement parallel to bore
B = Measurement across bore
a = Upper return point of upper piston ring
b = Bottom dead centre of piston
c = Lower return point of oil control ring
1-3 = Measuring points

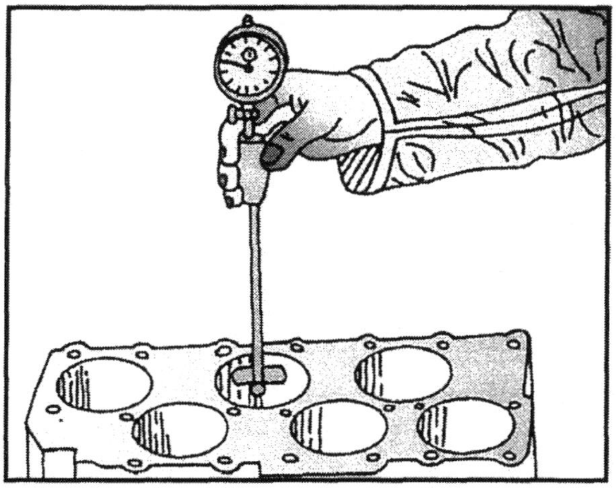

Fig. 1.31 – Measuring the diameter of a cylinder bore. The illustration shows a different engine but the principle is the same.

Note that all cylinder bores must be re-bored, even if only one of the bores is outside the diameter limit. A tolerance of 0.20 mm is permissible. If the wear is outside the limits, it is possible to have new cylinder liners fitted to the block. Your dealer will advise you what can be done.

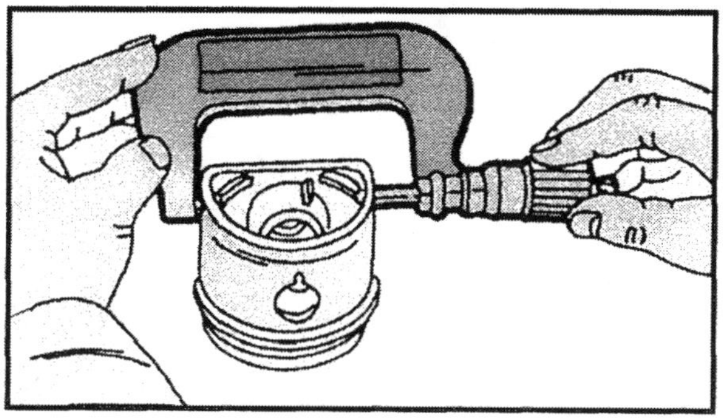

Fig. 1.32 – Measuring the diameter of a piston. Apply the micrometer at right angle to the piston pin approx. 4 mm away from the outer edge of the piston skirt. If in doubt have the pistons measured in an engine workshop.

The final cylinder bore diameter is determined after measuring the piston diameter. To measure the diameter, apply an outside micrometer 10 mm (0.4 in.) from the bottom edge and at right angle to the piston pin bore, as shown in Fig. 1.31. Add the piston running clearance to this dimension and 0.03 mm for the honing of the cylinders. The piston running clearance must not exceed 0.12 mm.

To measure the running clearance, determine the piston and cylinder bore diameters as described above and calculate the difference between the dimensions.

If the difference is more than 0.12 mm, have the cylinder bores re-bored to fit oversize pistons.

1.3.1.3. Checking Pistons and Connecting Rods

All parts should be thoroughly inspected. Signs of seizure, grooves or excessive wear requires the part to be replaced. Check the pistons and connecting rods as follows:

- Check the side clearance of each piston ring in its groove by inserting the ring together with a feeler gauge, as shown in Fig. 1.33. The grooves must be thoroughly cleaned before the check. If the wear limit exceeds the values in the technical data, either the rings or the pistons are worn.

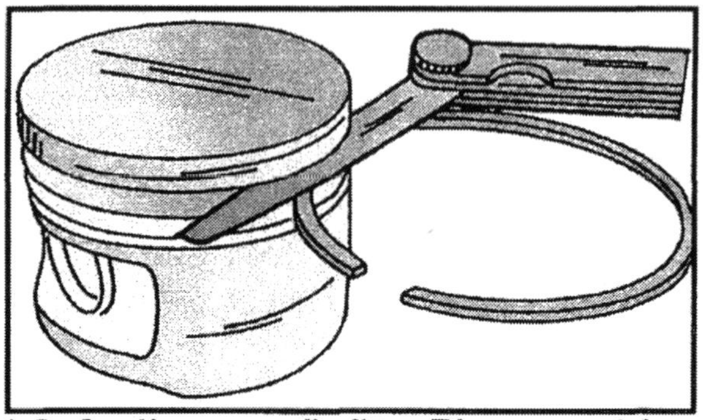

Fig. 1.33 – Checking the side clearance of piston rings.

- Check the piston ring gap by inserting the ring from the bottom into the cylinder bore. Use a piston and carefully push the piston ring approx. 1 in. further into the bore. This will square it up. Insert a feeler gauge between the two piston ring ends to check the ring gap, as shown in Fig. 1.34. Refer to Section 1.3.1.0. for the wear limits. Rings must be replaced, if these are exceeded. New rings should also be checked in the manner described.

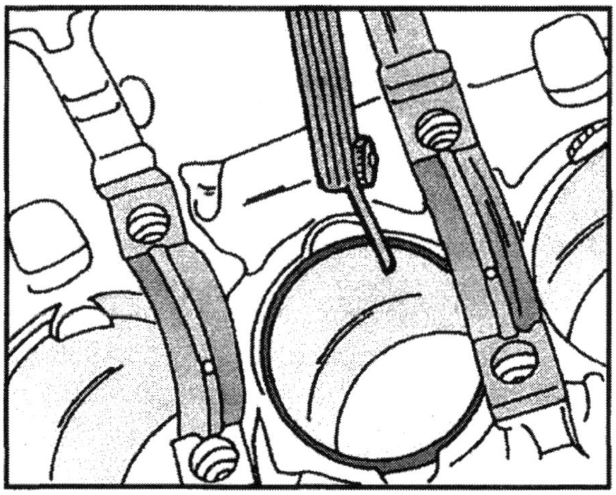

Fig. 1.34 – Checking piston ring gaps.

- Piston pins and small end bushes must be checked for wear or seizure. One individual connecting rod can be replaced, provided that a rod of the same weight group is fitted. Connecting rods are marked with either one or two punch marks and only a rod with the same mark must be fitted. The small end bushes can be replaced but we recommend to take them to an engine shop to have them replaced.
- Before re-using the connecting rod bolts check their length as shown in Fig. 1.35 . Replace any bolt that is longer the 48 mm.

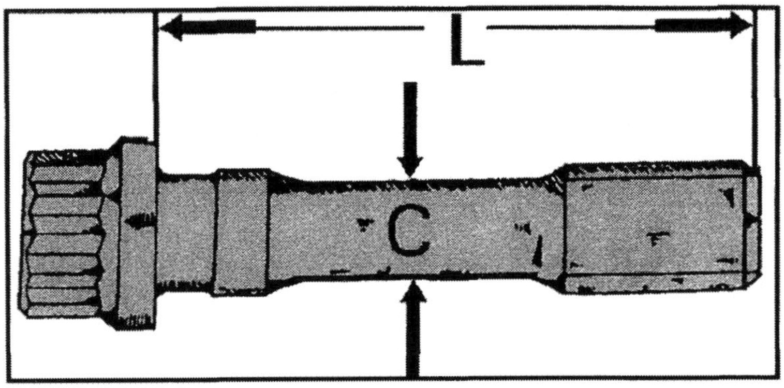

Fig. 1.35 – Connecting rod bolts can stretch in their length "L". By measuring the stretch neck "c", you will know if the bolts can be re-used (7.1 mm).

- Connecting rods should be checked for bend or twist, particularly when the engine has covered a high mileage. A special jig is necessary for this operation and the job should be carried out by an engine shop

The following information concern the connecting rods:

- Connecting rods which were over-heated due to bearing failure (bluish colour) must not be refitted.
- Connecting and bearing caps are matched to each other and must be fitted accordingly.

- New connecting rods are supplied together with the small end bearing bush and can be fitted as supplied.
- If the piston pin has excessive clearance in the small end bush, have a new bush fitted to obtain the correct piston pin running clearance and again we recommend to have the job carried out by an engine shop.

1.3.1.4. Checking the Big End Bearing Clearance

These operations are described in connection with the crankshaft (Section 1.4.3.).

1.3.1.5. Piston and Connecting Rods - Assembly

Fig. 1.26 shows a piston together with the connecting rod. The assembly can be carried out by referring to the illustration.

- Before fitting the pistons check the piston crowns to check the position of the arrow.

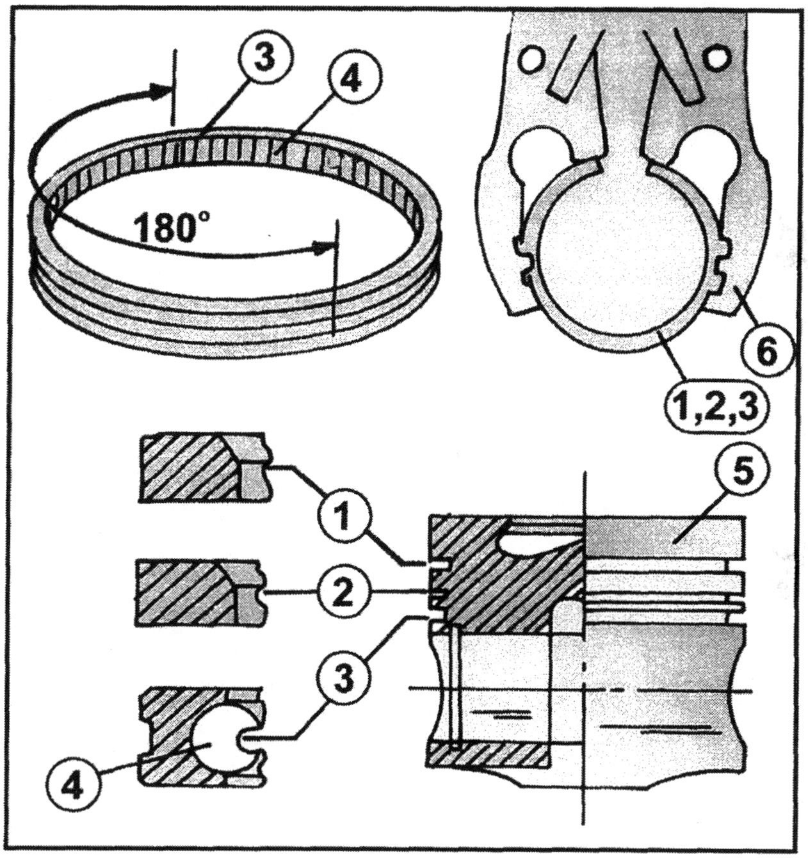

Fig. 1.36 – Sectional view of the piston. The section and arrangement of the piston ring gaps must be observed. The fitting of rings is shown on the top (R.H.). The L.H. view shows the oil control ring and the coil spring.

1 Straight compression ring
2 Tapered ring
3 Oil control ring
4 Coil spring
5 Piston

- Insert the connecting rod into the piston and align the two bores. Make sure that the arrow in the piston crown and locating lugs for the bearing shell location are facing the L.H. side of the engine, as shown in Fig. 1.26.
- Generously lubricate the piston pin with engine oil and insert it into the piston and connecting rod, using thumb pressure only. Never heat the piston to fit the piston pin. Fit the circlips to both sides of the piston, making sure of their engagement around the groove. Move the piston up and down to check for free movement. Fit the snap rings into the piston bores. Make sure they are located properly.
- Using a pair of piston ring pliers, fit the piston rings from the top of the piston, starting with the bottom ring. The two compression rings could be mixed up and Figs. 11.33 should be referred to avoid mistakes. Rings are either marked with "Top" or a paint mark which must point to the piston crowns. Arrange the piston ring gaps at 120° apart. The oil control ring must be fitted in accordance with the inset in Fig. 1.36. Under no circumstances mix-up the upper and lower compression rings.

1.3.1.6. Pistons and Connecting Rods – Installation

- Generously lubricate the cylinder bores with oil. Markings on connecting rods and bearing caps must be opposite each other. The arrows in the piston crowns must face towards the front of the engine, the marks in connecting rods and big end bearing caps must be opposite each other, both as shown in Fig. 1.26.
- Arrange the piston rings at equal spacings of 120° around the circumference of the piston skirt. The spring (4) in the oil control ring (3) must be arranged as shown in Fig. 1.36. Use a piston ring compressor to push the rings into their grooves, as shown in Fig. 1.37. Check that all rings are fully pushed in.

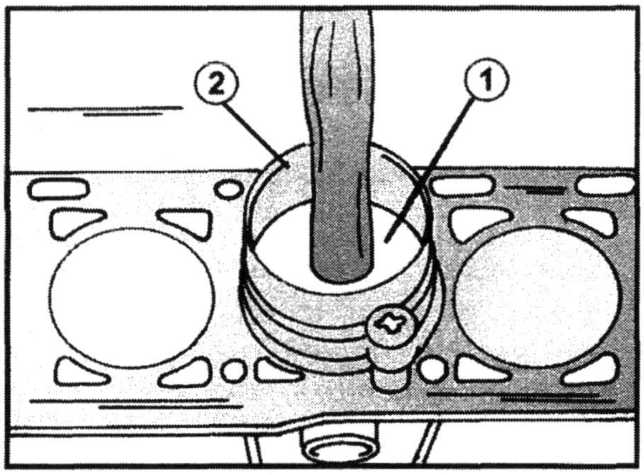

Fig. 1.37 – Fitting a piston (1) with a piston ring compressor (2).

- Rotate the crankshaft until two of the crankpins are at the bottom. Place the engine on its side to facilitate the installation. Bearing shell (5) in Fig. 1.26 should be in the con rod. Shells (5) and (6) must not be interchanged.
- Insert the second bearing shell (6) into the connecting rod bearing cap, with the locating tab on the L.H. side and fit the assembly over the connecting rod. Check that connecting rod/cap marks are facing each other.
- Coat the contact areas for the cap bolts with engine oil and fit and tighten the bolts to 2.5 kgm (18 ft.lb.) in several stages. From this position tighten each bolt by a further 90° (approx. 1/4 of a turn) without using the torque wrench. It is assumed that the stretch bolts have been measured as previously described.
- Rotate the crankshaft until the two remaining crankpins are at bottom dead centre and fit the two other or three other piston/connecting rod assemblies in the same manner.

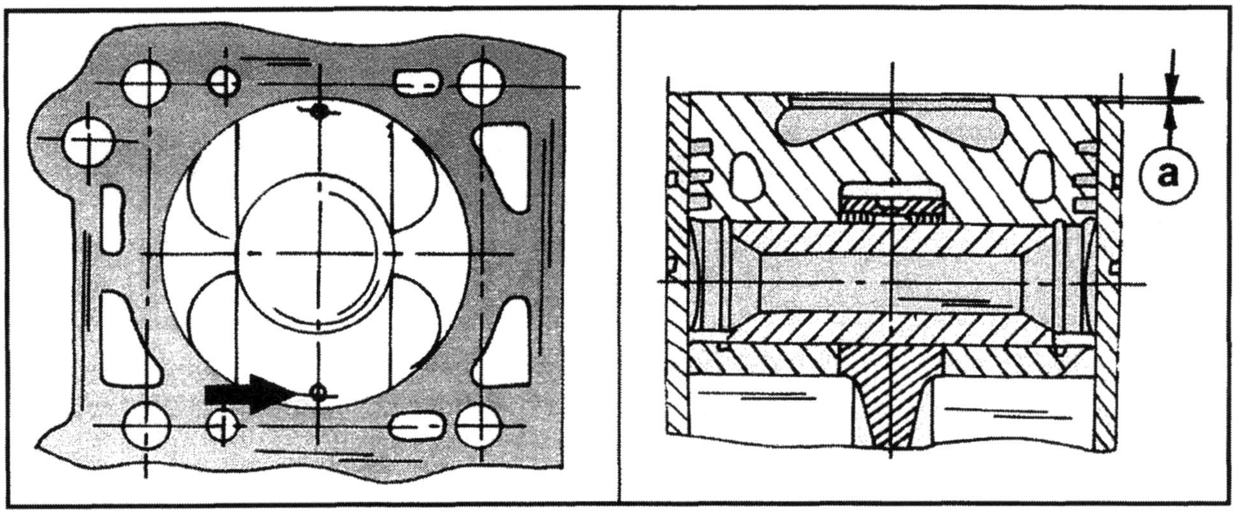

Fig. 1.38 – Checking the piston protrusion. The L.H. view shows where the measurement is carried out. The R.H. view shows with "a" where the piston protrudes above the cylinder block face.

- Check the pistons and connecting rods once more for correct installation and that each piston is fitted to its original bore, if the same parts are refitted.

- With a feeler gauge measure the side clearance of each big end bearing cap on the crankpin. The wear limit is 0.50 mm.
- If pistons have been replaced measure the protrusion of the pistons above the cylinder block face. This must be between 0.38 to 0.62 mm. To check the protrusion place the piston to the top dead centre position (must be exact) and measure the protrusion at the points shown by the arrows in the L.H. view of Fig. 1.38., i.e. in the direction of the piston pin. You measure the part of the piston above the cylinder block face as shown in the R.H. view of Fig. 1.38. The protrusion will determine the thickness of the cylinder head gasket during assembly, i.e. either a standard thickness gasket or a repair thickness gasket must be used. With the values obtained you will be able to obtain the correct gasket from your parts department.

1.4.1. TIMING CASE COVER

The timing case cover can be replaced with the engine fitted to the vehicle if for example work on the timing chain is required. We would like to point out that the operations are very comprehensive. Read the instructions in full before you start the work. The front of the engine must be accessible to gain access to the timing cover. The parts to be removed are shown in Fig. 1.39. Proceed as follows to remove and refit the cover. Some of the operations are only mentioned, but are described in the section in question. Note that on vehicles with A/C system can be treated in the same manner as only the compressor must be removed, with all lines/pipes attached.

- Disconnect the battery earth cable.
- Drain the cooling system and remove the complete radiator (section "Cooling System").
- Drain the engine oil.
- Remove the air cleaner housing, the fan and the fan shroud as described in section "Cooling System".
- Remove the vacuum pump and the high pressure pump.
- Remove the trim panel of the cylinder head cover, the injection pipes, the injectors and the cylinder head cover as described in connection with the removal of the cylinder head.
- The bottom engine compartment panel must be removed on some models (model 639.7... and 639.8 ... after VIN number 055347).
- Remove the Poly V-belt and the water pump.
- Remove the chain tensioner (described earlier on).
- Unscrew the cap closing the oil filter (section "Lubrication") and take out the filter element. This will enable the oil to return to the oil sump. The filter element and the sealing rings must be replaced during installation.
- Remove the steering pump (section "Steering"),
- Detach the compressor in the case of air conditioning from the timing cover and remove it from the cover with all refrigerant pipes/lines attached.
- Detach the generator from the timing cover and place it to one side with all cables attached to their terminals.
- Remove the starter motor.
- Remove the crankshaft pulley with the vibration damper.
- Remove the oil sump. The engine mountings removed during the operations, must be re-attached with bolts in order to lift the engine for the removal of the sump. The sump can now be removed as described.

- Remove the bolts (4) securing the cylinder head to the timing cover. Note that not all bolts have the same length. The bolts are tightened to 2.0 kgm (14.5 ft.lb.) during installation.

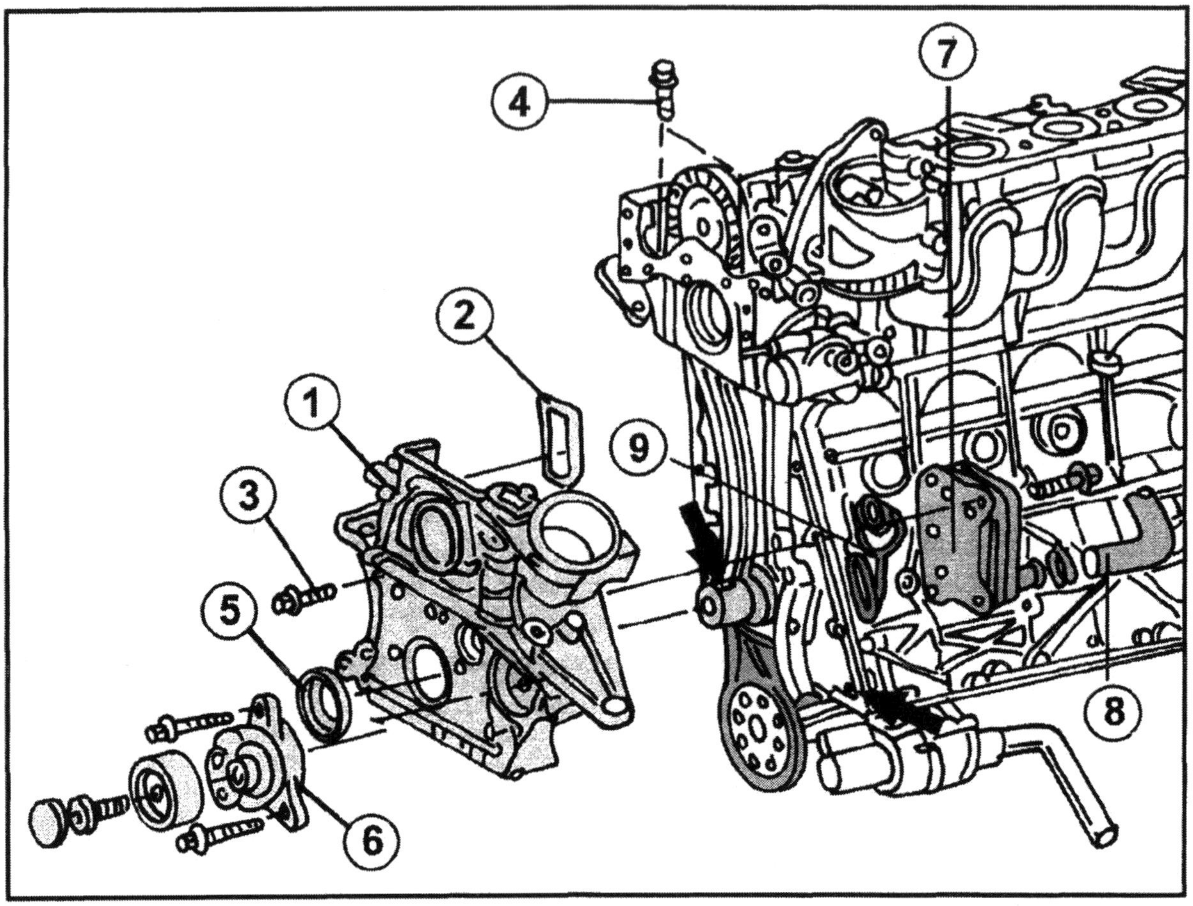

Fig. 1.39 – Removal and installation of the timing cover. The arrows show where the dowel pins are inserted.

1 Timing cover	4 M8 bolt, 2.0 kgm	7 Oil-water heat exchanger
2 Gasket	5 Crankshaft oil sea	8 Coolant hose
3 M8 bolt, 2.0 kgm	6 Poly belt tensioner	9 Gasket, for (7)

- Unscrew the bolts (3) and remove the timing cover (1). During installation clean the sealing surfaces. Replace the gasket (2) and push the timing cover over the dowel pins (arrows). The cover face must be coated with "Loctite 5970" before installation.
- With the cover removed extract the crankshaft oil seal (5). If a new cover is fitted transfer the tensioning device for the poly V belt from the old cover to the new cover. The same applies to the oil-water heat exchanger (7), but the gasket (9) must be replaced. The bolts are tightened to 1.4 kgm (10 ft.lb). The tensioning device is tightened to 3.0 kgm (22 ft.lb.) to the cover, the tensioning pulley, if removed, is tightened to 3.6 kgm (26 ft.lb.).

The installation if a reversal of the removal procedure, noting the tightening torques already given. Remember to fill the engine with oil and the cooling system.

1.4.3. CRANKSHAFT AND BEARINGS

1.4.3.0. Technical Data

All dimensions in metric units.

Machining tolerances:

Max. out-of-round of journals: 0.005 mm

Max. taper of main journals: 0.010 mm
Max. taper of crankpins: 0.015 mm
Max. run-out of main journals*:
Journals Nos. II and IV: 0.07 mm
Journal No. III: 0.10 mm

* Crankshaft placed with Nos. I and V journals in "V" blocks.

Main Bearing Journal Diameter:
Nominal: ....57.950 - 57.965 mm
1st repair size: ....57.700 - 57.715 mm
2nd repair size: ....57.450 - 57.465 mm
3rd repair size: ....57.200 - 57.215 mm
4th repair size: ....56.950 - 56.965 mm

Basic Bearing Bores:
For main bearings: ....62.500 - 62.519 mm
For big end bearings: .... 51.600 - 51.619 mm

Crankpin Diameter:
Nominal dimension: ....47.950 - 47.965 mm
1st repair size: ....47.000 - 47.715 mm
2nd repair size: .... 47.450 - 47.560 mm
3rd repair size: ....47.200 - 47.215 mm
4th repair size: .... 46.950 - 46.965 mm

Width of Crankpins:
Nominal Dimension: .... 27.960 - 28.044 mm
Repair sizes: .... Up to 28.30 mm

Bearing Running Clearances:
Main bearings: ....0.03 - 0.07 mm (best 0.055 mm)
Big end bearings: ....0.03 - 0.07 mm (best 0.050 mm)
Wear limit: ....0.080 mm

Bearing End Float:
Main bearings: ....0.10 - 0.25 mm
Big end bearings: .... 0.12 - 0.26 mm
Wear limit - Main bearings: ....0.30 mm
Wear limit - Big end bearings: ....0.50 mm

Bearing Shells:

	Main Bearings	*Big End Bearings*
Nominal Dimension:	2.25 mm	1.80 mm
1st repair size:	2.37 mm	1.92 mm
2nd repair size:	2.50 mm	2.05 mm
3rd repair size:	2.62 mm	2.17 mm
4th repair size:	2.75 mm	2.30 mm

Connecting Rod Bolts:
Thread: . . . . M8 x 1
Diameter of stretch neck: .7.4 mm
Min. diameter of stretch neck: . . . . 7.1 mm
Tightening torque – Big end bearing caps: . . . . 0.5 + 2.5 kgm (18 ft.lb) + 90°

Tightening torque – Main bearing caps:

- First stage:.. 5.0 kgm (36 ft.lb.)
- Second stage:... 90° more

1.4.3.1. Crankshaft - Removal and Installation

The engine must be removed to take out the crankshaft.

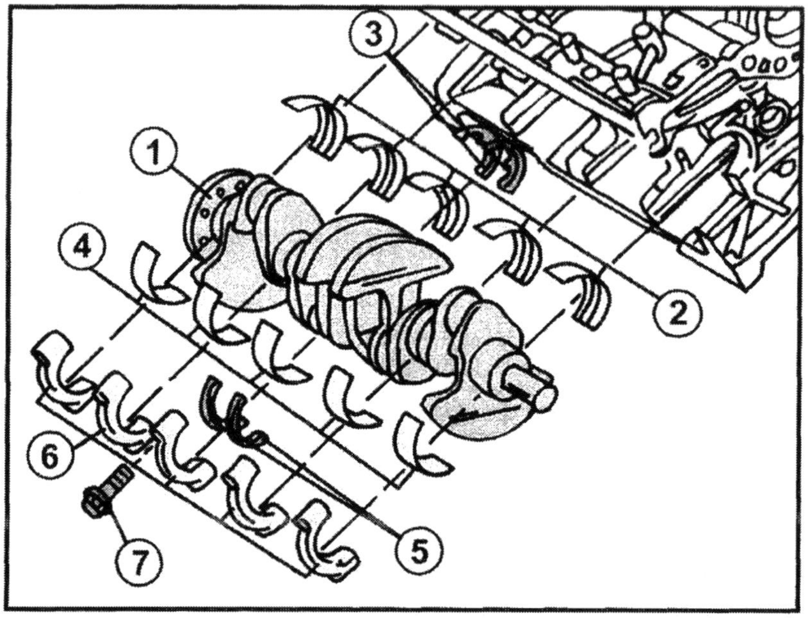

Fig. 1.40 – The crankshaft together with main bearing shells and thrust washers.

1 Crankshaft
2 Main bearing shells in crankcase
3 Thrust washers in crankcase
4 Main bearing shells in bearing caps
5 Thrust washers in caps
6 Main bearing caps
7 Main bearing cap bolts, 5.5 kgm + 90°

- Remove the transmission from the engine. Take care not to distort the clutch shaft.
- Counterhold the flywheel in a suitable manner and evenly slacken the clutch securing bolts. Use a centre punch and mark the clutch and flywheel at opposite points. Lift off the clutch plate and the driven plate. Immediately clean the inside of the flywheel and unscrew the flywheel. See also Fig. 1.41. Remove the drive plate for a torque converter of an automatic transmission in the same manner.

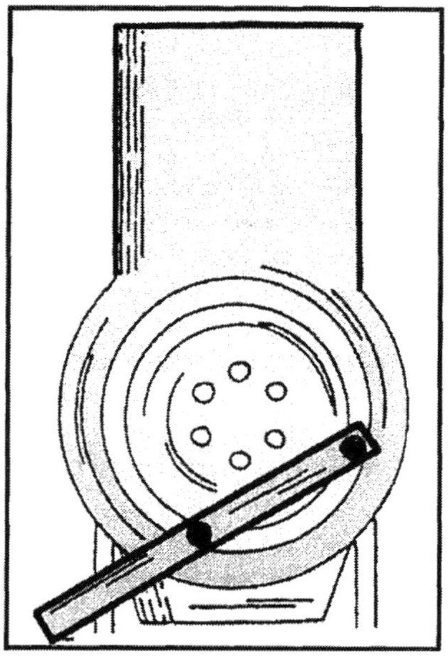

Fig. 1.41 – Flywheels and driven plates can be prevented from rotating as shown when the bolts are slackened. When tightening the bolts arrange the bar on the other side.

- With the flywheel still locked, remove the crankshaft pulley bolt and remove the crankshaft pulley/damper as described later on.
- Remove the cylinder head as described earlier on and the timing cover as described in Section 1.4.2. Remove the oil sump and oil pump.
- Remove the pistons and connecting rods as described in the section covering the pistons.
- The crankshaft end float should be checked before the crankshaft is removed. To do this, place a dial gauge with a suitable holder in front of the cylinder block and place the gauge stylus against the end flange of the crankshaft, as shown in Fig. 1.42.
- Use a screwdriver to push the crankshaft all the way to one end and set the gauge to "0". Push the shaft to the other side and note the dial gauge reading. The resulting value is the end float. If it exceeds 0.30 mm (0.012 in.) replace the thrust washers during assembly, but make sure to fit washers of the correct width. These are located left and right at the centre bearing. Note that only two washers of the same thickness must be fitted.

- Unscrew the oil seal flange from the rear of the cylinder block.

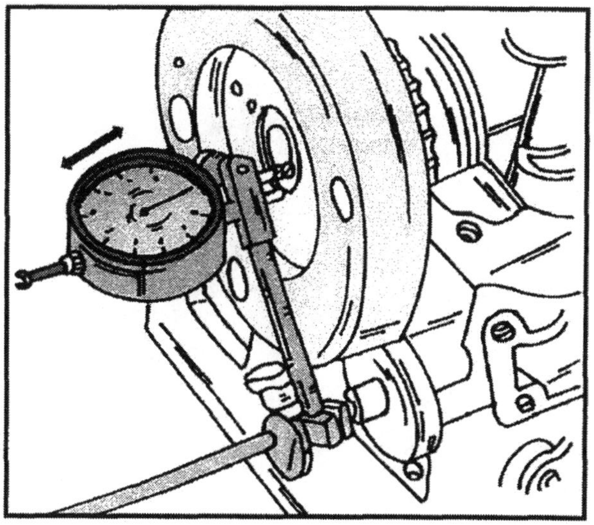

Fig. 1.42 – Checking the crankshaft end float.

- Unscrew the main bearing bolts (7) in Fig. 1.40 evenly across. The bearings caps are marked with the numbers 1 to 5. The numbers are stamped into the centre of the caps, as shown in Fig. 1.40. No. 1 cap is located at the crankshaft pulley side.
- Remove the bearing shells (4) from the bearing journals (they could also stick to the caps) and immediately mark them on their back faces with the bearing number. Also remove the thrust washers (5).

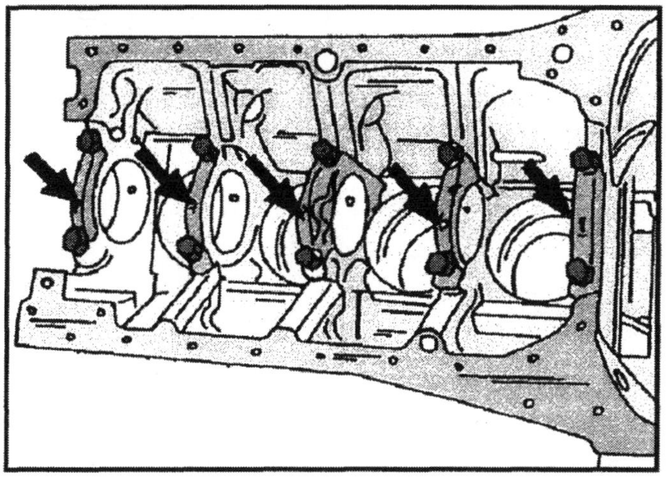

Fig. 1.43 – The arrows show the numbering of the main bearing caps. Caps must be fitted in their original order.

- Lift the crankshaft (1) out of the cylinder block and remove the remaining thrust washers from the centre bearing location and the remaining bearings shells. Keep the shells together with the lower shells and the bearing caps. These shells have an oil bore and a groove and must always be fitted into the crankcase when the crankshaft is installed.

1.4.3.1. Inspection of Crankshaft and Bearings

Main and crankpin journals must be measured with precision instruments to find their diameters. All journals can be re-ground four times and the necessary bearing shells are available, i.e. undersize shells can be fitted.

Place the crankshaft with the two end journals into "V" blocks and apply a dial gauge to the centre main journal. Rotate the crankshaft by one turn and read off the dial gauge. If the reading exceeds 0.06 mm, replace the crankshaft.

Check the main bearing and big end bearing running clearance as follows:

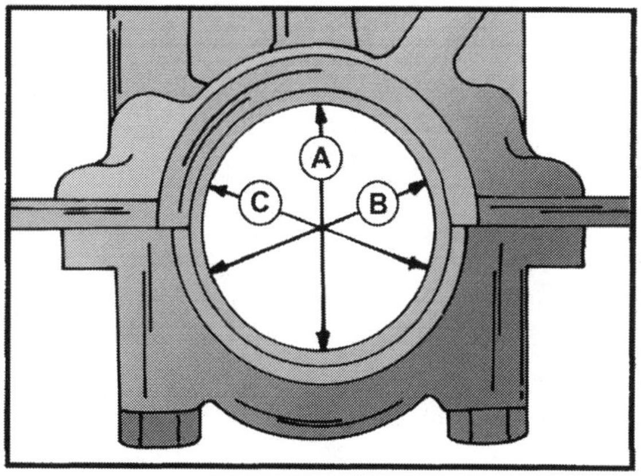

Fig. 1.44 – Measuring the inside diameter of the fitted bearing shells.

- Bolt the main bearing caps without shells to the crankcase, oil the bolt threads and fit each cap. Tighten the bolts to 5.5 kgm (39.6 ft.lb.) and then angle-tighten them a further 90° - 100°. Bearing caps are offset and can only be fitted in one position.
- Referring to Fig. 1.44 and measure the bearing bores in directions A, B and C and write down the results. If the basic

diameter is exceeded (see Section 1.4.3.0.), the bearing cap and/or the cylinder block must be replaced.

- Remove the bearing caps and refit them, this time with the well cleaned bearing shells. Re-tighten the bolts as specified.

Fig. 1.45 – Measuring the inside diameter of the fitted main bearing caps together with the shells.

- Measure the diameter of each bearing in accordance as shown in Fig. 1.45 and write down the results. Deduct the journal diameter from the bearing diameter. The resulting difference is the bearing running clearance, which should be between 0.031 - 0.073 mm, with a wear limit of 0.080 mm.
- Check the big end bearing clearances in a similar manner, but bolt the bearing caps to the connecting rods. Tighten the nuts to the value given in the technical data and angle-tighten as above. The bearing clearance should be between 0.031 - 0.073 mm, with the same wear limit.

Selection of bearing shells is rather complicated, and we advise you to take the cylinder block to an engine shop, if the above measurements have revealed that new bearing shells are necessary.

Fig. 1.46 – Correct fitting of the crankshaft thrust washers.

1.4.3.3. Crankshaft - Installation

Thoroughly clean the bearing bores in the crankcase and insert the shells (2) with the drillings into the bearing bores, with the tabs engaging the notches. Fit the thrust washers (3) to the respective bearing, with the oil grooves towards the outside, as shown in Fig. 1.46.

- Use the your forefingers as shown in Fig. 1.47 to hold the thrust washers against the bearing cap and fit the cap in position.

Fig. 1.47 – Fitting the main bearing cap together with the thrust washer.

- Lift the crankshaft in position and fit the bearing caps with the inserted shells (again shells well oiled and locating tabs in notches). Fit the two thrust washers to the centre bearing cap, again with the oil groove towards the outside. Place this cap in position, guiding the two thrust washers in order not to dislodge them. Use your forefingers to hold the washers as shown in Fig. 1.47. Quote the engine number when new washers are ordered. Fig. 1.48 shows the location of the bearing shells and thrust washers. The numbers refer to Fig. 1.40.

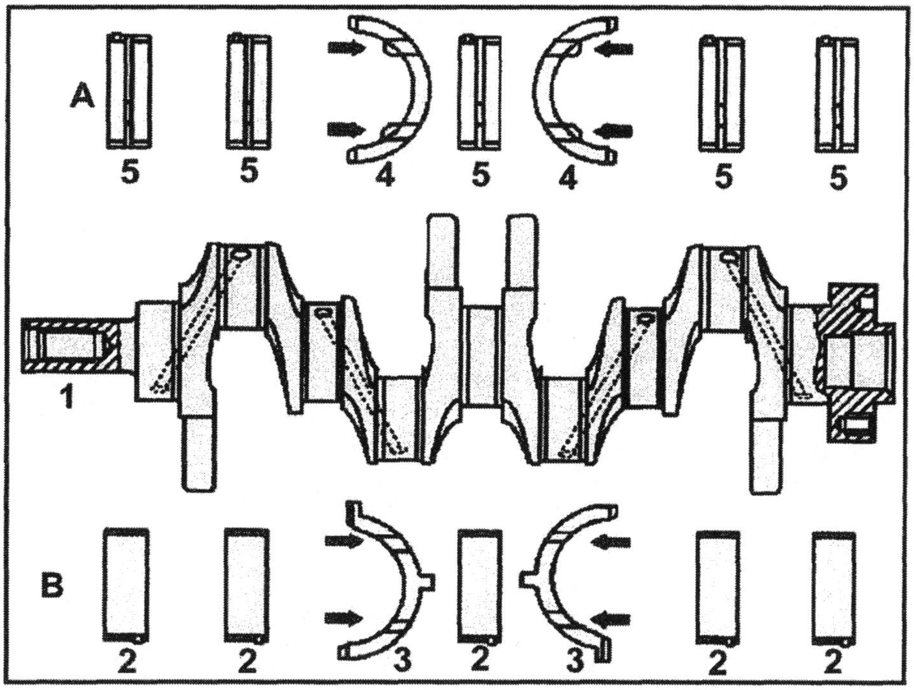

Fig. 1.48 – Main bearing shells and thrust washers as fitted to the crankcase (A) and main bearing caps (B).

- Check the numbering of the bearing caps and fit the well oiled bolts. A number is stamped into the bearing cap and indicates where cap No. 1 is fitted. Fit the remaining caps in accordance, always matching the number in the cap with the number in the crankcase.
- Measure the crankshaft cap bolts from the end of the thread to the underside of the bolt head. Bolts which are longer than 63.8 mm must not be re-used. Tighten the bolts from the centre towards the outside in several steps to a torque reading of to 5.5 kgm (39.5 ft.lb.) and from this position a further 90° (a quarter of a turn).

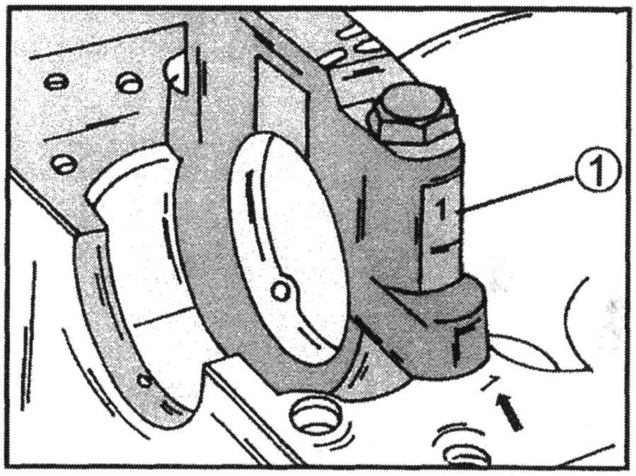

Fig. 1.49 – Crankshaft main bearing caps are marked as shown. The crankcase is marked with the same number, in this case with "1".

- Rotate the crankshaft a few times to check for binding.
- Re-check the crankshaft end float as described during removal. Attach the dial gauge to the crankcase as shown in Fig. 1.42. The remaining operations are carried out in reverse order to the removal procedure. The various sections give detailed description of the relevant operations, i.e. piston and connecting rods, rear oil seal flange, timing mechanism, flywheel and clutch or drive plate, oil pump, oil sump and cylinder head.

1.4.3.4. Flywheel or Drive Plate (Automatic)

The engine can be fitted with a conventional flywheel or a dual-mass flywheel. Always check the height of the old flywheel before fitting a new one.

Both flywheel and drive plate can be replaced with the engine fitted without re-balancing of the crankshaft. Proceed as described:

- Remove the transmission (Section 3.1.).
- Counterhold the flywheel in suitable manner and remove the clutch after having marked its relationship to the flywheel. Remove the drive plate in a similar manner. 8 bolts are used to secure the flywheel. A hole has been drilled between two of the bores and a similar hole is drilled into the crankshaft. These two bolts

must be aligned when the flywheel or the drive plate is fitted. Fig. 1.50 shows the alignment bore in the case of the flywheel. The drive plate has a similar hole.

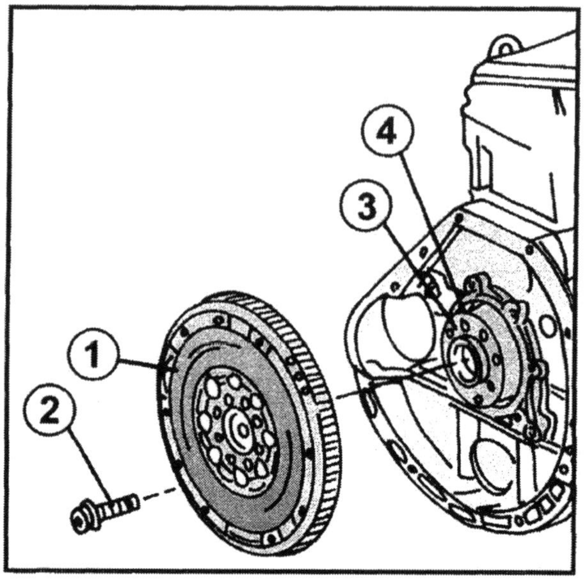

Fig. 1.50 – View of the flywheel with the location of the dowel pin.
1 Flywheel
2 Flywheel bolt, 4.5 kgm + 90°
3 Cylindrical dowel pin
4 Crankshaft

- Remove the flywheel or the drive plate. Distance washers are used in the case of the drive plate, which can also be removed. Measure the diameter of the mounting bolts at their smallest section (stretch neck). If less than 8.1 mm, replace the bolts. The measurement is carried out as shown earlier on for the connecting rod bolts.
- If the flywheel or the starter ring looks worn, take the wheel to your dealer to have the flywheel re-machined and/or the ring gear replaced.

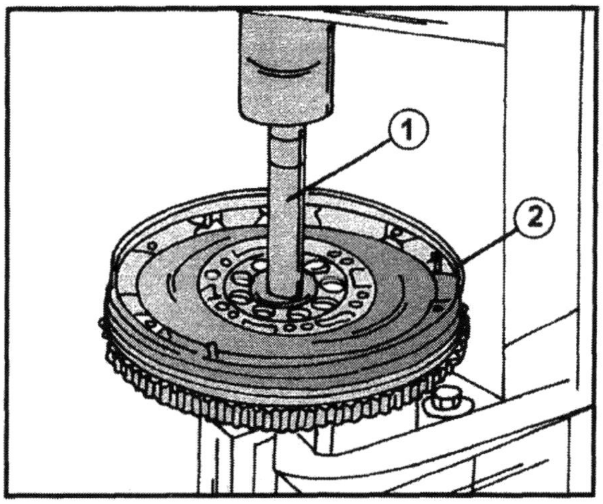

Fig. 1.51 – Pressing the guide bearing inside the flywheel (2). A press mandrel (1) must be used.

Fit the flywheel or the drive plate with the alignment bores in line. Fit a distance washer underneath and on top of the drive plate. Tighten the bolts evenly across to 4.5 kgm (32 ft.lb.) and from this position a further 90°. The angle is important to give the stretch bolts their correct tension.

Engines for manual transmissions are fitted with a ball bearing inside the flywheel. The bearing must be pressed out of the flywheel, i.e. you will have to have access to a press. Place the flywheel onto a press table as shown in Fig.1.51, insert the bearing and press the bearing in position. Grease the bearing after installation.

1.4.3.5. Crankshaft Pulley and Vibration Damper

The engine is fitted with a one piece crankshaft pulley/vibration damper assembly. The damper is located on the end of the crankshaft by means of a Woodruff key and secured by the central bolt in the end of the shaft. The tightening torque of the bolt is very high. Fig. 1.52 shows the attachment of the parts.

Remove the parts as follows, noting that a puller may be necessary to withdraw the hub:

- Remove the sound-proofing panel from underneath the engine compartment on some models (models 639.7/8 from VIN number 055347 only).
- Release the tension of the drive belt, as described later on.
- Engage a gear and apply the handbrake to lock the engine against rotation. In the case of a vehicle with automatic transmission remove the starter motor and lock the starter motor ring gear in suitable manner. The same can be carried out

when a manual transmission is fitted and the bolt cannot be removed by engaging a gear.

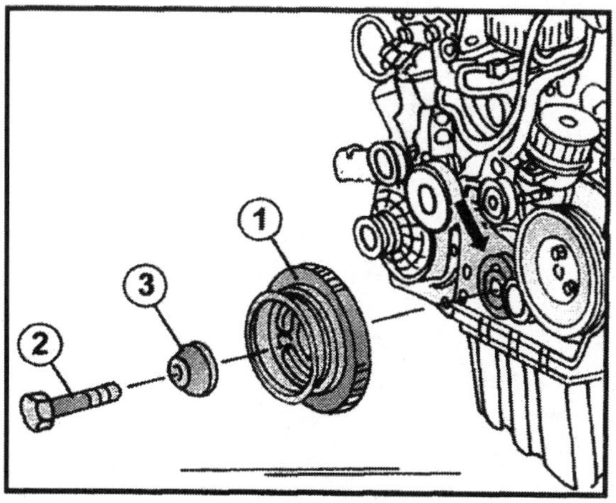

Fig. 1.52 – Details for the removal and installation of the crankshaft pulley/vibration damper assembly. Note the different tightening torque of the bolt (see below).

1 Crankshaft pulley/vibration damper
2 Central bolt
3 Washer

- Unscrew the crankshaft pulley/vibration damper bolt (2) in Fig. 1.52 and take off the washer (3).
- Withdraw the assembly from the end of the crankshaft. A tight vibration damper can be removed with a suitable puller. You can also try two tyre levers, applied at opposite points of the damper.

The crankshaft pulley has a certain diameter. If replaced, quote the engine type and number.

The installation of the crankshaft pulley and the vibration damper is carried out as follows:

- Rotate the crankshaft until the Woodruff key is visible and slide the crankshaft pulley/vibration damper with the key way over the key and the shaft. Make sure that the Woodruff key has engaged and has not been dislodged.
- Place the washer (3) over the centre bolt (2), coat the bolt threads with engine oil and fit the bolt. Tighten the bolt to 32.5 kgm (234 ft.lb.) and from the final position a further quarter of a turn. The crankshaft must still be locked against rotation.
- Fit the drive belt as described later on.
- The remaining operations are carried out in reverse order.

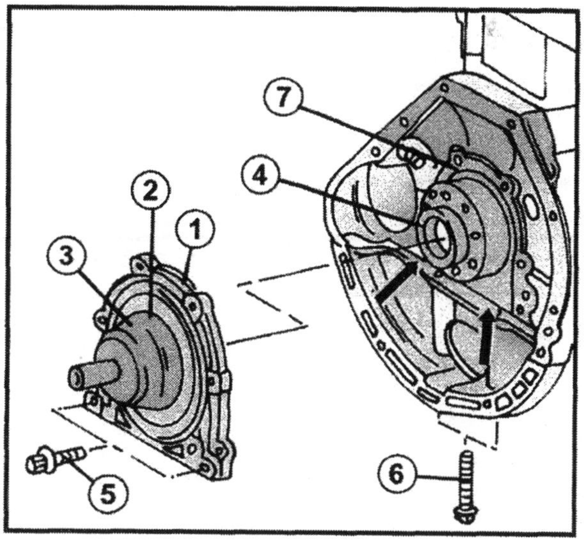

Fig. 1.53 – Removal and installation of the rear oil seal cover together with the oil seal. The faces marked with the arrows must be coated with "Loctite".

1 Cover with oil seal
2 Fitting sleeve (normally in repair kit)
3 Oil seal in cover
4 Crankshaft
5 Screws, inserted from front, 0.9 kgm
6 Screws, inserted from below, 0.9 kgm

1.4.3.6. Rear Crankshaft Oil Seal and Oil Seal Carrier

The rear crankshaft oil seal is located inside a flange which is bolted to the rear of the crankcase. Flange and oil seal are made together and cannot be replaced separately. Two dowels locate the flange correctly in relation to the crankshaft centre. The flange is fitted with sealing compound ("Loctite"). Fig. 1.53 shows the attachment of the oil seal flange (cover) and the integral oil seal.

Transmission and flywheel and/or drive plate must be removed to replace the cover/oil seal assembly. Also drain the engine oil.

- Place a jack underneath the engine (wooden block between jack head and engine) and lift the engine approx. 5 cm at the front axle in order to reach the bolts (6) in Fig. 1.53.
- Remove the bolts (5) from the front and the two bolts (6) from below.
- Insert two screwdrivers as shown in Fig. 1.54 and lever off the cover from the crankcase (7) without damaging the flange.

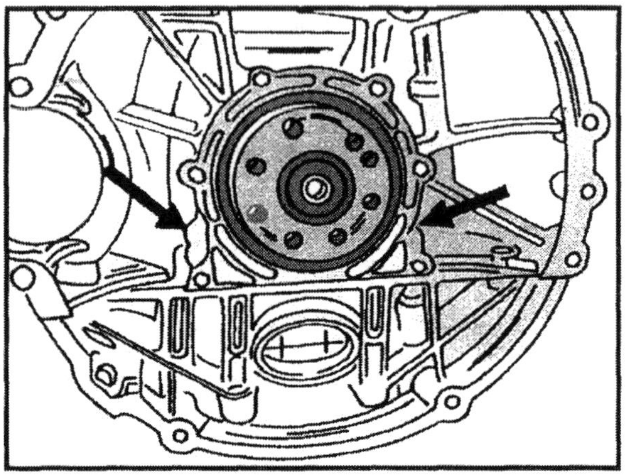

Fig. 1.54 – To remove the oil seal cover insert two screwdrivers carefully underneath two lugs shown by the arrows.

- Thoroughly clean the sealing faces and coat the cover face where it rests against the oil sump gasket. Push the cover against the crankcase. You should have a fitting sleeve (2) in Fig. 1.53 which will aid the fitting. Otherwise take care not to damage the seal. After the cover is flush against the crankcase push it upwards until the upper face is against the crankcase and remove the fitting sleeve. The oil seal (3) must have a snug fit.
- Fit the two bolts (6) from below and then the remaining bolts from the front. First tighten the lower bolts and then the remaining bolts. All are tightened to 0.9 kgm (7 ft.lb.).
- All other operations are carried out in reverse order.

1.4.3.7. Front Crankshaft Oil Seal

The front crankshaft oil seal is located in the timing cover. Oil leaks at this position can also be caused by a leaking timing cover gasket. Check before replacing the oil seal.

Remove the front crankshaft oil seal located in the timing cover. Oil leaks at this position can also be caused by a leaking timing cover gasket. Check before replacing the oil seal. Remove the crankshaft pulley/vibration damper as already described before the oil seal can be replaced. The seal can be carefully removed with a screwdriver (see Fig. 1.55). Screw a self-tapping screw into the outside of the seal and apply the screwdriver plate under the screw head. It is also possible to unscrew the oil seal cover to replace the seal.

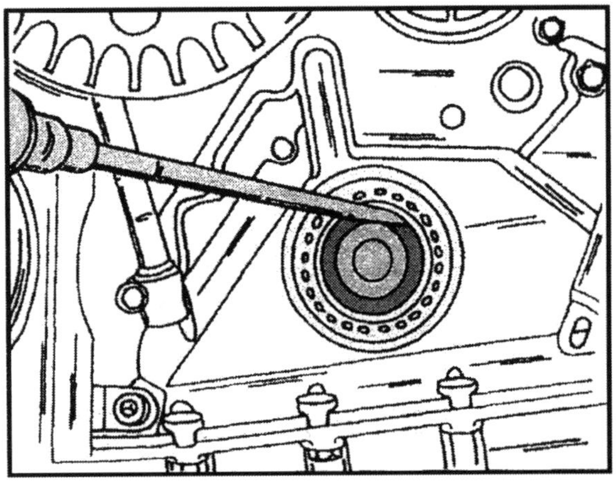

Fig. 1.55 – Removal of the front crankshaft oil seal.

Fill the space between sealing lip and dust protection lip with grease and carefully drive a new oil seal into the timing cover and over the crankshaft until the ring outer face is flush. Fit the two bolts from below and then the remaining bolts. All bolts are tightened to 7.2 kgm (7.2 ft.lb.).

Refit the vibration damper as described in Section 1.4.3.5. and carry out the remaining operations in reverse order.

1.4.4. TIMING MECHANISM

The following description deals with the timing mechanism of the engine. The removal and installation of the two camshafts has already been described in conjunction with the cylinder head. The endless timing chain is engaged with the camshaft sprockets, the high pressure pump sprocket and the crankshaft sprocket. The chain is guided by slide rails. The tension of the chain is ensured by means of a hydraulic chain tensioner, which is located in the timing cover and pushes onto a tensioning rail. The camshaft sprockets are fitted by means of a bolt and located by a Woodruff key.

A second, smaller chain is used to drive the oil pump. The chain is fitted around a second sprocket on the crankshaft and around the pump drive sprocket and has its own chain tensioner.

1.4.4.0. Chain Tensioner – Removal and Installation

The chain tensioner is fitted into the R.H. side of the cylinder head as shown in Fig. 1.56. The tensioning force of the chain tensioner is a combination of the fitted compression spring and the pressure of the engine oil. The oil contained inside the tensioner also absorbs shock loads from the timing chain. A chain tensioner cannot be repaired, i.e. must be replaced if suspect. A hand press is required (see below). To remove the tensioner you will have to know the engine number of a 646.982 or 646.983 engine. Remove the tensioner as follows:

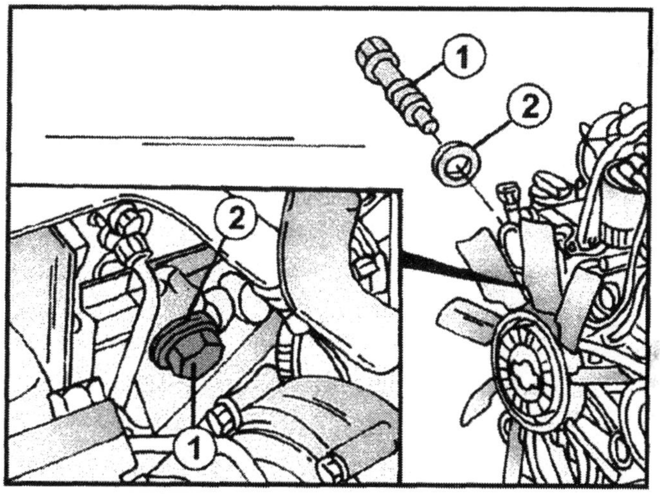

Fig. 1.56 – The chain tensioner (1) is fitted at the position shown and sealed off by means of sealing ring (2).

Several operations are required to gain access to the chain tensioner:

- Remove the air cleaner housing.
- Remove the air hose between the turbo charger and the air cleaner housing.
- Remove the oil dipstick guide tube from the side of the cylinder head on the engines mentioned above before engine end number 007337. Remove the securing bolt and press the guide tube carefully to one side. On these engines also remove a hose clamp immediately below the bolt and turn it to one side if necessary to create more space.
- The chain tensioner (1) in Fig. 1.56 can now be unscrewed from the side of the engine. The sealing ring (2) must be replaced.
- Fit the chain tensioner with the new seal and tighten it to 8.0 kgm (58 ft.lb.). The pressure plunger of the tensioner will push against the tensioning rail.
- Before installation fill the chain tensioner with engine oil. To do this, position the tensioner with the thrust pin facing down into the engine oil until it is over the collar at the hexagon. The tensioner is now placed under a hand press and depressed fully between 7 to 10 times. After the tensioner is filled it must be possible to compress the chain tensioner slowly and evenly by using considerable force.
- All other operations are carried out in reverse order.

1.4.4.1. Removal and Installation of Timing Chain

The replacement of the timing chain is practically impossible without using the special tools available to Mercedes dealers. As the engine must be removed in any case you will be able to replace the chain in accordance with the instructions given during the removal of the cylinder head, camshafts, etc. and the instructions given below. The timing chain can then be removed in one piece and refitted. The workshop separates the chain and rivets are used to assemble it.

Removal and Installation of tensioning rail and slide rail in timing cover

The location of the tensioning rail and the slide rail in the timing cover is shown in Fig. 1.57. If it is necessary to replace the tensioning rail (1) you will also have to replace the slide rail (3) and the tensioning lever (5).

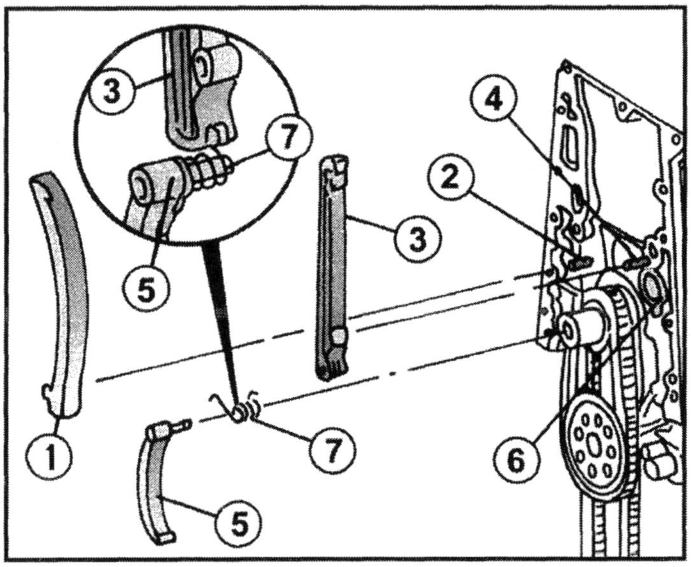

Fig. 1.57 – Removal and installation of the tensioning rail and the slide rail in the timing cover.
1 Tensioning rail
2 Bearing pin
3 Slide rail
4 Bearing pin
5 Tensioning lever
6 Bearing pin
7 Spring

Although the slide rail parts can be replaced with the engine in the vehicle, it may be of advantage to remove the engine, although the operations are more comprehensive and time consuming to replace the parts shown in the illustration. Also remove the cylinder head and the timing cover as described earlier on. Proceed as follows:

- With the engine removed, remove the cylinder head and the timing cover as described before.
- Pull the tensioning rail (1) off the bearing pin (2).
- Pull the slide rail (3) together with the tensioning lever (5) off the bearing pins (4) and (6) respectively.
- Release the spring (7) at the slide rail (3) and take it off the tensioning lever.

The installation is a reversal of the removal procedure. Ensure that the spring (7) is correctly installed to the tensioning lever.

Removal and Installation of slide rail in cylinder head

The location of the slide rail, located in the cylinder head, is shown in Fig. 1.58. The front cover of the cylinder head must be removed in order to gain access to the slide rail. Proceed as follows:

- Rotate the crankshaft in the direction of rotation until the piston of cylinder No. 1 is at top dead centre. The mark in the crankshaft pulley must be aligned.
- Lock the camshafts to prevent them from rotating. This has already been explained during the removal of the camshaft. Otherwise make sure that the shafts cannot move.
- Remove the front cover from the cylinder head as described in connection with the cylinder head.

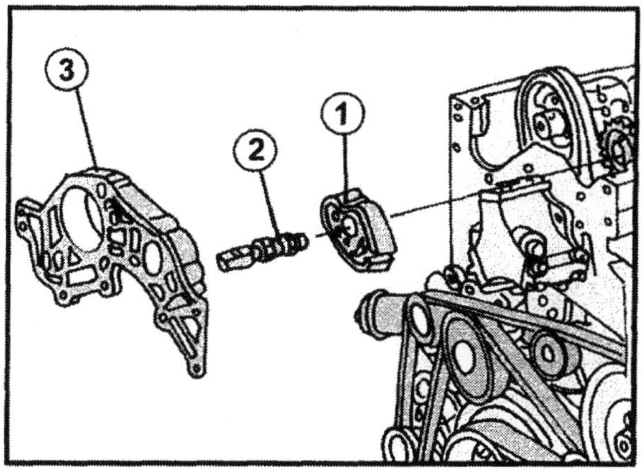

Fig. 1.58 – Location of the slide rail (1) in the cylinder head. A 17 mm hexagon 3/8 in. square wrench is required to unscrew the driver (2) after removal of the cylinder head cover (3).

- Using the wrench mentioned above unscrew the driver (2) out of the inlet camshaft and take off the upper slide rail (1).

The installation is a reversal of the removal procedure. The driver is tightened to 5.0 kgm (36 ft.lb.). Refit the front cover to the cylinder head as described earlier on and refit the engine.

Removal and installation of crankshaft timing sprocket

Various parts must be removed to gain access for the removal of the sprocket. These are the oil sump and the high pressure pump drive gear and the oil pump (refer to "Lubrication System"). After the parts above have been removed you will be able to remove the sprocket in accordance with Fig. 1.59. Note that a puller is required to withdraw the sprocket, but the special tool shown in the illustration can be replaced by a two-arm puller.

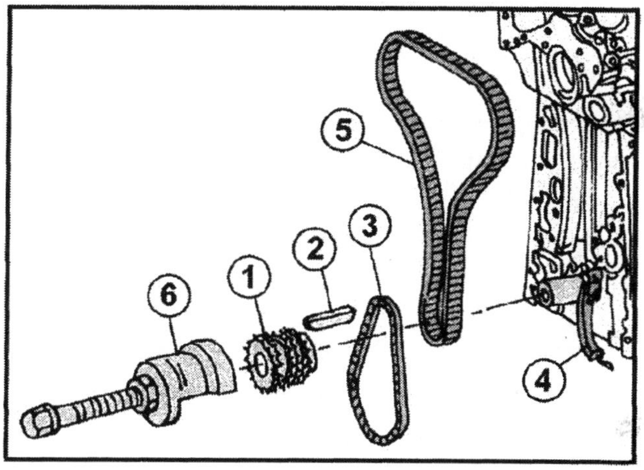

Fig. 1.59 – Details for the removal and installation of the crankshaft sprocket. The installation is the same on both engines.

1 Crankshaft sprocket
2 Woodruff key
3 Oil pump drive chain
4 Tensioning lever
5 Timing chain
6 Puller

- Push the tensioning lever (4) down against the force of the spring and take off the drive chain for the oil pump (3).
- Lift the timing chain (5) off the crankshaft pulley. Immediately check the condition of the timing chain and replace it if necessary.
- Use the puller (6) shown in the illustration or one operating in a similar manner and withdraw the sprocket (1) from the end of the crankshaft. The tensioning lever (4) must again be pushed against the tension of the spring to facilitate the operation. Check the condition of the Woodruff key (3) and replace it if necessary. A side cutter can be used to remove the old key.

The installation is a reversal of the removal procedure. Use a hammer to tap in the Woodruff key. The flat face must be parallel with the crankshaft end. Use a piece of tube to drive the sprocket over the crankshaft end. A tight sprocket can be heated up if necessary (wear leather gloves in this case).

Rotate the crankshaft a few times, re-check the timing marks and assemble the engine.

Removal and installation of camshafts

The camshafts are not the same on all engines. A number stamped into the camshaft indicates to which engine the particular shaft is fitted.

The shape of the cams has been changed and therefore also the marking of the camshafts. Fig. 1.60 shows where the numbers are located and explains where to check them. The engine in question must also be fitted with the correct camshaft.

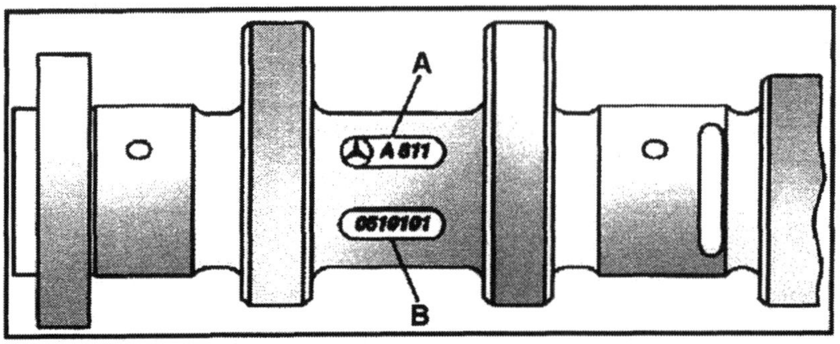

Fig. 1.60 – Marking of the camshafts after modification (A) and the marking of the inlet camshaft with the old shape (B). The marking can be found between the 1st and 2nd inlet cams.

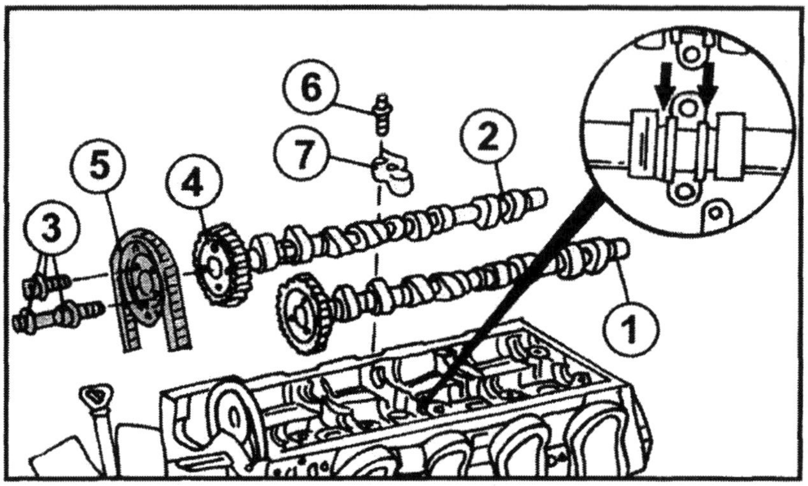

Fig. 1.61 – Camshaft installation. The arrows show where the alignment is checked at the camshaft bearing.
1 Inlet camshaft
2 Exhaust camshaft
3 Camshaft sprocket bolts (1.8 kgm)
4 Location of down pin
5 Camshaft sprocket
6 Bearing cap bolt (0.9 kgm)
7 Camshaft bearing cap

The camshafts are fitted to the upper end of the cylinder head and held in position by their bearing caps. The lower bearing bores are machined into the cylinder head. The camshafts are lifted out towards the top after removal of the camshaft bearing caps and the camshaft sprockets. The removal and installation has already been described in conjunction with the cylinder head. Fig. 1.61 shows the fitting of the camshafts.

1.4.5. POLY V-BELT – REMOVAL AND INSTALLATION

A single drive belt, also known as poly V-belt, is fitted to the front of the engine, but the layout of the belt is not the same on all engines and depends on a fitting of an air conditioning system or a vehicle without A/C system. If a compressor is fitted it will also be driven by the same belt. The same belt drive system is, however, fitted to all engines and is held in its correct tension by means of an automatic tensioning device. There is a difference also in the length of the belt, i.e. in the case of engines without A/C system or in the case of a fitted A/C system. If a new belt is fitted make sure your obtain the correct one.
The removal and installation of the belt has already been described in section "Cooling System". Follow the instructions for the diesel engine.

1.5. Tightening Torque Values

Note: Most tightening torque values are given during the removal and installation of the various engine parts. The following is a list of the most important values:

Cylinder Head Bolts (see Fig. 1.17):
- Tighten bolts (B): .... 1.5 kgm (11 ft.lb.)
- Tighten bolts (B): .... 6 kgm (43 ft.lb.)
- Tighten bolts (A): .... 2.0 kgm (14.5 ft.lb.)
- Tighten bolts (B): .... 90° angle-tightening (quarter of a turn)
- Re-tighten bolts (A): .... 2.0 kgm (14.5 ft.lb.)

- Re-tighten bolts (B): .... 90° angle-tightening (quarter of a turn)
Cylinder head bolts (in timing case) (A in Fig. 1.17): .... 2.0 kgm (14.5 ft.lb.)
Connecting Rod Bearing Caps:
- 1st stage .... 0.5 kgm (3.5 ft.lb.)
- 2nd stage .... 2.5 kgm (18 ft.lb.)
- 3rd stage .... Angle-tighten 90° (quarter of a turn)
Main Bearing Cap Bolts
- 1st stage .... 5.5 kgm (40 ft.lb.)
- 2nd stage .... Angle-tighten 90° (quarter of a turn)
Camshaft sprocket bolts .... 1.8 kgm (13 ft.lb.)
Chain tensioner .... 8.0 kgm (58 ft.lb.)
Cylinder head cover bolts .... 1.0 kgm (7.2 ft.lb.)
Upper part of charge air distribution pipe .... 0.9 kgm (7 ft.lb.)
Support strut to distribution pipe ....1.4 kgm (10 ft.lb.)
Support strut to crankcase .... 2.0 kgm (14.5 ft.lb.)
Injectors into cylinder head ...... 0.7 kgm (5 ft.lb.), then angle-tighten (1/4 of a turn)
Injector pipes union nuts .... 2.3 kgm (16.5 ft.lb.)
Front cover to cylinder head .... 1.4 kgm (10 ft.lb.)
Pre-delivery pump to cylinder head .... 1.0 kgm (7.2 ft.lb.)
Top cover of cylinder head .... 1.1 kgm (8 ft.lb.)
Chain tensioner to cylinder head .... 8.0 kgm (58 ft.lb.)
Vacuum pump housing .... 1.0 kgm (7.2 ft.lb.)
Centre bolt of crankshaft pulley (bolt head marked "8.8")
- 1st stage .... 20 kgm (145 ft.lb.)
-2nd stage .... Angle-tighten 90°
Centre bolt of crankshaft pulley (bolt head marked "10.9")
- 1st stage .... 32.5 kgm (234 ft.lb.)
-2nd stage .... Angle-tighten 90°
Flywheel or driven plate
- 1st stage .... 4.5 kgm (32.5 ft.lb.)
- 2nd stage .... Angle-tighten by 90°
Rear oil seal flange .... 0.9 kgm (7 ft.lb.)
Front oil seal cover .... 1.0 kgm (7.2 ft.lb.)
Alternator mounting bracket to cylinder block
- Bolt inserted from front .... 2.5 kgm (18 ft.lb.)
- Bolt inserted from the side .... 4.5 kgm (32.5 ft.lb.)
Alternator to mounting bracket .... 4.5 kgm (32.5 ft.lb.)
Bolt for T.D.C. transmitter (sensor) .... 1.0 kgm (7.2 ft.lb.)
Oil drain plug in sump .... 3.0 kgm (21 ft.lb.)
Oil sump to cylinder block
- M6 bolts .... 1.0 kgm (7.2 ft.lb.)
- M8 bolts .... 2.5 kgm (18 ft.lb.)
Exhaust manifold to cylinder head .... 2.5 kgm (18 ft.lb.)
Centre bolt, fan to water pump .... 4.5 kgm (32.5 ft.lb.)
Engine mounting, bracket, to engine mounting, front .... 8.3 kgm (60 ft.lb.)
Engine mounting, side, to gearbox .... 4.5 kgm (32 ft.lb.)
Engine mounting, rear, to mounting bracket .... 4.5 kgm (32 ft.lb.)
Exhaust manifold to exhaust pipe flange .... 3.0 kgm (22 ft.lb.)
Exhaust pipe to transmission bracket .... 2.0 kgm (14.5 ft.lb.)
Oil pipes to automatic transmission .... 2.5 kgm (18 ft.lb.)
Earth strap to chassis .... 2.3 kgm (17.0 ft.lb.)
Coolant drain plug (block) .... 2.8 kgm (20 ft.lb.)
Oil filter to crankcase .... 3.0 kgm (22 ft.lb.)

1.6. Lubrication System (all engines)

Petrol Engine

The lubrication system is a pressure-feed system. A gear-type oil pump is driven via the crankshaft timing sprocket by means of a separate single roller chain and is kept under tension with its own chain tensioner with torsion spring. Fig. 1.62 shows the arrangement of the chain drive.

The oil filter is fitted in vertical position on the side of the cylinder block. The filter element is held in the filter housing by means of a bolt, fitted to the top of the filter housing. The workshop uses a special wrench to undo this bolt. The filter consist of the lower part and a throw-away oil filter. An oil pressure switch is fitted to the side of the filter lower housing.

The engine is fitted with a light-metal oil sump, containing an oil level sensor. The oil sump can be removed with the engine fitted to the vehicle.

Diesel Engine

The lubrication system is a pressure-feed system. A gear-type oil pump is driven via the crankshaft timing sprocket by means of a separate single roller chain and is kept under tension with its own chain tensioner. The fitted position is shown in the section dealing with the diesel engine. The oil pump can be removed with the engine fitted to the vehicle. There are no provision to repair the oil pump, i.e. it must be replaced if defective.

A warning light in the instrument panel will light up, when the oil level is approaching the lower limit of the oil dipstick. At least 1 litre of oil should be filled in as soon as possible.

Oil pressure indication is electronic by means of a contact switch, fitted to the lower part of the oil filter. Increased oil pressure increases the resistance in the switch and changes the reading in the instrument accordingly. The oil filter is fitted in an upright position to the cylinder block. The oil flows through the filter element from the outside to the inside. A return shut-off valve prevents oil from flowing back through the oil pump into the oil sump when the engine is switched off. A by-pass valve opens when the pressure differential between the dirty and the clean end of the filter exceeds a certain value. The oil is then directed to the oil gallery without being cleaned. It should be noted that the oil filter element must be changed after 600 - 1000 miles if the engine has been overhauled.

The pistons are cooled with oil splash jets. The jets are fitted with a screw and a dowel pin to the cylinder block.

1.6.0. TECHNICAL DATA

Oil Capacities:...See Page 6
Oil pressure at idle speed: ....7 psi.
Oil pressure at 3000 rpm:...43 psi.

1.6.1. OPERATIONS ON THE PETROL ENGINE

1.6.1.1. Oil Pump – Removal and Installation

The oil sump must be removed from the vehicle to remove the oil pump. Fig. 1.63 shows details of the removed pump together with the oil sump.

- Remove the oil sump as described in Section 1.6.1.2.

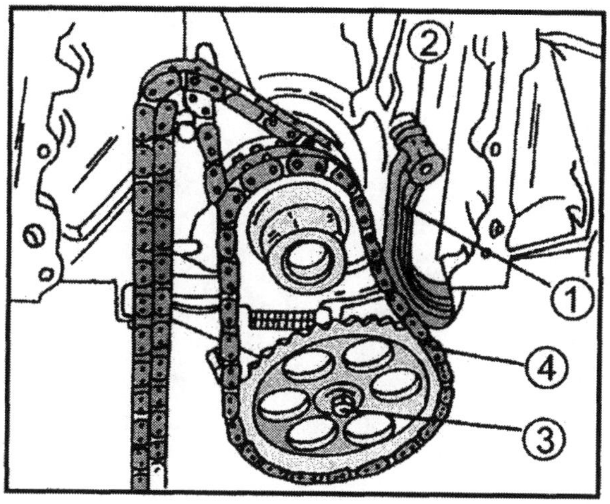

Fig. 1.62 – View of the fitted oil pump drive (petrol).

1 Chain tensioner blade
2 Torsion spring
3 Bolt and washer
4 Drive chain

- Undo the bolt (3) in Fig. 1.62 securing the pump chain sprocket and remove together with the washer. Remove the oil pump sprocket together with the drive chain (4) from the crankshaft. When removing the sprocket you will notice that one side is curved. This side must face the oil pump during installation. Fig. 1.64 shows this side with the arrow. The oil pump sprocket can only be fitted in one position.
- Remove the pump securing bolts from the bottom of the crankcase and remove together with the washers. The mounting bracket will also come away. The pump is now withdrawn towards the bottom. Note that two dowels locate the pump.

The pump sprocket and drive chain can be replaced as a set. An overhaul of the pump is not possible.

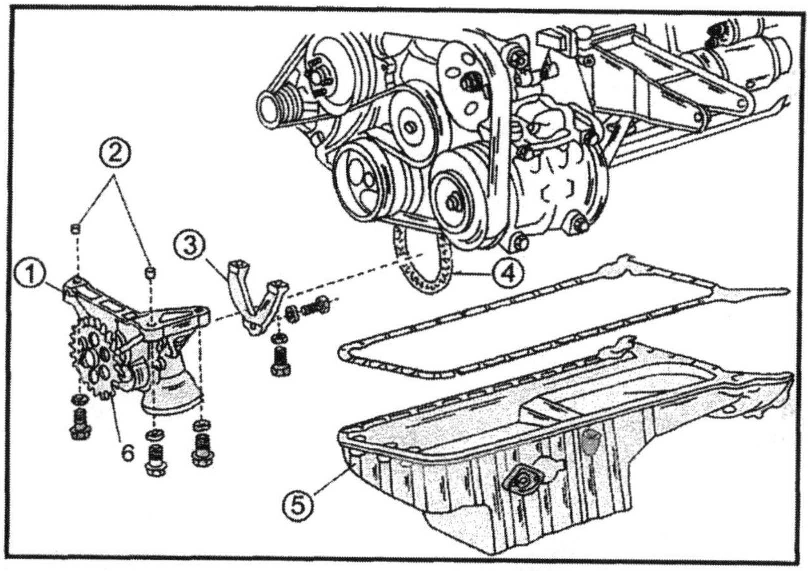

Fig. 1.63 – Attachment of oil pump and oil sump (petrol). The oil sump of this engine consists of two parts as shown in Fig. 1.65.

1 Oil pump assembly
2 Dowel sleeves
3 Reinforcement bracket
4 Oil pump drive chain
5 Oil sump
6 Oil pump sprocket

The installation is a reversal of the removal procedure. Observe the following points:

- Fill the pump with engine oil before installation. Fit it to the crankcase, ensuring that the dowels (2) in Fig. 1.63 engage. Fit and tighten the bolts to 2.3 kgm (16.5 ft.lb.). Fit the bracket (3) and tighten the two bolts to 1.0 kgm (7.2 ft.lb.).

Fig. 1.64 – Attachment of the oil pump sprocket to the oil pump shaft. Engage as shown.

- Fit the oil pump sprocket in accordance with Fig. 1.64. Fit the bolt with the washer and tighten the bolt to 3.2 kgm (23ft.lb.).
- The remaining operations are now carried out in reverse order. After releasing the chain tensioner (1) in Fig. 1.62 the chain will be correctly tensioned.

1.6.1.2. Oil Sump – Removal and Installation

The oil sump can be removed with the engine fitted to the vehicle, but the engine must be lifted out of its mountings to remove the engine mountings on the L.H. and R.H. sides. A suitable lifting device with the necessary chains or ropes must therefore be available before the operation is commenced. The oil sump consists of an upper and a lower part, as shown in Fig. 1.65. The following instructions describe the removal of the complete oil sump. The lower part can be removed without removing the upper part, if necessary. The removal of the lower sump part can be carried out without lifting the engine. If necessary follow the instructions given below. The engine oil should be drained immediately whichever operation is carried out.

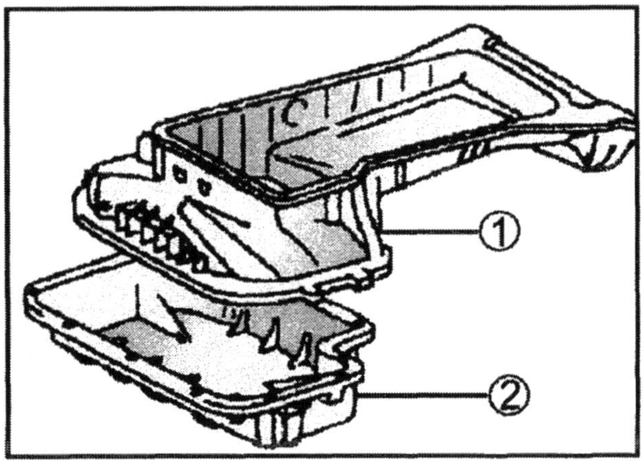

Fig. 1.65 – The oil sump consists of an upper part (1) and a lower part (2). Both can be removed individually as required.

Removal and Installation of the complete Oil Sump

- Remove the air cleaner housing and partially drain the cooling system (the cooling system must be drained as a coolant hose must be disconnected).
- Remove the oil dipstick guide tube as described later on. The tube is withdrawn in an upward direction. Before installation check the sealing ring and replace if necessary.
- If an automatic transmission is fitted disconnect the fluid lines for the oil cooling.
- Lift the engine out of its mountings with a suitable hoist until the engine mountings are no longer under tension and remove the bolt securing the engine mounting to the mounting bracket on the L.H. and R.H. side of the engine. The mounting bolts are tightened to 5.8 kgm (42 ft.lb.).
- Lift the engine as far as possible, but take care that the rear of the engine does not touch the body in order not to damage the hot film air mass sensor and the fuel line.
- Remove the lower oil sump part (2) in Fig. 1.65.

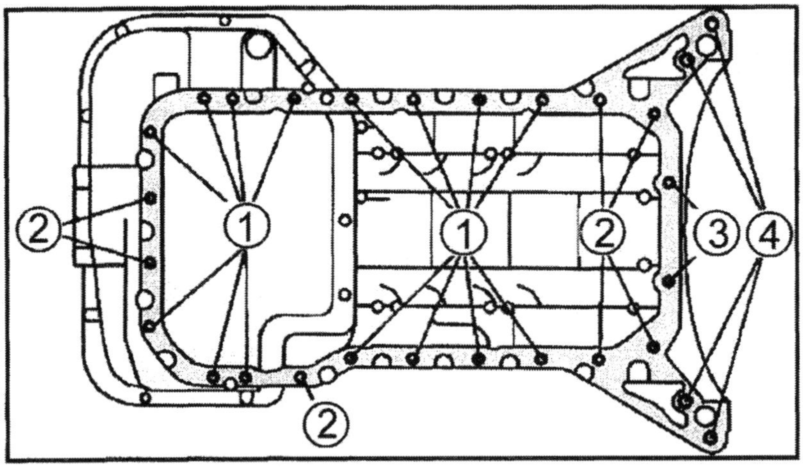

Fig. 1.66 – Location of the oil sump securing bolts of the petrol engine.

1 M6 x 20 mm bolts
2 M6 x 40 mm bolts
3 M6 x 90 mm bolts
4 M8 x 30 mm bolts

- At the transmission end remove two bolts securing the oil sump to the clutch bell housing. Tighten the M8 bolts (4) to 2.0 kgm (14.5 ft.lb.) during installation.
- Detach a coolant hose from the coolant pipe.
- Unscrew the bolts securing the oil sump. The bolts are not all of the same length and diameter. Fig. 1.66 shows the location of the bolts around the oil sump. The

end of the oil sump must be flush with the transmission mounting face. Fit and tighten the different bolts in accordance with their diameter. M6 bolts are tightened to 0.9 kgm (7 ft.lb.), M8 bolts to 2.0 kgm (14.5 ft.lb.). All bolts are tightened evenly and gradually until the specified torque has been obtained.

- Fit the oil drain plug (2.5 kgm/18 ft.lb.) and fill the engine with oil.

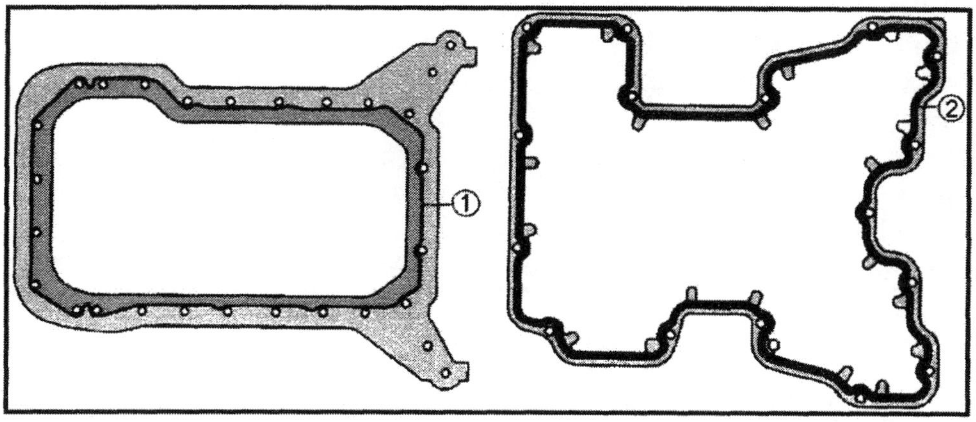

Fig. 1.67 – Application of the sealing compound. In the L.H. view to the upper part of the oil sump (1), in the R.H. view to the lower part of the sum (2). Follow the instructions given below.

The installation is a reversal of the removal procedure, noting the tightening torques given above. Before installation clean the sealing faces of oil sump and crankcase. The sealing faces in Fig. 1.67 must be coated with "Loctite 5970" sealing compound and the sump must be fitted within 10 minutes. The bead of the sealing compound must not exceed 2.0 mm and not "flattened".

Fit the oil sump upper part, followed by the lower part and fit the bolts in accordance with Fig. 1.66, noting the diameter, length and torque of the individual bolts.

Removal and Installation of the Oil Sump bottom Section

- Remove the lower engine compartment panelling and drain and collect the engine oil.
- Disconnect the electrical connector plug from the oil level sensor.
- Remove the clamp securing a coolant pipe from the bottom section of the oil sump. In a similar manner remove the clamp securing the automatic fluid line to the sump part.
- Remove the thermostatic valve (self-tapping screw) for the automatic transmission after removal of the securing screws. The fluid lines must not be disconnected. The screw must be tightened to 1.4 kgm (10 ft.lb.) during installation.
- If the bottom oil sump section is to be replaced, remove the sensor for the oil level, temperature, etc. Otherwise leave it in position.
- Unscrew the bolts securing the oil sump part and lower the sump towards the bottom. The fluid lines and the thermostatic valve for the automatic transmission must be pushed to one side to obtain more space.

The installation is a reversal of the removal procedure. Before installation clean the sealing faces of oil sump and crankcase. The sealing faces in Fig. 1.67 must be coated with "Loctite 5970" sealing compound and the sump must be fitted within 10 minutes. The bead of the sealing compound must not exceed 2.0 mm and not "flattened". Tighten the oil sump screws to 1.4 kgm (10 ft.lb.). Fit the oil drain plug, tighten it to 3.0 kgm (22 ft.lb.) and fill the engine with the correct amount of oil.

1.6.1.3. Oil Level Sender Unit – Removal and Installation

The oil level sender unit is inserted into the underside of the oil sump, i.e. the front of the vehicle must be placed on chassis stands to remove the sensor. The sensor is monitoring the oil level, the oil temperature and the oil quality. From below you will recognise the sensor, attached by three securing bolts, both on petrol and diesel engines. After draining the engine oil withdraw the cable connector plug from the sender unit (switch) and remove the screws. The sensor can now be withdrawn. Remove the oil seal below the sender unit (always replace).

Install in reverse order. Tighten the screws, but note the different tightening torques. Tighten the sensor of a petrol engine to 1.1 kgm (8 ft.lb.), the sensor of a diesel engine to 1.4 kgm (10 ft.lb.). Finally fill the engine with oil with the quantity specified on page 6. The tightening torque of the oil drain plug for all engines is 3.0 kgm (22 ft.lb.).

Start the engine and check for oil leaks when the engine is warm.

1.6.1.4. Oil Pressure Switch – Removal and Installation

The switch is located at the front end of the engine, L.H. side above one of the belt pulleys. The switch can be removed after withdrawing the cable connector. The switch is tightened to 2.0 kgm (14 ft.lb.).

1.6.1.5. Oil Filter – Removal and Installation

The filter can be removed with a filter wrench. The workshop uses a special tool, shown in Fig. 1.68, which is placed over the screw cap at the top of the filter and engages with the ribs. A suitable socket, extension and ratchet complete the arrangement. Replace the oil filter as follows:

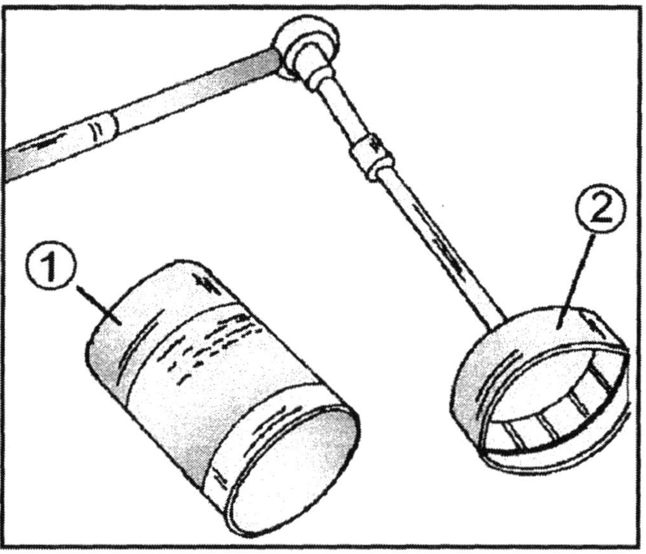

Fig. 1.68 – The screw cap on top of the oil filter (1) is removed with the special tool (2) as shown. Alternatively follow the description below.

- Disconnect the battery. If the engine oil is to be changed (in most cases) remove the sound-proof panel underneath the engine.

Fig. 1.69 – Location of the oil filter of a petrol engine. The numbers are referred to in the text.

- Use a filter chain wrench and unscrew the screw cap (1) in Fig. 1.68 (unless you have the special tool). Fully unscrew the screw cap and remove it. The filter element will be attached to the cap (catch dripping oil). In Fig. 1.69 you can see where the filter cap (1) and the filter element (2) are located.

- Fit the new filter element to the screw cap, replace the sealing ring (3) in Fig. 1.69 and fit the cap (1) to the filter housing. Tighten the cap to 2.5 kgm (18 ft.lb.). You will have to estimate the torque without the filter wrench.
- Finally check and correct the oil level or fill the engine with oil.

If the complete oil filter requires replacement (rarely), free the oil pipe from the retaining clips and disconnect them from the filter housing. Replace the sealing rings. Installation is a reversal.

1.6.1.6. Oil/Water Cooler/Oil Dipstick Guide Tube

The removal and installation of the oil/water cooler is a complicated operation. Replacement is an operation for a workshop. Dipstick guide tube see description earlier on.

1.6.1.7. Oil Change

Changing the engine oil is a straight-forward operation after removal of the sound-proofing panel underneath the engine. The oil filter should always be replaced at the same time. Tighten the oil drain plug to 3.0 kg (22 ft.lb.).

1.6.2. OPERATIONS ON THE CDI DIESEL ENGINE

Most operations described for the petrol engine also apply to the diesel engine. The following sections only refer to operations which are different.

1.6.2.0. Oil Pump – Removal and Installation

Fig. 1.70 shows the parts involved during the removal and installation.

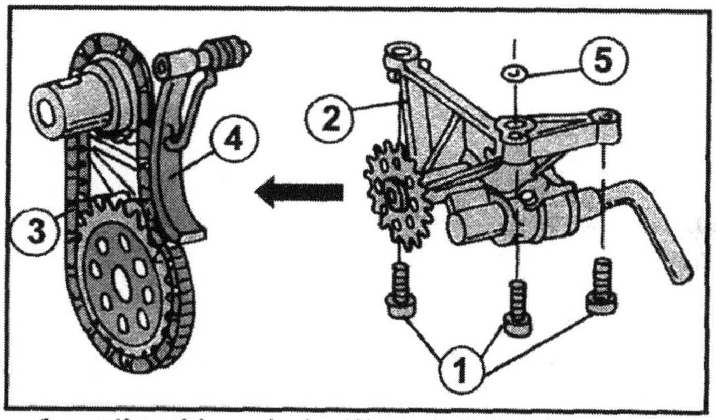

Fig. 1.70 – Removal and installation of the oil pump.
1 Pump securing bolts, 1.9 kgm
2 Oil pump
3 Oil pump drive chain
4 Tensioning lever
5 Pump oil seal

- Remove the oil sump as described below.
- Remove the bolts (1) from below.
- Press the tensioning lever (4) away from the drive chain (3) and disengage the oil pump from the drive chain and remove the pump (2) towards the bottom.

The installation is a reversal of the removal procedure, noting the following points:

- The pump oil seal (5) must always be replaced. Thoroughly clean the pump strainer before the pump is refitted. If a new pump is fitted fill it with engine oil before installation to prevent it from seizing up during the starting up of the engine.
- Refit the pump to the cylinder block and insert and tighten the bolts (1) to 1.9 kgm (14 ft.lb.)
- Refit the oil sump as described below and fill the engine with oil. Start the engine and check all sealing faces for oil leaks.

1.6.2.1. OIL PUMP - REPAIRS

Oil pumps should not be dismantled and/or repaired. If a new pump is fitted check the drive chain and the chain sprocket to prevent fitting a new pump together with new drive components.

1.6.2.2. Oil Sump – Removal and Installation

The oil sump can be removed with the engine fitted to the vehicle, but the engine must be lifted out of its mountings to remove the engine mountings on the L.H. and R.H. sides. A suitable lifting device with the necessary chains or ropes must therefore be available before the operation is commenced. We suggest to read the instructions in full before commencing with the operations, as the job is fairly complicated. The engine oil should be drained immediately.

- Remove the air cleaner housing.
- Lift the engine out of its mountings with a suitable hoist until the engine mountings are no longer under tension and remove the bolt securing the engine mounting to the mounting bracket on the L.H. and R.H. side of the engine. The mounting bolts are tightened to 5.8 kgm (42 ft.lb.).
- Lift the engine as far as possible, but take care that the rear of the engine does not touch the body in order not to damage any parts of the engine.
- Place a suitable container underneath the oil drain plug. Remove the oil drain plug and allow the oil to drain into the container. Remove the oil filler cap (see description later on) to speed up the draining of the oil. It also helps, if the engine oil is warm, i.e. drive the vehicle a few miles before the oil is drained.
- Remove the lower engine compartment panelling in the case of models 639.7/8 after VIN number 055347.
- From the front of the oil sump detach the automatic transmission fluid line if an automatic transmission is fitted. To do this unscrew a union nut from the thermostatic valve and the bracket from the oil sump (located below the belt pulleys). Then unscrew the thermostat from the oil sump. Tighten the union nut to 1.8 kgm (13 ft.lb.). The valve screw (M6) is tightened to 1.4 kgm (10 ft.lb.).
- Remove the bolts securing the charge air pipe and separate the pipe from the oil sump.
- The oil filler pipe (oil filler neck) attached to the oil sump must be removed. Remove one bolt from the engine support and a second bolt from the oil sump. The pipe can now be withdrawn from the sump. The two bolts are tightened to 1.2 kgm (8.5 ft.lb.).
- Disconnect the cable plug from the oil level/temperature/quality sensor from the bottom of the oil sump.

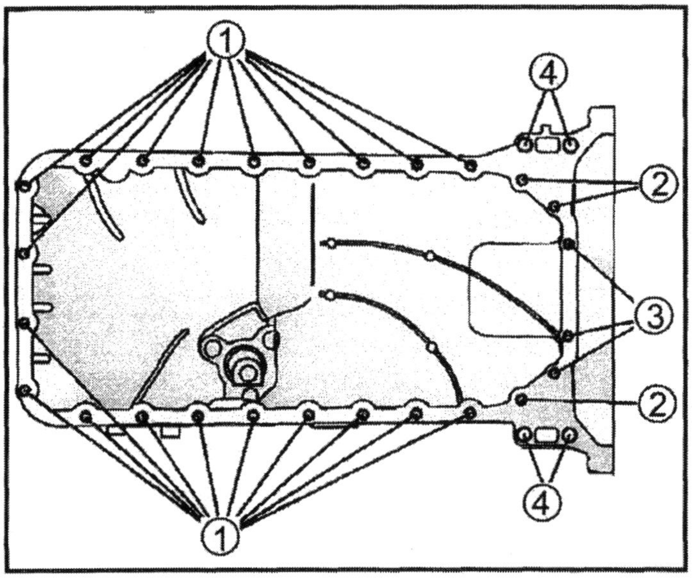

Fig. 1.71 – Bolt locations and length of the oil sump securing bolts.
1 M6 bolt, 20 mm
2 M6 bolts, 33 mm
3 M6 bolt, 85 mm
4 M8 bolts, 40 mm

- At the rear of the oil sump remove two bolts securing the oil sump to the transmission case (clutch bell housing). These are the large bolts and are tightened to 4.0 kgm (29 ft.lb.) during installation.
- Unscrew the bolts securing the oil sump. The bolts are not all of the same length and diameter. Fig. 1.71 shows the location of the bolts around the oil sump. The end of the oil sump must be flush with the transmission mounting face. Fit and tighten the

different bolts in accordance with their diameter. M6 bolts are tightened to 0.9 kgm (7 ft.lb.), M8 bolts to 2.0 kgm (14.5 ft.lb.). All bolts are tightened evenly and gradually until the specified torque has been obtained.

- The oil sump can now be detached from the engine but at this stage cannot be removed. A helper will be required to take out the sump. Slowly raise the engine, observing that the viscous fan is not damaged. It is also required to lift up the transmission with a mobile or hydraulic jack. After the transmission is securely supported remove the rear engine mount (bolts are tightened to 2.0 kgm). Ask the helper to lift the transmission a little further until the oil sump can be seen to be free. Remove two bolts securing the large intake tube and then lift out the sump. The sealing ring for the intake tube must always be replaced.

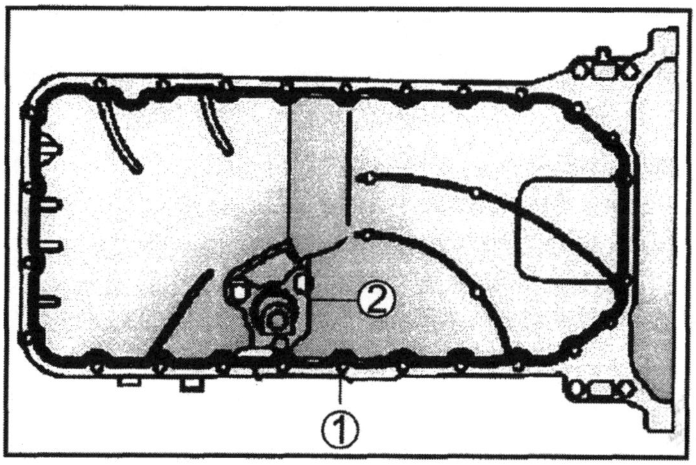

Fig. 1.72 – Coat the oil sump face as shown by (1) with the sealing compound specified below. (2) shows the location of the sensor.

The installation of the oil sump is a reversal of the removal procedure, noting the following points:

- Before installation clean the sealing faces of oil sump and crankcase. The sealing faces in Fig. 1.72 must be coated with "Loctite 5970" sealing compound and the sump must be fitted within 10 minutes. The bead of the sealing compound must not exceed 2.0 mm and not "flattened". Tighten the oil sump bolts to the crankcase to 0.9 kgm (M6 bolts) or 2.0 kgm (M8 bolts). Fig. 1.71 shows where the bolts are located. Note their length.
- Tighten the bolts securing the clutch bell housing to the oil sump to 4.0 kgm and the bolts securing the oil sump to the timing case cover to 0.9 kgm. The same torque applies to the bolts on the rear oil seal cover.
- Tighten the oil drain plug to 3.0 kgm.
- Tighten the oil filler fitting to 1.2 kgm to the oil sump.
- Refit the engine mountings and tighten them to the torque values given above.
- Finally fill the oil sump with the correct quantity of the recommended oil (Page 6).

Fig. 1.73 – Fitted oil filter (diesel engine).
1 Filter screw cap, 2.5 kgm
2 Filter insert (element)
3 Sealing ring

1.6.2.3. Oil Filter

Fig. 1.73 shows where the oil filter is located. The following text describes the removal and installation of the filter. We would like to point out that a special filter wrench is used in the workshop. The hexagon fitted to the top of the oil filter cover on earlier and similar engines has now been removed. If you intend to carry out regular servicing of your vehicle we recommend to obtain such a filter wrench, available under Parts No. 103 589 02 09 00.

Otherwise it may be possible to use a large pair of pliers or grips, but remember that access is limited. Remove the filter as follows:

- Remove the oil filter screw cap (1) in Fig. 1.73 from the top of the filter.
- Remove the oil filter screw cap together with the oil filter insert (2) out of the oil filter housing and take off the sealing ring (3).
- Thoroughly clean the inside of the filter housing with a clean rag.
- Fit the new filter insert (2) together with a new oil seal (3) and refit the filter screw cap. The cap is tightened to 2.5 kgm (18 ft.lb.).
- Check the oil level in the engine and correct if only the oil filter element has been replaced. Otherwise fill the engine with the correct amount of oil.
- Start the engine and allow it to run for a while. Then switch off and check the filter surrounding for oil leaks.

1.6.2.4. Oil Dipstick Guide Tube

Diesel Engine

The oil dipstick tube is located next to the large air intake hose. It can be removed as follows, noting that some other operations are necessary before the guide tube can be removed. We also must mention that a special support is used in the workshop to guide the guide tube in position during installation.

- Withdraw the oil dipstick.
- Remove the oil feed line to the turbo charger in the following order. Remove the large intake hose from the turbo charger after slackening the hose clamps. Remove the bolt securing the feed line from the turbo charger at one end and the bolt securing the feed line to the cylinder head. And take out the oil feed line. During installation clean the sealing surfaces and replace the gaskets. Tighten the turbo charger connection bolt to 1.8 kgm (13 ft .lb.) and the bolt on the cylinder head to 0.9 kgm (7 ft.lb.).
- Remove the bolt securing the dipstick guide tube. The mounting clamp must be removed.
- On models 639.7/8 after VIN number 055347 remove the bottom engine compartment panelling.
- Remove the poly V-belt as described later on.
- Remove the charge air hose from the turbo charger. This is the hose leading from below into the turbo charger, above the alternator.
- Disconnect the cable connector plug from the alternator, remove the four alternator mounting bolts and push the alternator to one side with the line connected. Secure the alternator with a piece of wire so it cannot drop down.
- The oil dipstick guide tube can now be withdrawn from the crankcase. It will be a tight fit and must be forced out with a suitable tool.

During installation insert the guide tube into the crankcase and force it down. Fit a bolt into the mounting clamp and bring the guide tube into the correct position (this is when the workshop uses the support 111 589 12 31 00). Then tighten the mounting clamp.

The remaining operations are carried out in reverse order. The alternator bolts are tightened to 2.0 kgm (14.5 ft.lb.).

Petrol Engine

Fig. 1.74 shows the attachment of the dipstick tube on these engines. The tube is removed as follows:

- Withdraw the oil dipstick (1) from the tube (3).

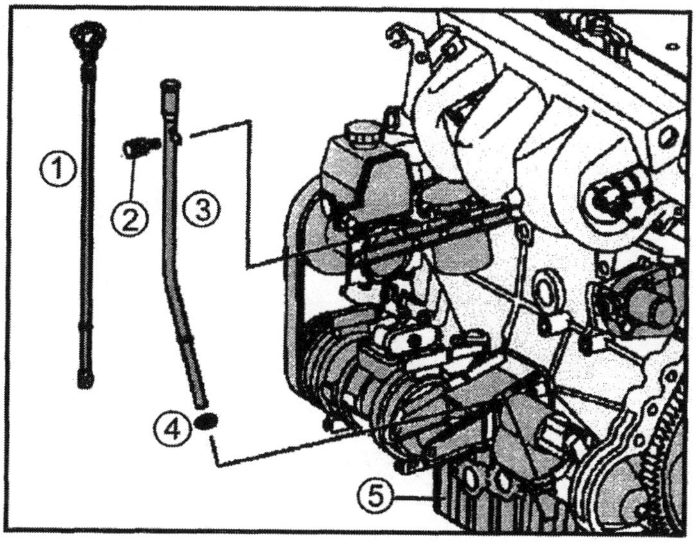

Fig. 1.74 – Details for the removal and installation of the oil dipstick guide tube of a petrol engine.

1 Oil dipstick
2 Combi screw
3 Dipstick guide tube
4 Retainer
5 Oil sump

- Unscrew the combi screw (2) at the upper end and withdraw the tube (3) from the oil sump (5). Some oil may run out, collect it. The retainer (4) at the lower end must always be replaced.

The installation is a reversal of the removal procedure. The bolt (2) is tightened to 1.0 kgm.

1.6.2.5. Engine Oil Change

The engine oil should be changed every 10,000 miles. Remember that there are a few litres of engine oil to handle and the necessary container to catch the oil must be large enough to receive the oil. Dispose of the old oil in accordance with the local laws. Never discharge the engine oil into a drain. Drain the oil as follows, when the engine is fairly warm:

- Jack up the front end of the vehicle and place the container underneath the oil sump. Unscrew the oil drain plug (ring spanner or socket). Take care, as the oil will "shoot" out immediately. Remove the oil filler cap to speed-up the draining.
- Check the plug sealing ring and replace if necessary. Clean the plug and fit and tighten to 3.0 kgm (22 ft.lb.).
- Fill the engine with the necessary amount of oil. Make sure that the oil is suitable for diesel engines.
- Refit the oil filler cap and drive the vehicle until the engine operating temperature is reached. Jack up the vehicle once more and check the drain plug area for oil leaks.

1.6.2.6. Oil Pressure Switch – Removal and Installation

The switch is located at the front end of the engine, L.H. side above one of the belt pulleys. The switch can be removed after withdrawing the cable connector. The switch is tightened to 2.0 kgm (14 ft.lb.).

1.6.3. ENGINE OIL PRESSURE

The oil pressure can only be checked with an oil pressure gauge, which is fitted with a suitable adapter in place of the oil pressure switch. We recommend to leave the oil pressure check to a workshop. Low oil pressure can also be caused through a low oil level in the sump.

1.6.4. OIL FILLER NECK

The arrangement of the oil filler neck is constructed in an unusual manner, as shown in Fig. 1.75. The oil is filled into the oil filler (1) at the upper end of the engine and then directed via a hose (2) and the oil filler pipe (3) to the oil sump. The oil filler (1) is fitted to the air cleaner housing. To remove the individual items proceed as follows:

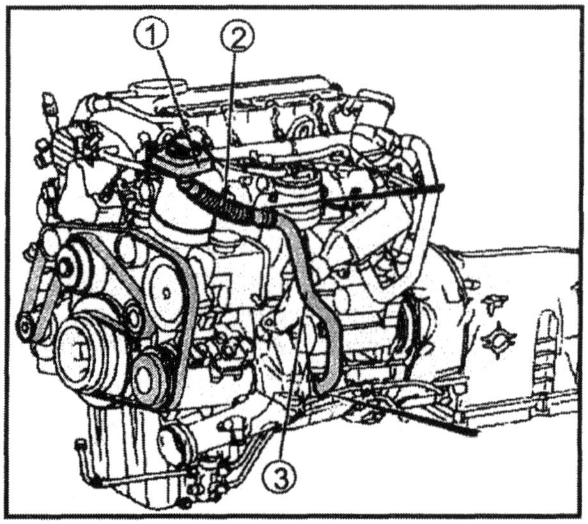

Fig. 1.75 – Details for the removal and installation of the oil filler neck (diesel) the numbers are referred to in the text.

- Remove the oil filler device (1) from the air cleaner housing by pressing the retaining spring towards the rear and removing the oil filler upwards.
- Disconnect the hose (2) from the filler neck (1).
- Remove a bolt from the fuel filter bracket, located at the position shown by the upper arrow.
- On models 639.7/8 after VIN number 055347 it will be necessary to remove the engine compartment panelling from underneath the engine.
- Remove a bolt located at the position shown by the lower arrow, remove a second bolt from the oil sump and withdraw the oil filler tube (3) towards the bottom. A sealing ring is inserted which must always be replaced.

The installation is a reversal of the removal procedure. The bolt securing the filler tube (3) to the oil sump is tightened to 1.2 kgm (9 ft.lb.).

1.7. Cooling System

The cooling system operates with an expansion tank, on one side of the engine compartment. A coolant level indicator is fitted into the expansion tank. If the level drops below the "Min" mark for any reason, the switch contacts will close and light up a warning light in the instrument panel. A check of the coolant level is therefore redundant.

The water pump is fitted to the front at the bottom of the cylinder block, is sealed off by means of an "O" sealing ring and secured by means of four bolts. The thermostat is mentioned later on.

1.7.0. TECHNICAL DATA

Type: .... Water pump-assisted thermo-siphon system with
.... impeller-type water pump
Filling Capacity: .... See Page 6
Anti-freeze amount: .... See Section 1.7.1.0
Thermostat:
- Opens at: .... 87° C
- Fully opens at: .... 102° C

1.7.1 COOLANT - DRAINING AND REFILLING

- If the engine is hot open the expansion tank cap to the first notch and allow the pressure to escape. The coolant must have a temperature of less than 90° C. but we recommend to wait until the temperature has dropped to 50° C. In all cases use a thick rag to cover the expansion tank cap to protect your hand.
- Remove the noise dampening panel from underneath the vehicle on models 639.7/8 after VIN number 055347.
- Unscrew the coolant drain plug at the bottom of the radiator. Collect the coolant in a suitable container. If the anti-freeze solution is in good condition, your can re-use it. Poke a piece of wire into the bore of the drain hole to dislodge sludge, if

the coolant flow is restricted. On a diesel engine there is also a plug in the side of the cylinder block at the position shown in Fig. 1.76, on a petrol engine there is a drain plug on both sides of the cylinder block. From the surrounding details shown in Fig. 1.77 you will be able to located the drain plugs. Remove the plug(s) with a suitable spanner. The plug(s) are refitted and tightened to 1.2 kgm (petrol engine) or 3.0 kgm (diesel engine) after the coolant has drained.

Fig. 1.76 – The cylinder block drain plug of a diesel engine is located at the position shown. The engine is removed for clarity.

To ensure that the cooling system is filled without an air lock, proceed as follows when filling in the coolant. Refer to Section 1.7.1.0 for the correct anti-freeze amount to be added. Anti-freeze marketed by Mercedes-Benz should be used, as this has been specially developed for the engine.

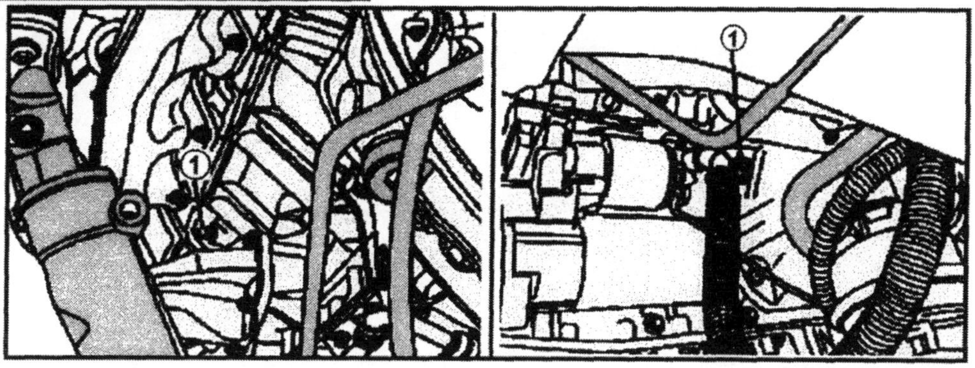

Fig. 1.77 – The location of the coolant drain plugs of a petrol engine. On the L.H. side is the plug on the L.H. side of the block. The R.H. view shows where the plug on the other side is located.

- Set both heater switches to the max. heating capacity, by moving the controls. If an automatic climatisation system is fitted, press the "DEF" button.
- Fill the pre-mixed anti-freeze solution into the expansion tank filler neck until it reaches the "Cold" mark. Do not fit the expansion tank cap at this stage.
- Start the engine and run it until the operating temperature has been reached, i.e. the thermostat must have opened. Fit the cap when the coolant has a temperature between 60° to 70° C. A thermometer can be inserted into the radiator filler neck to check the temperature. If the system has been filled correctly the coolant level should be approx. 1 cm above the level mark on the reservoir.
- Check the coolant level after the engine has cooled down and correct if necessary once more.
- Refit the noise dampening panel underneath the vehicle if removed.

1.7.1.0. Anti-freeze Solution

The cooling system is filled with anti-freeze when the vehicle leaves the factory and the solution should be left in the system throughout the year. When preparing the anti-freeze mixture, note the following ratio between water and anti-freeze solution. We recommend to use the anti-freeze supplied by Mercedes-Benz. It may cost you a little more, but your engine will thank you for it. The following ratios should be observed:

Engine – Diesel – Total Capacity 8.6 litres:

To –37° C:

- 4.3 litre anti-freeze, 4.3 litre of water

To –45° C:

- 4.7 litre anti-freeze, 3.9 litre of water

Engine – Diesel – Total Capacity 10.5 litres:

To –37° C:

- 5.3 litre anti-freeze, 5.2 litre of water

To –45° C:

- 5.8 litre anti-freeze, 4.7 litre of water

Engine – Petrol – Total Capacity 8.1 litres:

To –37° C:

- 4.1 litre anti-freeze, 4.0 litre of water

To –45° C:

- 4.5 litre anti-freeze, 3.6 litre of water

Engine – Petrol – Total Capacity 9.9 litres:

To –37° C:

- 5.0 litre anti-freeze, 4.9 litre of water

To –45° C:

- 5.5 litre anti-freeze, 4.4 litre of water

1.7.2. RADIATOR AND COOLING FAN

1.7.2.0. Checking Radiator Cap and Radiator

The cooling system operates under pressure. The expansion tank cap is fitted with a spring, which is selected to open the cap gasket when the pressure has risen to the value engraved in the cap. If the cap is replaced, always fit one with the same marking, suitable for the models covered.

To check the radiator cap for correct opening, a radiator test pump is required. Fit the pump to the cap and operate the pump until the valve opens, which should take place near the given pressure (1.2 kg/sq.cm. = 17 psi.). If this is not the case, replace the cap. Fig. 1.78 shows the working principle of such a test pump.

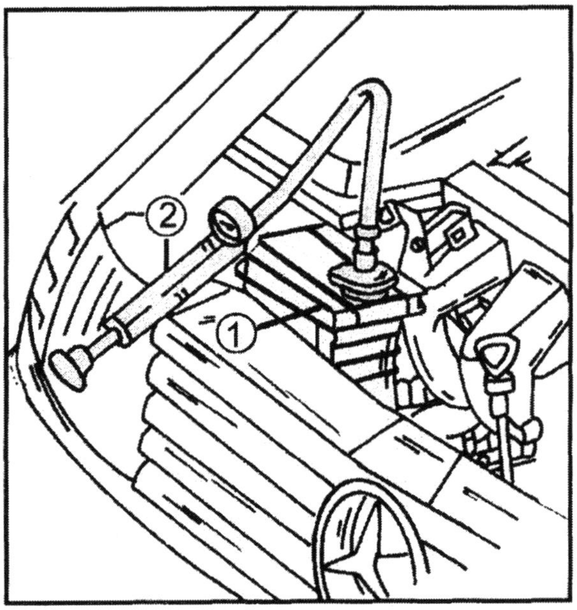

Fig. 1.78 – A radiator test pump (2) is used to check the cooling system for leaks and the expansion tank cap for correct opening. The pump is connected to the expansion tank (1).

The same pump can also be used to check the cooling system for leaks. Fit the pump to the expansion tank filler neck and operate the plunger until a pressure of 1.0 kg/sq.cm. is indicated. Allow the pressure in the system for at least 5 minutes. If the pressure drops, there is a leak in the system.

1. 7.0.1. Radiator – Removal and Installation

The removal and installation of the radiator is a time consuming operation as the complete front end of the vehicle must be removed to gain access to the radiator. The pressure of a fitted air conditioning system must be released to take out the radiator. ***The following description therefore only refers to vehicles without an air conditioning system, as the pressure cannot be released by ordinary means.***

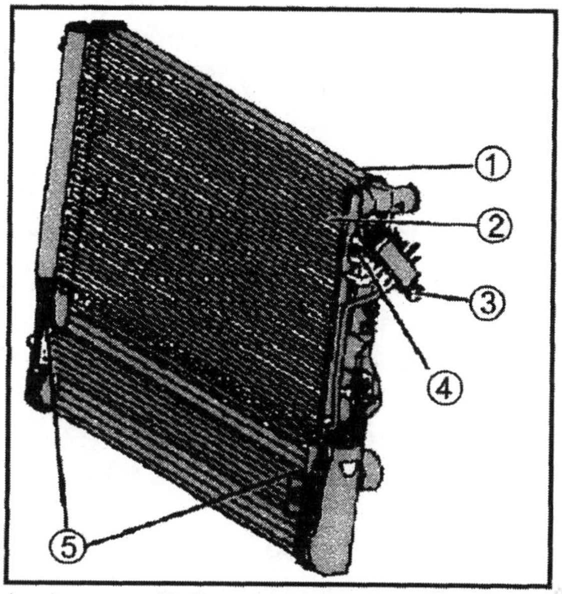

Fig. 1.79 – View of the radiator with the location of some of the parts.

1 Radiator
2 A/C condenser
3 Pressure sensor (A/C only)
4 Clip
5 Bracket

There are differences between petrol and diesel models which must be taken note of as the operations are carried out. Fig. 1.79 shows the radiator assembly as it is removed, but note this includes the parts fitted if an A/C system is installed, but should be ignored.

- Disconnect the battery negative cable.
- Drain the cooling system as described in Section 1.7.1. On a model with automatic transmission, use clamps and clamp off the two hoses for the oil cooler. Disconnect the hoses from the R.H. side of the radiator. Some fluid will drip out. Immediately plug the hose ends and connections in a suitable manner.
- Remove the fan shroud as described below for the petrol and diesel engine. This involves the removal of the fan in the case of a diesel engine or the removal of the air cleaner housing and the fan from the viscous clutch in the case of the petrol engine.
- Remove the front grille. Looking at the top of the grille (bonnet open) you will see three bolts (one on the outside on each end and one in the centre). After removal of the bolts, lift the grille up and out.
- Disconnect all coolant hoses from the radiator (top and bottom) and from the water pump.
- In the case of a diesel model unclip the wiring harness running along the top of the radiator and remove the side radiator covers upwards.
- From the bottom of the vehicle remove the radiator mounting bolts at the left and at the right and lift the radiator up and out.

The installation is a reversal of the removal procedure. Take care that the two guide pins enter the locating bores in the radiator crossmember. Tighten the cylinder block drain plug to 3.0 kgm (21 ft.lb.) and the radiator drain plug to 1.5 kgm (11 ft.lb.).

Refit the front grille in reverse order.

After installation of the radiator check that the gap between the fan blades and the radiator air baffle is the same on all sides. If necessary remove the securing clips and centre the baffle.

If an automatic transmission is fitted, re-connect the hose connections. Check the fluid level in the automatic transmission, if you think some fluid has drained. Finally refill the cooling system as described in Section 1.7.1. Start the engine and check all cooling hose and drain points for leaks.

1.7.3. FAN SHROUD

Diesel Engine

The fan must be removed to take out the fan shroud. Some of the items must be located before they can be removed. Proceed as follows:

- Remove the air cleaner housing and drain the cooling system.
- Detach the upper coolant hose from the radiator connections, unclip the coolant hoses at the upper end of the fan shroud and push them towards the rear.
- Remove the top radiator cover. To do this, take off the expanding clips and pull off the cover towards the rear.
- Looking at the front of the engine compartment disconnect the coolant hoses from the radiator.
- Unlatch the side radiator covers at the top and tilt them to the back.
- Disconnect the charge air hose on the R.H. and L.H. side of the charge air cooler.
- Unlatch the side radiator covers at the bottom and tilt them to the back.
- The radiator must be protected from damage. To do this, prepare a sheet of metal or plastic, 1 mm thick, and with a dimension of 400 x 680 mm and insert it in front of the radiator. Make up two little brackets to attach the plate to the radiator so it cannot move during the next operations.
- Unscrew the bolt in the centre of the viscous clutch in the centre of the water pump and carefully lift the fan upwards. The viscous is covered under a separate heading later on.
- The fan shroud can now be removed in an upwards direction.

The installation is a reversal of the removal procedure. The bolt securing the viscous clutch to the water pump is tightened to 4.5 kgm (32.5 ft.lb.).

Petrol Engine

The viscous fan clutch must be removed to take out the fan shroud. Some of the items must be located before they can be removed. Proceed as follows:

- Remove the air cleaner housing and drain the cooling system.
- Detach the two coolant hoses from the top of the fan shroud and push them towards the rear.
- Below the coolant hoses remove the electrical wiring harness and place the harness towards the rear.
- Remove the top radiator cover. To do this, take off the expanding clips on the left and on the right and pull off the cover towards the rear.
- Looking at the front of the engine compartment disconnect the coolant hoses from the radiator on the left and on the right.
- Unlatch the side radiator covers at the top and tilt them to the back (on both sides).
- Unscrew the bolt in the centre of the viscous clutch in the centre of the water pump and carefully lift the fan upwards. The viscous is covered under a separate heading later on.
- The fan shroud can now be removed in an upwards direction.

The installation is a reversal of the removal procedure. The bolt securing the viscous clutch to the water pump is tightened to 4.5 kgm (32.5 ft.lb.).

1.7.4. WATER PUMP (Petrol Engine)

The removal and installation of the pump is not a very difficult operation, apart from the fact that the front of the engine must be exposed to gain access to some of the parts. These include the electric air pump, the viscous fan and the Poly V-Belt. Fig. 1.80

shows the bolts securing the pump and you will see that bolts of different length are used.

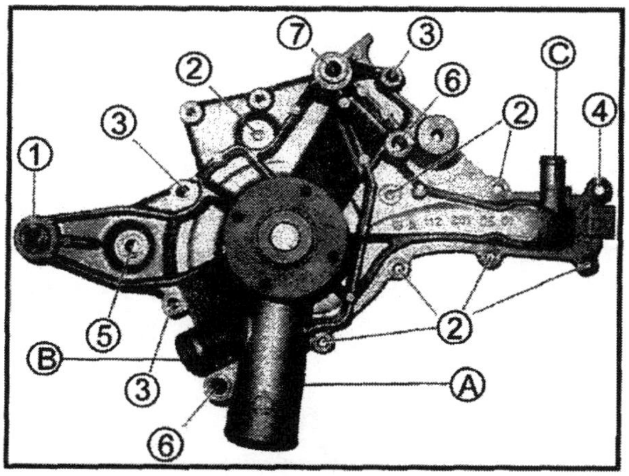

Fig. 1.80 – Details for the removal and installation of the water pump (petrol). Note the different bolt length.

1 Bolt, alternator to timing case
2 Bolt, M6 x 19 mm
3 Bolt, M6 x 43 mm
4 Bolt, M8 x 80 mm
5 Bolt, M8 x 90 mm
6 Bolt, M8 x 110 mm
7 Bolt, pulley to coolant pump

- Remove the air cleaner housing and the vibration damper for the crankshaft.
- Drain the cooling system as described earlier on. Only drain the radiator. Remember the engine must be cold.

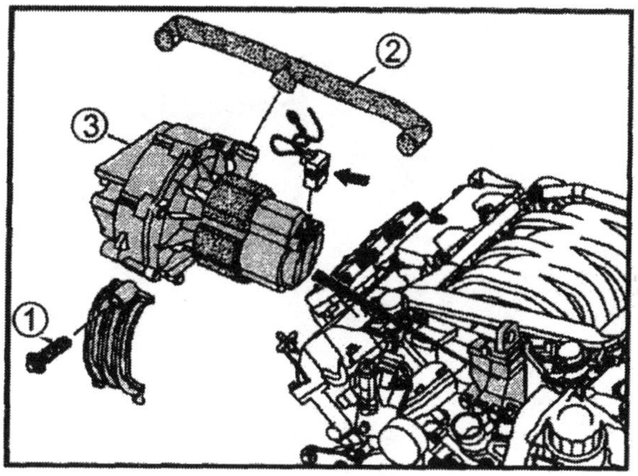

Fig. 1.81 – Removal of the air pump. The numbers are referred to bellow.

- Remove the air pump as follows: Fig. 1.81 shows where the air pump is located. Disconnect the cable connector plug (arrow), remove the bolt (1), take off the securing clamp and place the earth cable to one side (bolt torque 0.8 kgm), disconnect the air hose (2) and remove the electric pump (3). The installation of the pump is carried out in reverse order.
- Remove the coolant temperature sensor, located just above the pump. A spring clip must be withdrawn to take out the sensor.
- Remove the viscous fan as described later on.
- Remove the Poly V-belt as described later on.
- Remove the water pump pulley. Hold the pulley against rotation by inserting a screwdriver blade between one of the bolts and the centre hub and remove the bolts securing the water pump pulley and withdraw the pulley. Tighten the pulley bolts to 0.9 kgm (7 ft.lb.) during installation. Also remove the belt guide pulley from the coolant pump. The pulley is secured by bolt (7) in Fig. 1.80 but is hidden by a cover plug which must be removed with a small screwdriver. Tighten the bolt to 0.9 kgm (7 ft.lb.).
- Remove the coolant hoses (A) and (B) in Fig. 1.80 and a further hose (C) after undoing the hose clamps from the water pump.
- At the upper left-hand side of the pump remove a bracket. Place the bracket with the connected lines to one side.
- Unscrew the water pump from the cylinder block and remove it towards the bottom. Note the length of the different bolts, given in Fig. 1.80.

The water pump cannot be overhauled and must be replaced in case of damage or wear.

The installation is a reversal of the removal procedure. Make sure that the sealing faces of coolant pump and timing case cover are free of oil or grease. Attach the new gasket with two or three spots of sealing compound to the water pump housing.

Tighten the bolts securing the pump to the timing case to 2.0 kgm (14.5 ft.lb.). The water pump pulley is tightened also to 0.9 kgm.

Finally refill the cooling system, start the engine and check all parts belonging to the cooling system for leaks.

1.7.5. WATER PUMP (Diesel Engine)

The water pump is fastened to a light-alloy housing which is bolted to the timing case cover. Note that two V-belt guide pulleys must be removed as two of the water pump securing bolts are located behind the guide pulleys. The removal and installation can be carried out as follows:

- The front of the engine can be exposed to remove the pump. This means you will have to remove the front grille as described earlier on during the removal of the radiator.
- Drain the cooling system. It is enough to open the drain tap in the radiator. Never open the expansion tank if the coolant temperature exceeds 90° C.
- Remove the fan shroud (see earlier on).
- The radiator must be protected from damage. To do this, prepare a sheet of metal or plastic, 1 mm thick, and with a dimension of 400 x 680 mm and insert it in front of the radiator. Make up two little brackets to attach the plate to the radiator so it cannot move during the next operations.

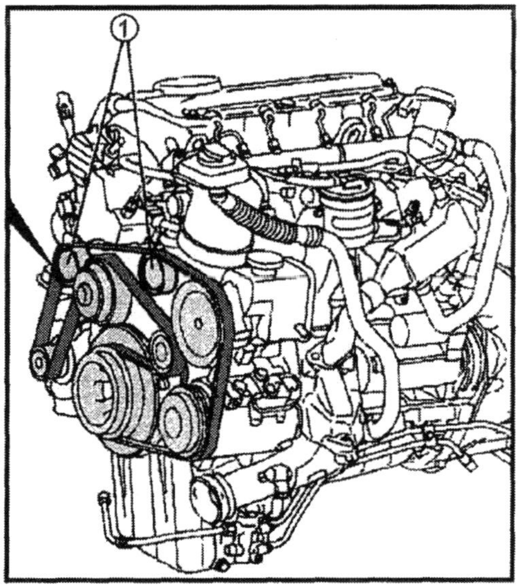

Fig. 1.82 – View of the Poly V-belt drive. The two guide pulleys (1) must be removed after the belt has been taken off. Both pulleys are fitted with an end cover in the centre which must be removed.

- Remove the Poly V-belt as described later on and then remove the two guide pulleys shown in Fig. 1.82. An end cover is placed over the centre of each pulley which must be removed with a small screwdriver. Prevent the pulley from rotating and remove the pulley securing bolt. The bolts are tightened to 3.5 kgm (25 ft.lb.) during installation. One pulley is fitted to the water pump, the second one to the timing case cover. As already mentioned above, two of the pump securing bolts are located behind the guide pulleys.
- Detach the coolant hoses from the water pump after slackening of the respective hose clamps. Check the condition of hoses and clamps before re-using them.
- Unscrew the bolts from the pump. Withdraw the pump and take off the gasket. The gasket must always be replaced.
- If the coolant (water) pump is replaced remove the belt pulley from the old pump. Hold the pulley against rotation by inserting a screwdriver blade between one of the bolts and the centre hub and remove the bolts securing the water pump pulley. Withdraw the pulley. Tighten the pulley bolts to 1.0 kgm (7.2 ft.lb.) during installation.
- The installation of the water pump is a reversal of the removal procedure. Clean the sealing surfaces before installation and replace the gasket, as already mentioned above. Tighten the bolts to 1.4 kgm (10 ft.lb.). Finally refit all removed parts and fill the cooling system.
- Start the engine and check all connections of the cooling system for leaks.

1.7.6. DRIVE BELTS AND DRIVE BELT TENSION (Petrol)

A single flat belt, so-called "poly belt" is used to drive the units. The layout of the belt depends on the equipment. Figs. 1.83 shows the two versions. One of the layouts will be fitted to your engine.

The tension of the belt is held under constant tension by means of a tensioning device. A fitted shock absorber prevents vibration in the belt system.

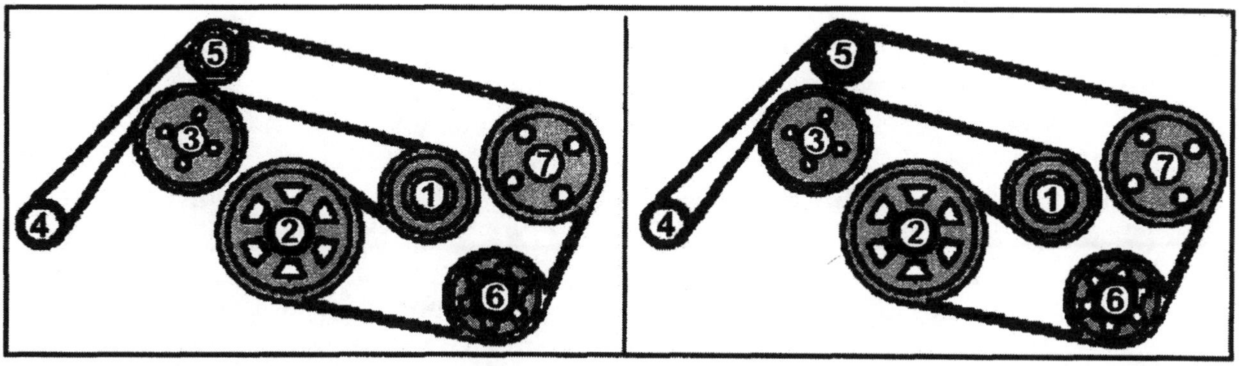

Fig. 1.83 – The layout of the drive belt of a petrol engine. On the left for vehicles without A/C system, on the right with A/C system. Note that two guide pulleys are fitted to models without A/C.

On the left:
1 Tensioning roller
2 Crankshaft pulley
3 Water pump pulley
4 Alternator pulley
5 Guide pulley
6 Guide pulley
7 Power steering pump pulley

On the right:
1 Tensioning roller
2 Crankshaft pulley
3 Water pump pulley
4 Alternator pulley
5 Guide pulley
6 A/C compressor pulley
7 Power steering pump pulley

The belt drive should be checked every 12,000 miles. To do this, mark an easy accessible part of the belt with chalk and crank the engine by at least one turn by applying a socket at the centre bolt of the crankshaft pulley. Check the whole length of the belt, i.e. until the chalk mark can be seen once more. From the condition of the belt you will be able to judge if a new belt is required.

Remove the belt as follows:

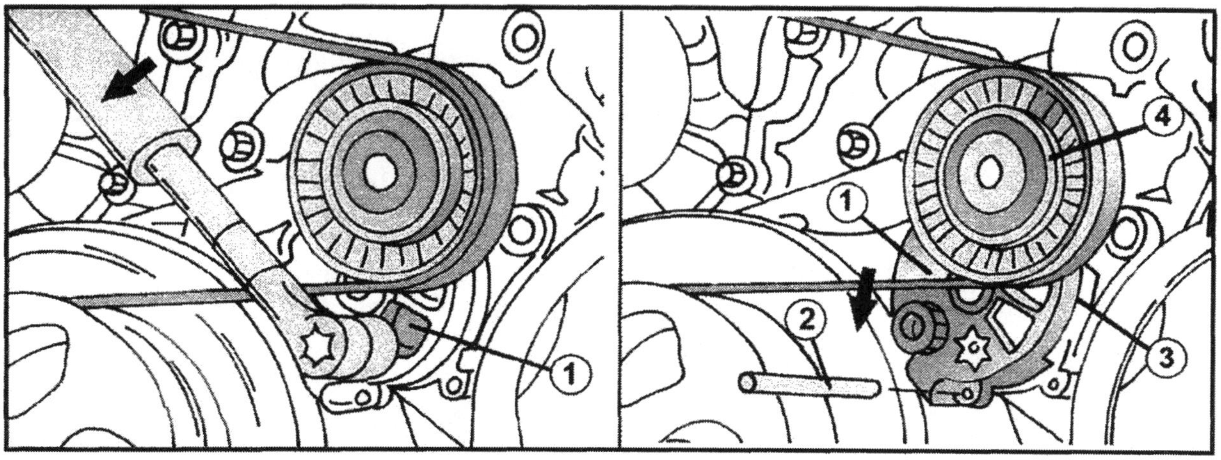

Fig. 1.84 – Removal of the Poly V-belt of a petrol engine. The numbers are referred to on the text.

- Remove the air cleaner housing.
- Unclip the coolant hose between the radiator and the thermostat housing from the bracket and the clamp behind the oil filter housing and place the coolant hose to one side.

- Refer to the L.H. view of Fig. 1.84 and turn the front part (1) of the pulley tensioning device against the spring force, i.e. turn it in the direction of the arrow. Then refer to the R.H. view and insert the pin (2) of 4 mm diameter (or drift, or drill shank) into the holes, positioned in line in the front part and in the housing (3) in order to lock the tensioning pulley (4) in this position. The V-belt can now be removed from the various pulleys.

The installation is carried out as follows:

- Check the removed V-belt for damage and traces of wear before installation. Replace the belt if necessary, fitting the correct belt.
- Fit the V-belt over the various pulleys, in all cases first over the tensioning pulley (1) in Fig. 1.83. Then depending on the version fit it over the remaining pulleys shown in Fig. 1.83. Make sure the belt engages with the grooves in the pulleys.
- Refer to Fig. 1.85 and turn the front part (1) of the tensioning device sufficiently against the force of the spring (direction of the arrow, R.H. view) until it is possible to withdraw the inserted pin (2) or drift/drill bit, depending what has been used out of the holes in the front part (1) and the housing (3) of the tensioning device (3).

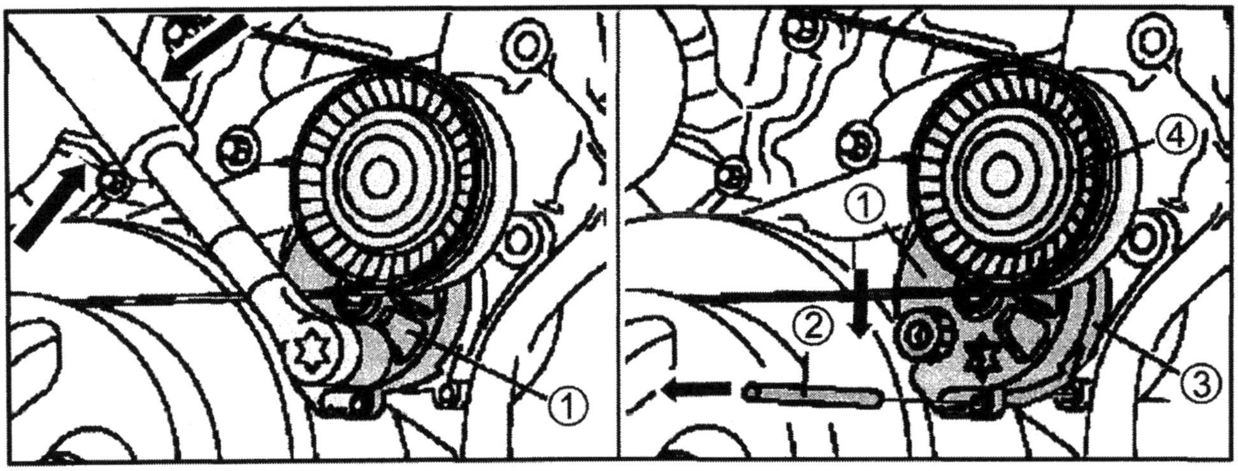

Fig. 1.85 – Details for the installation of the poly V-belt of a petrol engine. The numbers refer to the description.

- Slowly turn the front part (1) of the tensioning device in the L.H. view in the direction of the arrow until the V-belt is automatically tensioned by the force of the spring. The belt is now tensioned and no further operations are necessary. Rotate the crankshaft a few times before and re-check the belt drive before the removed parts are refitted.

1.7.7. DRIVE BELTS AND DRIVE BELT TENSION (Diesel)

A single drive belt, also known as Poly V-belt, is fitted to the front of the engine, but the layout of the belt is not the same on all engines and depends on a fitting of an air conditioning system or a vehicle without A/C system. If a compressor is fitted, the length of the belt is different. The same belt drive system is, however, fitted to all engines and is held in its correct tension by means of an automatic tensioning device. The length of the belt is also different, i.e. different for vehicles without A/C system and vehicles with A/C system. If a new belt is fitted make sure you obtain the correct one. The removal of the belt requires a drift or pin of 4 mm diameter or a 4 mm drill, the use of which is described in the following instructions. After the front of the engine has been laid bare of all obstructing parts remove the belt as follows. Fig. 1.86 shows details for the removal and installation.

- Disconnect the battery earth cable.
- Remove the air cleaner housing.

Fig. 1.86 – Details for the removal and installation of the Poly V-belt.
1 Poly V-belt
2 Tensioning device
3 4 mm pin
4 Tensioning pulley

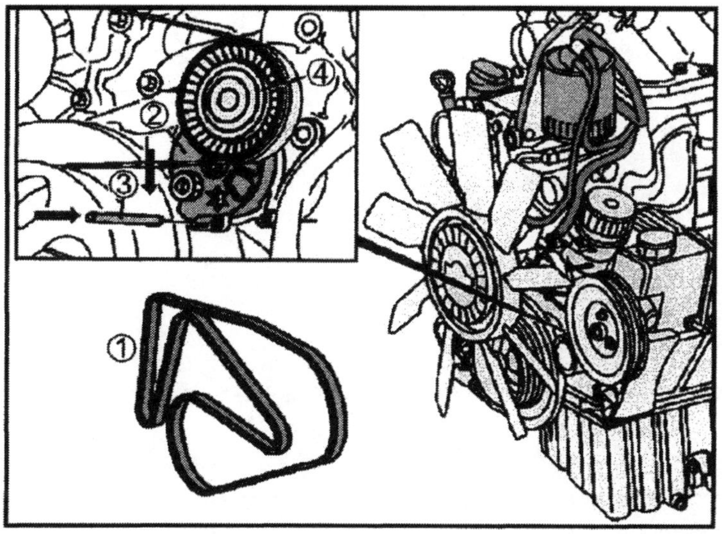

- Refer to the L.H. view of Fig. 1.84 and turn the front part (1) of the pulley tensioning device against the spring force, i.e. turn it in the direction of the arrow. The refer to the R.H. view and insert the pin (2) of 4 mm diameter (or drift, or drill shank) into the holes, positioned in line in the front part and in the housing (3) in order to lock the tensioning pulley (4) in this position. The V-belt can now be removed from the various pulleys.

The installation is carried out as follows:

- Fit the V-belt over the various pulleys, in all cases first over the tensioning pulley (4) in Fig. 1.86. Then depending on the version fit it over the remaining pulleys shown in Fig. 1.87. It should be noted that the pulley (6) drives the compressor on a vehicle with air conditioning whereas a guide pulley is used on a vehicle without A/C system. Make sure the belt engages with the grooves in the pulleys.
- Refer to Fig. 1.85 and turn the front part (1) of the tensioning device sufficiently against the force of the spring (direction of the arrow, R.H. view) until it is possible to withdraw the inserted pin (2) or drift/drill bit, depending what has been used out of the holes in the front part (1) and the housing (3) of the tensioning device (3).

Fig. 1.87 – The poly V-belt is fitted over the various pulleys as shown. The layout is the same on all engines. Note the comment about pulley (6) given above.
1 Tensioning pulley
2 Crankshaft pulley
3 Coolant pump pulley
4 Alternator pulley
5 Guide pulley
6 Compressor/Guide pulley
7 Steering pump pulley
8 Guide pulley

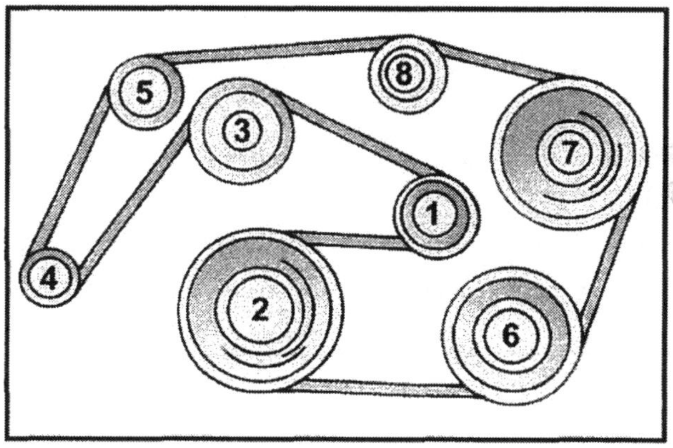

Check the removed V-belt for damage and traces of wear before installation. Replace the belt if necessary, fitting the correct belt.

- Pull out the pin (2).
- Slowly turn the front part (1) of the tensioning device in the L.H. view in the direction of the arrow until the V-belt is automatically tensioned by the force of the spring. The belt is now tensioned and no further operations are necessary. Rotate the crankshaft a few times more and re-check the belt drive before the removed parts are refitted.

1.7.8. VISCOUS FAN CLUTCH

Petrol Engine

The removal and installation of the viscous fan clutch (fluid coupling) presents no particular problems after access to the front of the engine has been created. The air cleaner housing must be removed. The fan clutch is attached as shown in Fig. 1.88. The workshop uses a special open-ended spanner to apply to the securing nut to enable the application of a torque wrench, as shown in Fig. 1.89. Otherwise you will have to estimate the torque during installation. Proceed as follows:

Fig. 1.88 – Removal and installation of the cooling fan and viscous fan clutch.
1 Torque wrench
2 Special open-ended spanner
3 Coolant pump pulley
4 Viscous fan

- Disconnect the battery earth cable, observing the necessary precautions.
- Counterhold the water pump pulley in a suitable manner. The workshop uses a special counter holder which has two pins and can be inserted into two holes in the pulley.
- Undo the union nut at the rear of the viscous fan as shown in Fig. 1.88 and lift out the fan wheel (4)from the engine compartment.
- If necessary remove the screws and take the fan wheel off the fluid clutch (4).

Fig. 1.89 – The water pump pulley can be prevented from rotating as shown with the special counter-holder (1).

The installation is a reversal of the removal procedure. The installation position relative to the viscous fan clutch is fixed by the rib on the fan wheel. Attach the fan wheel with 1.0 kgm to the clutch and tighten the nut to 4.5 kgm (32.5 ft.lb.). The water pump drive wheel must again be prevented from rotating as described above.

Diesel Engine

Follow the instructions in section 1.7.3 (fan shroud) until the protective plate is inserted as explained. The fan clutch is secured as shown in Fig. 1.90.

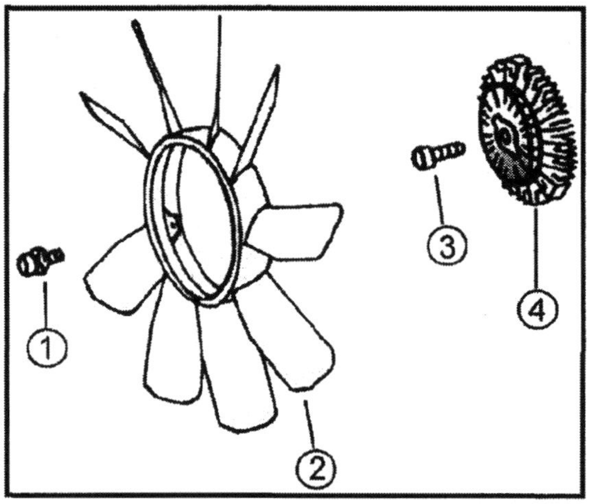

Fig. 1.90 – Fan and viscous fan clutch of a diesel engine.
1 Fan blade bolt
2 Fan blade
3 Bolt, 4.5 kgm
4 Viscous fan clutch

- Undo the centre bolt (3) securing the fluid coupling at the coolant pump and lift out the fan wheel (2) in an upwards direction.
- If necessary remove the screws (1) and take the fan wheel off the fluid clutch (4).

The installation is a reversal of the removal procedure. The installation position relative to the viscous fan clutch is fixed by the rib on the fan wheel. Attach the fan wheel with 1.0 kgm to the clutch and the centre bolt (3) to 4.5 kgm (32.5 ft.lb.). The water pump drive wheel must again be prevented from rotating as described above.

Fig 1.91 – Details for the removal and installation of the thermostat of a petrol engine.

1 Cover screw
2 Hose to expansion tank
3 Thermostat housing cover
4 Hose to radiator
5 Thermostat
6 Sealing ring
7 Thermostat housing

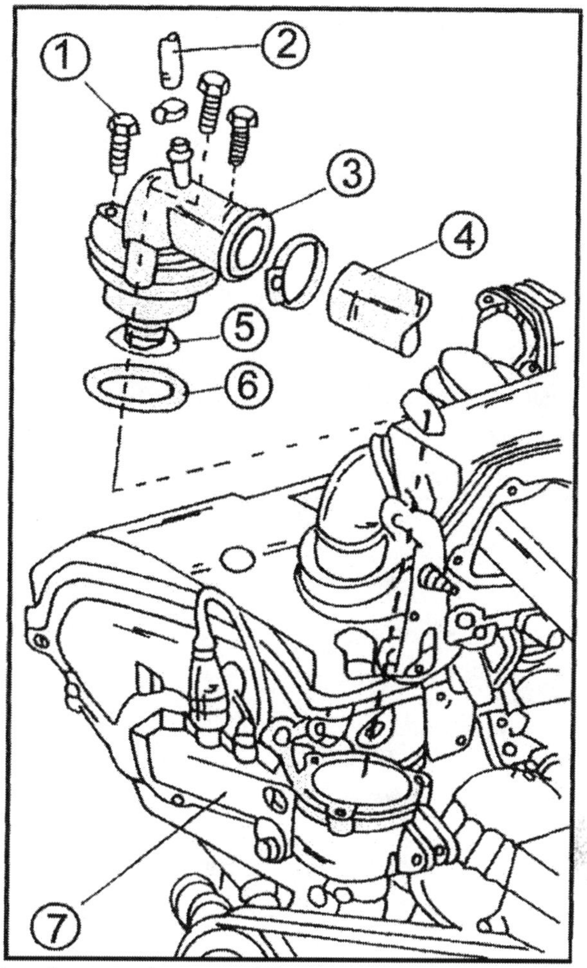

1.7.9. THERMOSTAT

The thermostat of a petrol engine is fitted into a cover, where the large cooling hose is connected, as can be seen in Fig. 1.91. It should be noted that the thermostat cannot be removed from its cover, i.e. a new thermostat means that a complete cover must be fitted. The thermostat of a diesel engine can only be removed together with the thermostat housing.

Petrol Engine

The thermostat can be removed after draining the cooling system. Then refer to Fig. 1.91:

- Remove the air cleaner housing.
- Disconnect the hoses (2) and (4) from the housing cover (3) after slackening the hose clamps. The cover can now be removed together with the thermostat.

The installation is a reversal of the removal procedure. The "O" sealing ring (6) must be replaced. Tighten the thermostat housing cover bolts to 1.4 kgm (10 ft.lb.). The thermostat cannot be repaired and must be replaced as specified above.

Diesel Engine

The thermostat is fitted inside a housing to the cylinder head and can be found at the position shown in Fig. 1.92. The thermostat itself is an integral part of the thermostat housing i.e. a defective thermostat can only be replaced together with the thermostat housing. Remove the thermostat assembly as follows:

Fig. 1.92 – Thermostat location (all models).

- Remove the air filter housing.
- Drain the cooling system. Always wait until the temperature is below 90° C. Even then remove the expansion tank cap very slowly. Use a thick rag to protect your fingers and hands. Only turn the cap to its first stop to allow all vapour to escape. It is enough to drain the cooling system to the level of the thermostat.
- Disconnect the large coolant hose after slackening of the hose clamp from the thermostat housing. The hose clamp must be replaced during installation. The coolant hose must be replaced if no longer in good condition.

- Remove the retaining bracket securing the fuel pipe after unscrewing the screw.
- Unscrew the thermostat housing securing bolts and take the housing off the cylinder head and disconnect the plug from the coolant temperature sensor which is fitted to the thermostat housing. The gasket underneath the housing must always be replaced.

The installation is a reversal of the removal procedure observing the points given already above. The sealing faces of thermostat housing and cylinder block must be thoroughly clean before the gasket is positioned. Tighten the securing bolts to the cylinder head with 0.8 kgm (6.5 ft.lb.).

A thermostat cannot be repaired, but can be tested by immersing it in a container of cool water and gradually raising the temperature to check the opening temperature (also applies to the thermostat of a petrol engine).

- Suspend the thermostat on a piece of wire so that it does not touch the sides or the bottom of the container. Suspend a thermometer in a similar manner. Fig. 1.93 shows the arrangement, but remember that the complete housing must be immersed, not only the thermostat as shown.

Fig. 1.93 – Testing a thermostat.

- Gradually heat the water and observe the thermometer. The thermostat should begin to open at 87° C and should be fully open at 105° C.
- Otherwise replace the thermostat. Allow the thermostat to cool down and check if it closes properly.

The installation is a reversal of the removal procedure. When fitting the thermostat use a new sealing ring. Finally fill the cooling system and check the whole system for leaks.

1.8. Diesel Fuel Injection System

1.8.0. INTRODUCTION

Absolute cleanliness is essential during any repairs or work on the diesel fuel injection system, irrespective of the nature of the work in question. Thoroughly clean union nuts before unscrewing any of the injection pipes.

The high pressure injection pump, the injectors and the injection pipes are the main components of the fuel injection system.

The operations described on the following pages are headed accordingly. In all cases we recommend to read the instructions in full before you commence with any removal work. In general there are no special tools required, but you will need some experience working on the engine. Many of the removal operations are only mentioned and you will have to refer to the relevant chapter or section for a full description.

PRECAUTIONS WHEN WORKING ON DIESEL INJECTION SYSTEMS

Whenever repairs are carried out on a diesel fuel injection system, whatever the extent, observe the greatest cleanliness.

- Only carry out work on diesel injection systems under the cleanest of conditions. Work in the open air should only be carried out when there is no wind, to prevent dust entering open connections.
- Before removal of any union nut clean all around it with a clean cloth.
- Removed parts must only be deposited on a clean bench or table and must be covered with a sheet of plastic or paper. Never use fluffy shop rags to clean parts.
- All open or partially dismantled parts of the injection system must be fully covered or kept in a cardboard box, if the repair is not carried out immediately.
- Check the parts for cleanliness before Installation.
- Never use an air line to clean the exterior of the engine when connections of the injection system are open. With the availability of air compressors which can be plugged into a cigar lighter socket, you may be tempted to use air for cleaning.
- Take care not to allow diesel fuel in contact with rubber hoses or other rubber parts. Immediately clean such a hose if it should happen accidentally.
- Do not use naked lights anywhere near the engine. Although diesel fuel is less inflammable, there is always the risk of fire.

1.8.1. HIGH PRESSURE INJECTION PUMP – REMOVAL AND INSTALLATION

The precautions given above must be observed at all times.

Fig. 1.94 shows the parts to be disconnected or removed during the removal of the pump.

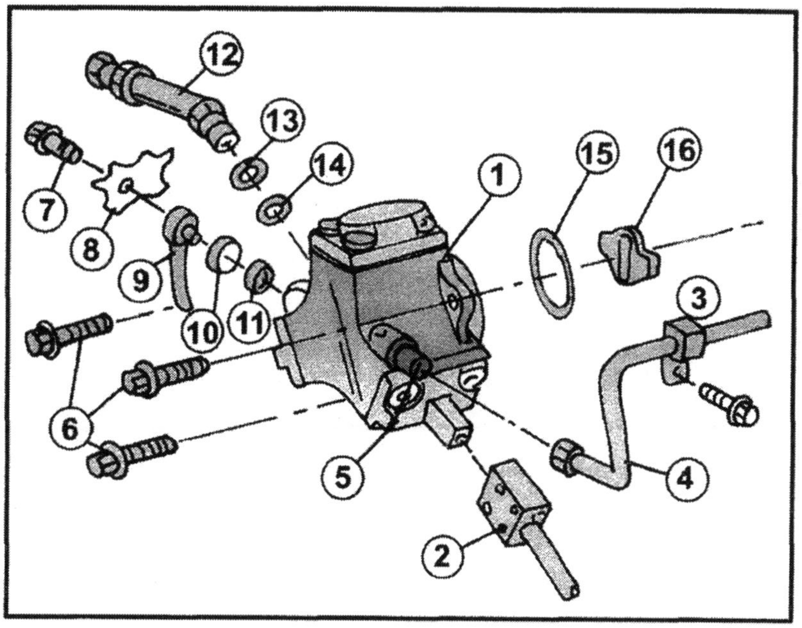

Fig. 1.94 – Details for the removal and installation of the high pressure pump. The numbers are referred to below.

- Disconnect the battery earth cable.
- Remove the vacuum pump (section "Brakes").
- Take the tension off the Poly V-belt as described in "Cooling System" section and lift it off the guide pulley. The lower bolt securing the high pressure pump (6 in Fig. 1.94) can only be removed when the belt is removed from this area.
- Withdraw the connector plug (2) from the high pressure pump (1). A second plug is located below and must also be withdrawn. One plug is connected to the fuel temperature sensor, the other one to the quantity control valve.
- Detach the bracket (3) and take off the pressure pipe (4) from the pump (1). Never slacken the threaded connection (5). Use an open-ended spanner to counterhold at the threaded connection when slackening or tightening the union nut. When tightening the union nut do not exceed a torque of 2.3 kgm (16.5 ft.lb.) as it is possible that the threaded connection will be slackened the next time the union nut is removed. Take care not to bend the pipe during removal.

- Take off the bracket (8) for the fuel lines at the high pressure pump and take off the fuel return line (9). Tighten the securing bolt (7) to 0.9 kgm (7 ft.lb.) during installation. The two seals (10) and (11) must always be replaced.

Note: If a new pump is fitted detach the intermediate pipe (12) from the pump. The sealing rings (13) and (14) must be replaced in this case.

Fig. 1.95 – The driver must engage correctly into the intermediate drive gear of the high pressure pump at the position shown by the arrow.

- Remove the high pressure pump (1) from the cylinder head after unscrewing the securing bolts (6). We would like to point out that it is not easy to reach the lower bolt of the three shown in the illustration without lifting off the V-belt. **Warning:** The high pressure pump must not be opened. The driver (16) can be taken off. The sealing ring (15) must be replaced during installation.
- During installation of the driver pay attention to the relative position of the driver to the intermediate gear. This can be checked at the position shown by the arrow in Fig. 1.95. If the driver (16) looks worn, also replace the intermediate drive gear.

The installation is a reversal of the removal procedure. In addition to the tightening torques already given above tighten the pump securing screws to 1.4 kgm (10 ft.lb.).

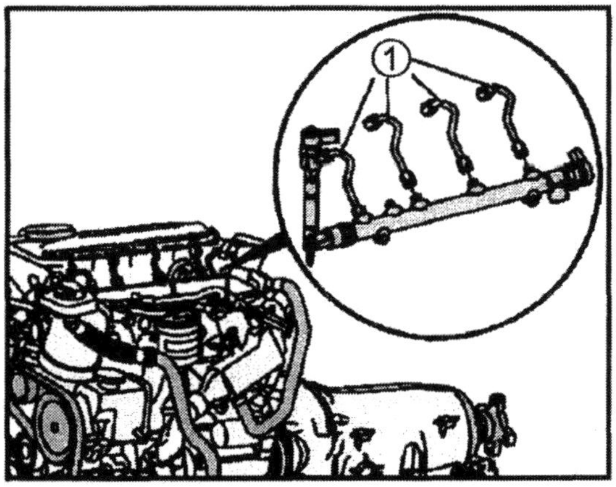

Fig. 1.96 – The injection pipes (1) can be found at the position shown.

Injection Pipes – Removal and Installation

Fig. 1.96 shows where the injection pipes are located. The pipes can be removed or replaced as follows:

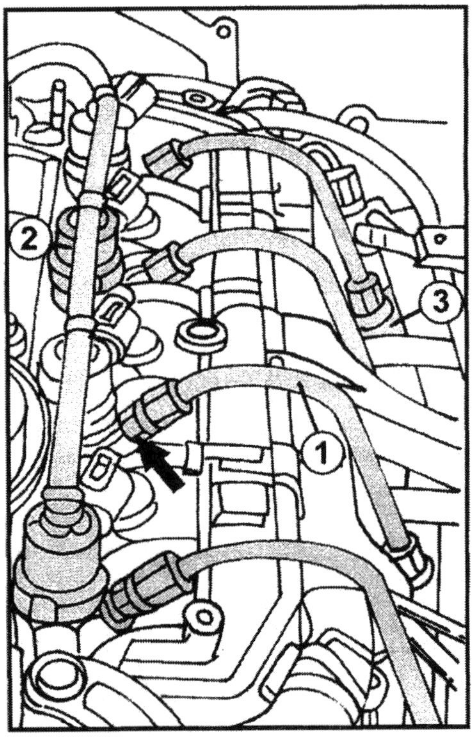

Fig. 1.97 – Counterhold the threaded fitting of the injectors (arrow) with an open-ended spanner at the injectors (2) and the injection rail (3) when removing the injection pipes (1).

- Remove the air cleaner housing.
- Remove the trim panel of the cylinder head cover as described during the removal of the cylinder head.
- Remove the union nuts securing the injection pipes (1) to the injectors (2) and to the injection rail (3) in Fig. 1.97. Counterhold the threaded fitting of the injectors (arrow) when slackening the union nut(s). Do not exceed the tightening torque of 2.3 kgm (16 ft.lb.) during installation. Otherwise the threaded fitting may unscrew during the next removal of the injector.

Fig. 1.98 – The injection pipes (1).

- Remove the injection pipes without bending or kinking them. Lay them out in the order shown in Fig. 1.98. Close the open ends of the pipes.

The installation is a reversal of the removal procedure, noting the tightening torque already given above and make sure the pipes seat properly. To install the pipes free of stress or tension, slacken the injection rail in order to avoid any risk of a pipe fracturing. Do not tighten the rail fully until all pipes have been secured.

1.8.2. INJECTORS – REMOVAL AND INSTALLATION

The injectors are a tight fit in the cylinder head. The workshop has a puller to remove a very tight injector. Fig. 1.99 shows the parts of an injector. An injector can be removed as follows. The battery should be disconnected and the ignition switched off:

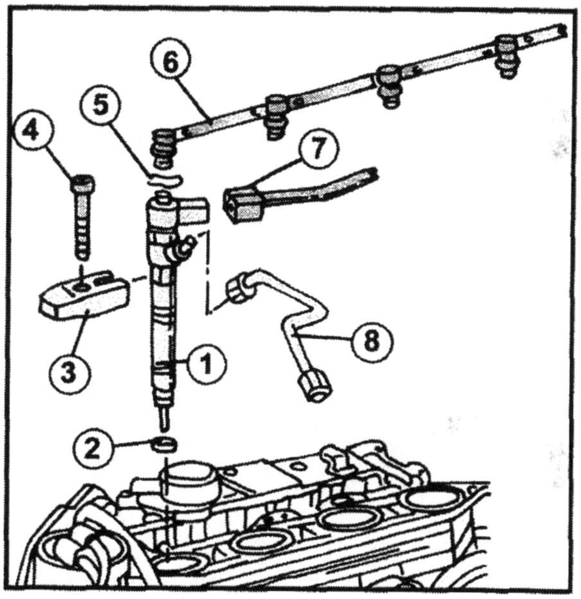

Fig. 1.99 – Details for the removal and installation of the injectors. The numbers are referred to in the text.

- Remove the trim panel of the cylinder head cover and the injection pipes as described above.
- Remove the connector plug from the camshaft Hall sensor and the plug (7) from the injector (1) or from all four injectors.
- Remove the securing clips (5) from the connection of the leak-off pipe (6) and push the pipe to one side.
- Remove the stretch bolts (4) and the tensioning clamp (3). The bolts must always be replaced. Tighten the bolts (4) to 0.7 kgm and from the final position a further 90°.

The injector can now be removed. As already mentioned, a tight injector is removed in the workshop with a special puller (611 589 01 33 00) which is applied in place of the tensioning clamp. If a new injector is to be fitted quote the engine type and model year as injectors are sometimes changed to improve their performance.

The installation is a reversal of the removal procedure. Note the comment already given for the tightening of the injection pipes. The sealing ring (2) must always be replaced.

1.8.3. FUEL FILTER – REMOVAL AND INSTALLATION

The high pressure injection pump, the injectors and the associated injection pipes are sensitive to contamination. A fuel filter is responsible to remove any foreign matter. The filter element should be replaced every in accordance with the intervals given in your service instructions (approx. 40,000 miles). Looking into the engine compartment you will find the fuel filter at the position shown in Fig. 1.100 on the R.H. side of the engine (direction of drive).

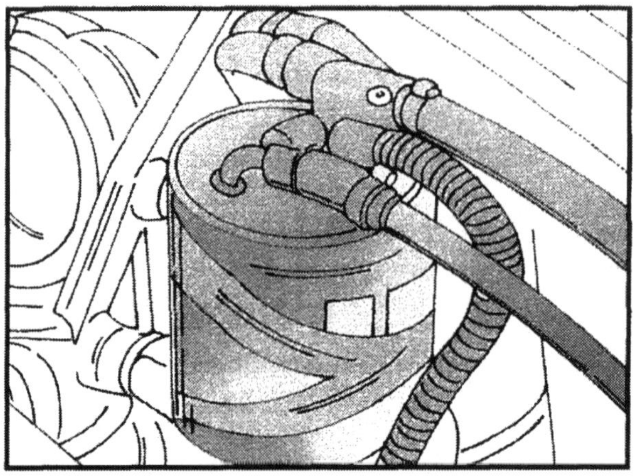

Fig. 1.100 –View of the fuel filter as fitted to the engine.

Fig. 1.101 shows the attachment of the filter. All precautions given previously must be observed. The air cleaner housing must be removed.

- Slacken the hose clamp of the fuel inlet line (9) and detach the line. Take care, as fuel will run out. Also note the installation position of the hoses on the filter.
- Dismantle the plug connection of the fuel output line (8) at the fuel filter (1). Again watch out for dripping fuel.

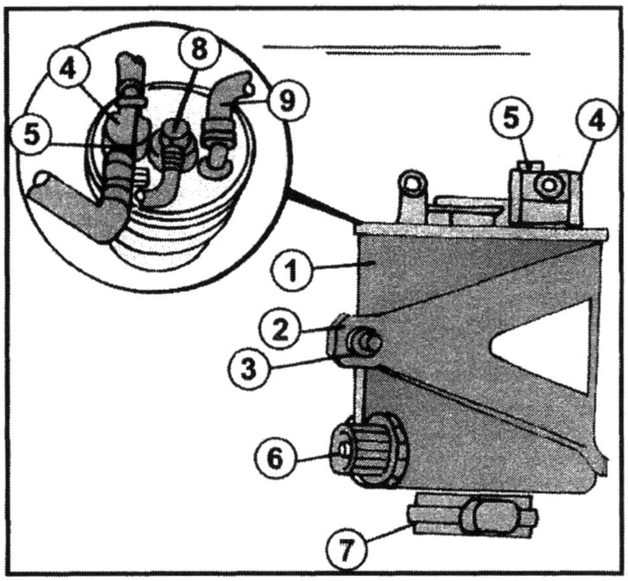

Fig. 1.101 – Details for the removal and installation of the fuel filter. The numbers are referred to in the text.

- Unscrew the bolts (5) at the pre-heater valve (4) and swivel the valve (4) out of the circlip and remove. Collect escaping fuel.
- Attach a drain hose with an inner diameter of approx. 4 mm to the drain connection of the drain plug (6). Fig. 1.101 will show you where the drain plug is located. Open the drain plug (6) and allow the water and fuel to run out. Dispose of the water/fuel mixture in a responsible manner.
- Disconnect the cable connector plug at the water level sensor. The cable connector has a release button which must be released. It should be noted that not all filters are fitted with a water separator.
- Slacken the clamp bolt (3) securing the filter housing clamp (2) and remove the filter (1) out towards the top.

The installation of the filter is carried out as follows:

- Assemble the fuel filter. To do this, press the water level sensor (7) against the spring force into the new fuel filter and rotate it into the installation position by turning it a quarter of a turn to snap in position at the bottom of the filter.
- Insert the filter (1) into the filter housing clamp bracket (2) and turn the filter to its correct position.
- Connect the cable connector plug to the water level sensor (7). The locking button of the water level sensor connector plug (7) must snap in position.
- Check that the fuel filter is in its correct position and tighten the clamp bolt (3) to 0.8 kgm, no more.
- Replace the sealing ring of the drain plug (6) and fit it in position. Tighten the plug to 0.6 kgm.
- The fuel filter must now be filled with fuel, using a small funnel. This takes place via the opening for the pre-heating valve (4). Make sure that the funnel is clean and only use clean diesel fuel. The location of the valve can be seen in Fig. 1.107. Then install the pre-heating valve (4).
- Connect the fuel input line (9) and the output line (8).

- Start the engine and let it run at idle speed. The workshop uses a low pressure bleeding instrument, but if the operations have been carried out carefully as described there should be no problems to start the engine.

1.8.4. VARIOUS PARTS OF THE SYSTEM

Inlet Manifold Pressure Sensor

The sensor is fitted downstream of the air cleaner. It can be replaced after withdrawing the cable connector plug. Pull out the sensor. An "O" sealing ring is fitted underneath the sensor and must be replaced.

Fig. 1.102 – Location of the quantity control valve (1) and the intake port shutoff switchover valve.

Quantity Control Valve

The valve is located at the position (1) shown in Fig. 1.102 on the L.H. side and is secured by means of three screws to the high pressure pump. Remove the air cleaner housing to obtain access to the valve. After withdrawing the plug connector clean the area of the high pressure pump thoroughly and undo by one to two threads at a time until all are removed.

Pull out the valve by hand rotating it as you pull it out. Do not use a screwdriver to pry it out and do not tilt it left and tight or up and down. The "O" sealing rings under the valve must be replaced.

The installation is a reversal of the removal procedure. Again rotate the valve as you push it in. Align the valve with the screw threads and fit and tighten the screws to 0.7 kgm.

Lambda Probe (O2 sensor)

The Lambda probe is fitted on the R.H. side of the lower connection of the large air intake duct and the duct must be removed from the air cleaner after slackening off the clamps.

Remove the connector plug for the sensor from a bracket and then withdraw the connector plug from the Lambda probe. The probe can now be unscrewed with a suitable open-ended ring spanner.

During installation coat the threads of the Lambda probe, not the body of the probe, with heat-resistant grease and tighten the Lambda probe to 4.5 kgm (32.5 ft.lb.).

Fuel Temperature Sensor

The sensor is located just above the quantity control valve shown in Fig. 1.102. To remove the sensor the air cleaner housing must be removed. The sensor can be removed after withdrawing the connector plug. Installation is carried out in reverse order. Replace the sealing ring under the sensor. Tighten the sensor to 2.5 kgm (18 ft.lb.).

Intake Port Shutoff Switchover Valve

This item can be found below the fuel filter at the position (2) in Fig. 1.102. The starter motor must be removed to gain access. After withdrawing the connector plug from the valve at the upper end, unscrew the three securing bolts and take out the valve towards the bottom. Install in reverse order.

Hot Film Mass Air Flow Sensor

The sensor cannot be seen from the outside as it is located inside the large air intake hose. Remove the hose from the air cleaner housing after slackening off the hose clamp. After withdrawing the cable connector plug locate and remove the two securing screws and withdraw the sensor. If necessary replace the "O" sealing ring. Fit the sealing ring with a little grease. Installation is carried in reverse order.

1.8.5. COMMON RAIL – REMOVAL AND INSTALLATION

Fig. 1.103 shows the items to be disconnected or removed to take out or replace the common rail fuel distribution tube. Observe all precautions regarding fuel when removing the rail.

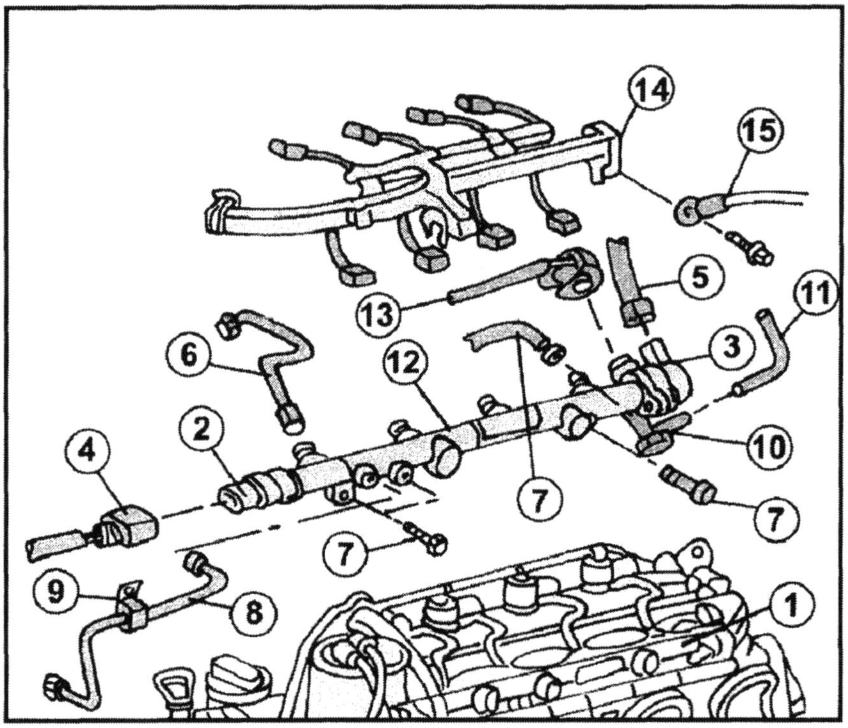

Fig. 1.103 – Details for the removal and installation of the common rail. Follow the numbers as given below.

- Disconnect the battery earth cable.
- Remove the trim panel of the cylinder head cover. Refer to section "Diesel Engine" later on in the manual.
- Remove the injection pipes (6) as described earlier on in this section.
- Unplug the cable connector plug (4) from the rail pressure sensor (2) on one side of the rail and the plug (5) from the pressure control valve (3) at the other end.
- Detach the retaining clamp (9) securing the pressure pipe (8) to the high pressure pump-rail and disconnect the pipe from the rail and the high pressure pump, following the instructions already given during the removal of the injection pipes. Tighten the union nuts to 2.3 kgm (16.5 ft.lb.).
- Disconnect the fuel return flow line (13) from the rail (1). Pay attention to the locking arm. Replace the seal.
- Disconnect the fuel return flow line (12) and the leak oil line (11) at the angled piece (10).
- Detach the cable duct (14), unplug the cable connector plugs from the injectors and the camshaft position sensor and lift the cable duct off the rail.

- Take off the fuel rail (1) and remove the fuel rail towards the rear.
- If the fuel rail is replaced, fit it onto the existing angle piece (10). Replace the seals.

The installation is a reversal of the removal procedure, noting the points already given above. The banjo bolt for the leak oil line must be tightened to 2.0 kgm (14 ft.lb.).

1.8.6. RAIL PRESSURE SENSOR – REMOVAL AND INSTALLATION

The pressure sensor is fitted to the common rail. Fig. 1.104 shows where it is located on the side of the high pressure pump. Observe all precautions regarding the fuel.

Remove the sensor as follows:

- Remove the air cleaner housing and the fuel filter (4) as already described.
- Unplug the cable plug connector (1) from the rail pressure sensor (2).

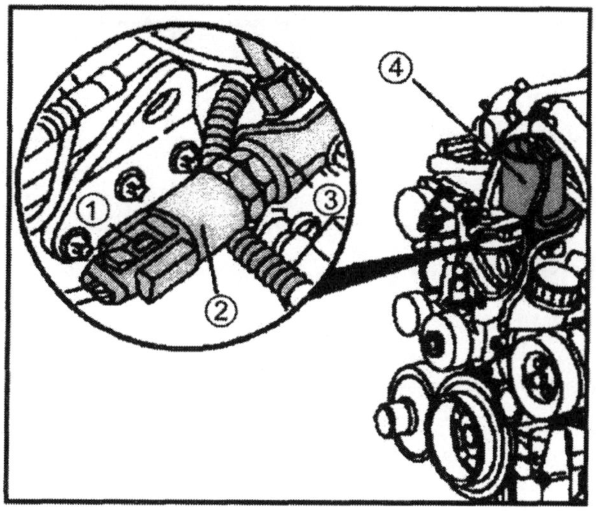

- Unscrew the pressure sensor (1) from the rail (3).

Fig. 1.104 – Details for the removal and installation of the rail pressure sensor.

The installation is a reversal of the removal procedure. Tighten the sensor to 7.0 kgm (50.5 ft.lb.).

1.8.7. AIR CLEANER

1.8.7.1. Removal and Installation of Air Cleaner Housing

The air cleaner is situated in the centre of the engine compartment. Remove and install the complete air cleaner housing as follows:

- Remove the large intake manifold hose from the air cleaner housing after slackening off the clamp.
- Remove the oil filler device at the upper end of the engine. To do this, press the retaining spring towards the rear and take out the oil filler towards the top.
- Pull the air cleaner housing out of its rear mounting and towards the front.
- Certain engines (depending on emission control) are fitted with a connector next to the air cleaner inlet bore. Withdraw the plugs.
- The air cleaner housing can now be removed.

The installation is a reversal of the removal procedure.

1.8.7.2. Replacing the Air Cleaner Element

The air cleaner element can be replaced after the removal of the air cleaner housing. Fig. 1.105 shows details of removal and installation. Remove the upper section of the air cleaner housing (1) in the L.H. view. The bolts remain in the upper section. During installation insert the tabs (A) in the R.H. view into the recesses (B) of the lower section of the air cleaner housing.

Remove the air cleaner element (2). Use a moist rag and wipe out dust or other matter.

The installation is a reversal of the removal procedure.

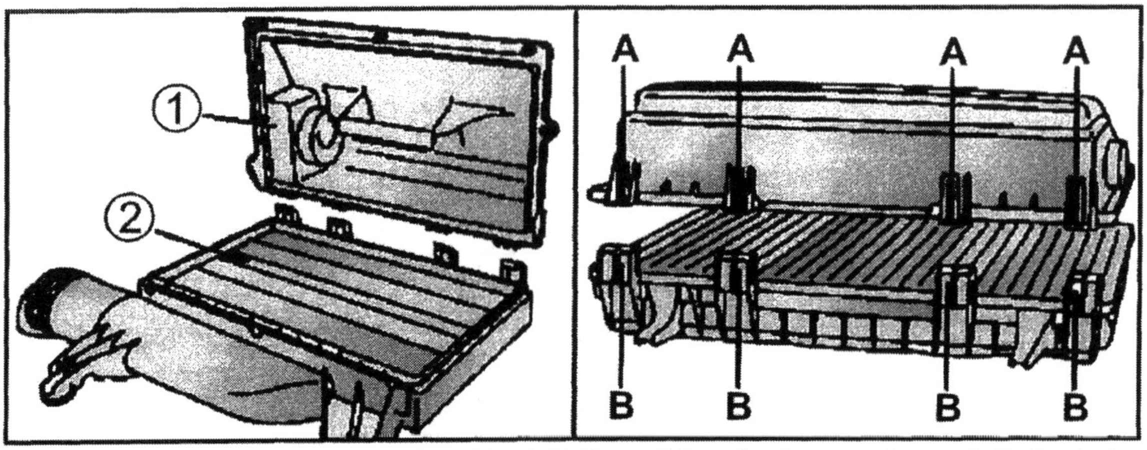

Fig. 1.105 – Details for the removal and installation of the air cleaner element. Refer to text.

Make sure that the element is correctly seated on all edges. It is a good practice to make a note when the element has been replaced.

1.8.8. GLOW PLUGS

The glow plug system consists of the glow plugs, the glow time relay and the warning lamp in the instrument panel.
When the ignition is switched on, the glow plug relay will receive current via terminal "15". The relay directs the current from the plus terminal "30" via a fuse to the glow plugs. The glow plugs receive a voltage of at least 11.5 volts and a current of 30 amps, which is however, reduced to 8 to 15 amps by means of a regulator, thereby preventing burning out of the plugs. The glow plugs heat up to 90° C within 10 seconds and can reach a temperature of 1180° C after 30 seconds.

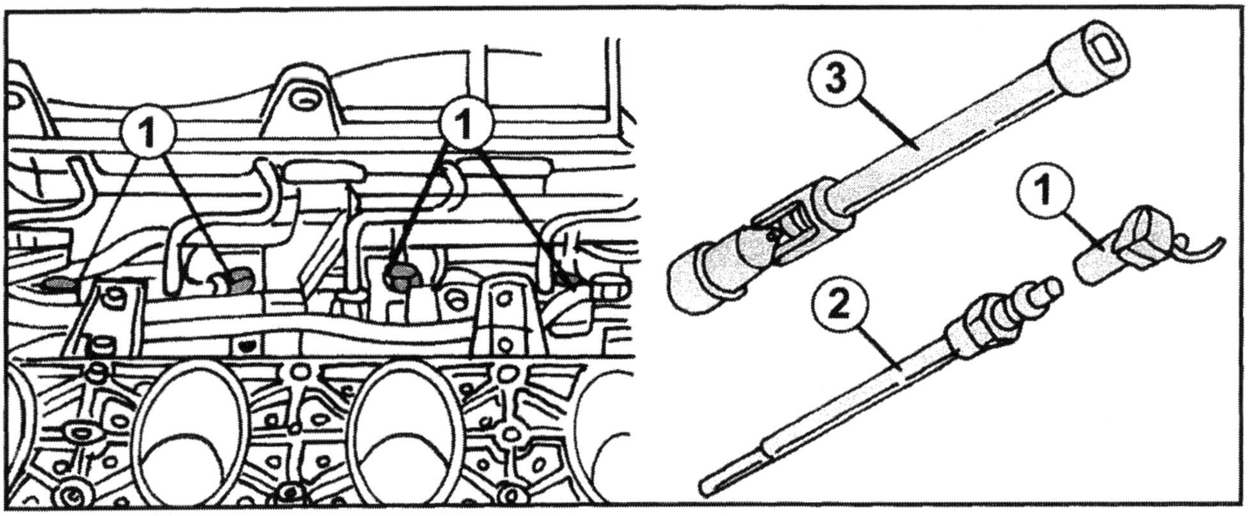

Fig. 1.106 – Removal and installation of glow plugs.

The glow plug time relay determines the operating time of the glow plugs. This relay senses the outside temperature, i.e. at very low temperatures, for example – 30°) C, the plugs can glow as long as 25 seconds. During the summer months, however, glowing time may be as little as 2 seconds. If the engine is not started immediately after the warning light has gone "off", the current feed will be interrupted through a safety circuit. Subsequent starting of the engine will switch in the glow plug circuit via starter motor terminal "50".

Glow plugs must again be obtained by quoting the engine number and model year. The plugs are sometimes modified and only your MB parts supplier will be able to sell you the correct plugs.

The glow plugs are fairly well hidden below as you can see in Fig. 1.106. A special socket with universal joint and extension are required to reach them. A ratchet is of an advantage. Additionally the air cleaner housing must be removed.

Use a pair of pliers and unplug the cable connector plugs (1) from the glow plugs (2). Using the special wrench mentioned above and shown in Fig. 1.106 unscrew the plug(s) out of the cylinder head. If a plug is difficult to remove refit the connector plugs, start up the engine and allow it to reach operating temperature. If you want to remove the plugs from a removed cylinder head and the plug is tight, you can heat the cylinder head to 90° C.

Note that not all engines have the same glow plugs. Quote the engine number when new plugs are purchased.

The installation of the glow plugs is a reversal of the removal procedure. Plugs are tightened to 2.0 kgm (14.5 ft.lb.). Do not over-tighten the cable securing nuts (0.4 kgm). Problems with glow plugs should always be investigated by a workshop dealing with diesel equipment as diagnosis is not a straight forward operation.

2. CLUTCH

2.0. Technical Data

Type: .... Single dry plate diaphragm clutch
Operation: .... Hydraulic system with central clutch operator
Clutch diameter: .... Refer to parts lists

2.1. Introduction

The clutch is operated by means of a hydraulic system. Pressure is generated by the depressing of the clutch pedal and transmitted via the clutch master cylinder to a so-called central clutch operator inside the clutch bell housing which takes the place of the clutch slave cylinder but incorporates the clutch release bearing. The diameter of the clutch is not the same on all engines and we refer you to your MB parts department if the clutch must be replaced, quoting the engine type and engine number.

2.2. Removal and Installation

To remove the clutch unit, it will be necessary to separate the transmission from the engine, with the assembly fitted to the vehicle. After the clutch has been removed you will have the parts shown in Fig. 2.1.

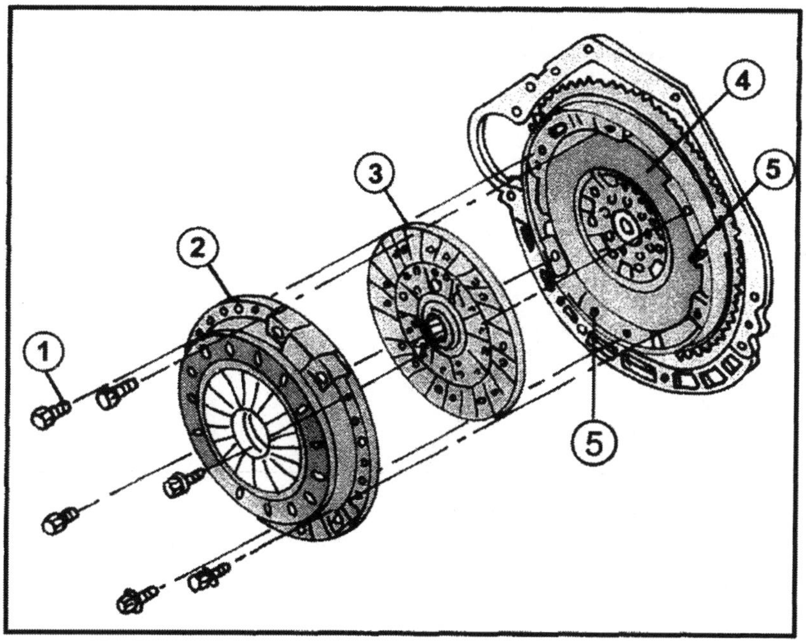

Fig. 2.1 – The component parts of the clutch.
1 Clutch securing bolts
2 Clutch pressure plate
3 Clutch driven plate
4 Flywheel
5 Dowel pins

- Remove the transmission (Section 3.1).
- Mark the clutch in its fitted position on the flywheel if there is a possibility that the clutch unit is to be re-used. To remove the clutch, unscrew the six bolts (1) in Fig. 2.1, securing the pressure plate (2) to the flywheel (4) and lift off the clutch unit and then the driven plate (3), now free. Before removing the driven plate, note the position of the longer part of the driven plate hub, as the driven plate must be refitted in the same way. Note that the clutch plate is located in position by means of the dowel pins (5).

Install in the reverse sequence to removal, noting the following points:

- If the old clutch unit is fitted, align the marks made before removal. A new clutch can be fitted in any position.

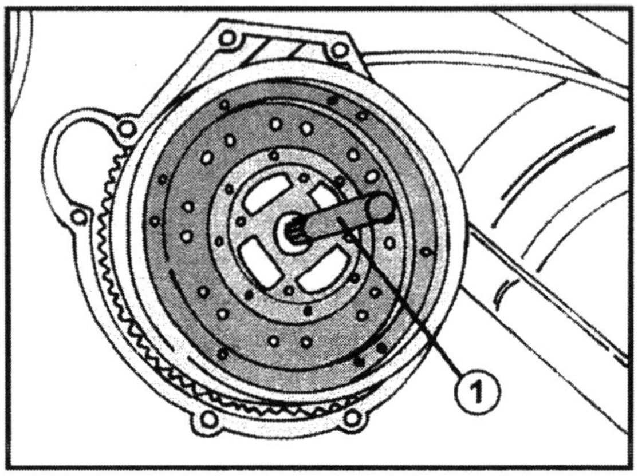

Fig. 2.2 – The clutch centering mandrel (1) is inserted into the driven plate and the flywheel before the clutch is fitted.

- A centering mandrel is required to centre the clutch driven plate inside the flywheel. Tool hire companies normally have sets of mandrels for this purpose. An old transmission (clutch) shaft for the transmission in question, which you may be able to obtain from a Mercedes workshop, can also be used. Experienced D.I.Y. mechanics will also be able to align the clutch plate without the help of a mandrel. Insert the mandrel as shown in Fig. 2.2. Fit the clutch pressure plate (2) over the dowel pins (5) in Fig. 2.1.
- Fit and tighten the six clutch to flywheel bolts (1) to a torque reading of 2.3 kgm (16.5 ft.lb.). The flywheel must be locked against rotation when the clutch bolts are tightened.

2.3. Servicing

The cover assembly - pressure plate and diaphragm spring must not be dismantled. Replace, if necessary with a complete assembly from your dealer or distributor.

Fig. 2.3 – To check the driven plate for run-out, clamp it between the centres of a lathe and check with a dial gauge.

Inspect the driven plate and the linings, replacing the complete plate if the linings are worn down close to the rivets. A driven plate with the linings contaminated with grease or oil cannot be cleaned successfully and should also be replaced. All rivets should be tight and the torsion springs should be sound and unbroken. Check the condition of the driven plate splines. Clamp the driven plate between the centres of a lathe and apply a dial gauge to the outside of the plate as shown in Fig. 2.3, at a diameter of approx. 175.0 mm (6.4 in.). The max. run/out of the driven plate should be no more than 0.5 mm (0.02 in.).

Check the rivet fastening of the clutch pressure plate and replace the plate, if loose rivets can be detected.

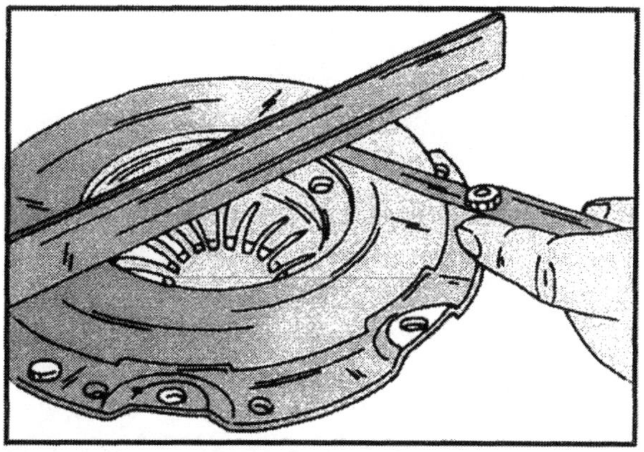

Fig. 2.4 – Checking the clutch pressure plate for distortion. The gap should not be more than given below.

Place a straight edge (steel ruler) over the friction face of the pressure plate and insert feeler gauges between the ruler and the surface. If the gap at the innermost spot of the friction face is no more than 0.03 mm (0.012 in.), the plate can be re-used. Fig. 2.4 shows this check.

Check the inner ends of the clutch plate springs at the position shown in Fig. 2.5. The fingers of the diaphragm spring must all be at the same height. It may be possible to bend them slightly with a pair of pliers, but renewal is always the best remedy.

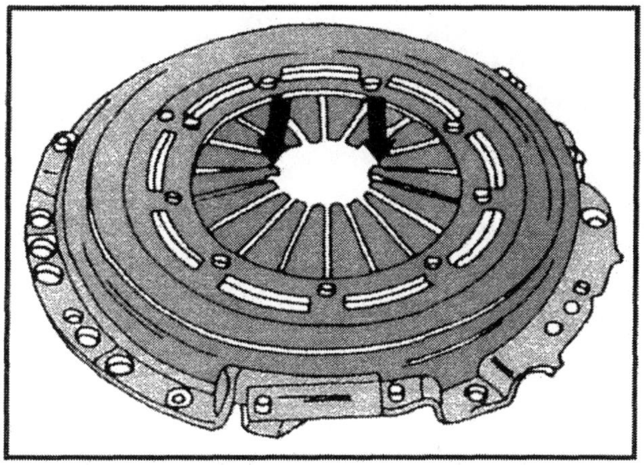

Fig. 2.5 – The "fingers" of the diaphragm spring can wear at the positions shown. All must be at the same height.

2.4. Clutch Release Mechanism

Engagement and disengagement of the clutch is by means of the central clutch operator inside the transmission housing with incorporated clutch release bearing. The system is free of clearance and any wear of the driven plate is automatically compensated by the system.

2.4.0. HYDRAULIC CLUTCH OPERATOR – REMOVAL AND INSTALLATION

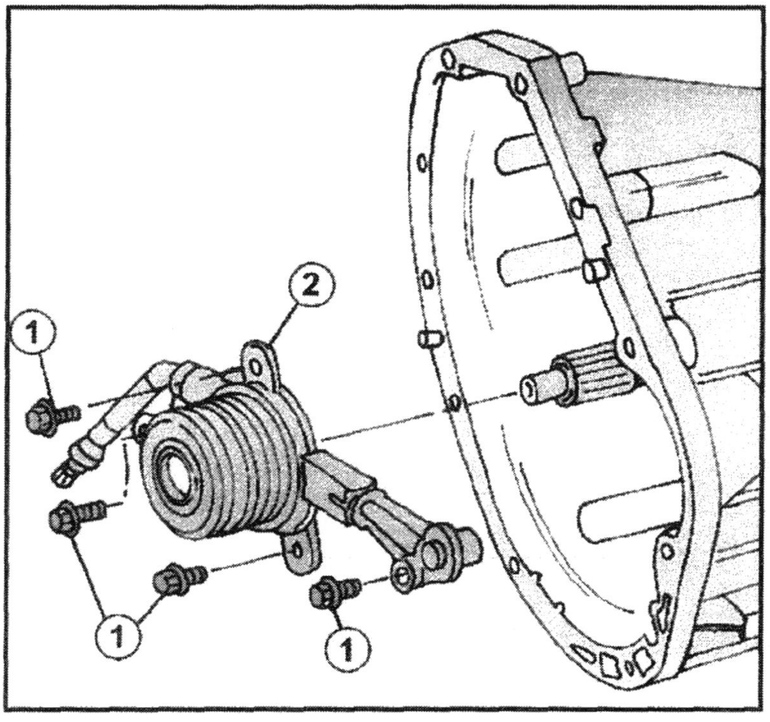

Fig. 2.6 – Attachment of the central clutch operator inside the gearbox housing. The numbers are referred to during the removal instructions.

Precautions when working with Brake Fluid

As the hydraulic system is filled with brake fluid read the notes below before any operations are carried out on the hydraulic clutch system.

- Only store brake fluid in a closed container, tin, etc. which must be marked accordingly. Brake fluid can damage you health if swallowed.

- Protect your hand with gloves and if possible your eyes with goggles. If your hands come into contact with brake fluid immediately wash them with water and soap.
- Take care not to drip brake fluid onto painted areas of the vehicle. Brake fluid will damage the paint. If it happens, wipe it off immediately.
- Brake fluid is able to absorb moisture from the air. This is the reason that it must be kept in a closed container.
- Even the smallest quantity of oil in the brake fluid can damage the rubber parts of the hydraulic system.
- Brake fluid released from the system, as is the case during the bleeding operation, must not be re-used to top-up the fluid reservoir.

The transmission must be removed to gain access to the hydraulic clutch operator. Fig. 2.6 shows the attachment of the parts in question. With the transmission removed unscrew the four bolts (1) and take off the central clutch operator (2) from the gearbox (3).
The installation is carried out in reverse order. Use new micro-encapsulated bolts (1). Tighten the bolts to 0.9 kgm (7 ft.lb.). Mercedes-Benz recommends to use a suitable thread cutter to re-cut the bolt threads inside the gearbox housing before the clutch operator is fitted. After installation of the gearbox bleed the clutch system as described further on in the text.

2.4.1. CLUTCH MASTER CYLINDER – REMOVAL AND INSTALLATION

Fig. 2.7 shows where the clutch master cylinder is fitted near the clutch pedal. The clutch pedal must be removed to replace the cylinder. Remove the cylinder as follows:

- Remove the master cylinder (7) from the clutch pedal (8) and from the pedal mounting bracket (1). The cylinder is secured on both sides with parts (2). On the push rod remove on one side the securing clip at the upper end of the clutch pedal (8) and remove the cylinder sideways out of its position. On the end of the push rod you will find a flange (12) which must be facing the clutch pedal during installation.

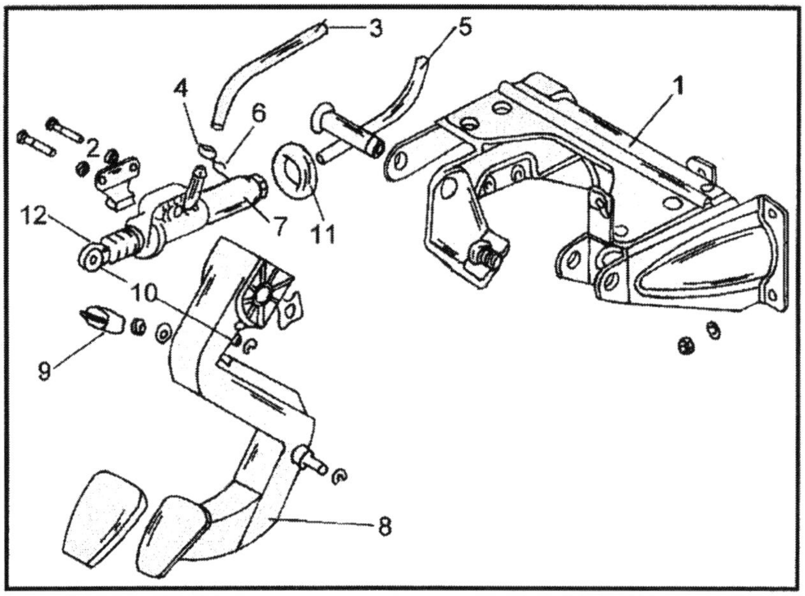

Fig. 2.7 – Details for the removal and installation of the clutch master cylinder.
1 Pedal bracket
2 Securing parts, cylinder
3 Fluid hose to cylinder
4 Hose clamp
5 Connecting to slave cylinder
6 Retaining clip
7 Clutch slave cylinder
8 Clutch pedal
9 Tempomat switch (if fitted)
10 Push rod attachment
11 Rubber protection cap
12 Flange and cylinder

- Use a syringe and draw enough fluid out of the fluid reservoir to bring the fluid level below the "Min." mark (see Fig. 2.8). Otherwise use a bleeder hose, push it over the slave cylinder bleeder screw, open the screw and operate the clutch pedal until the system is empty (collect the fluid in a glass jar).

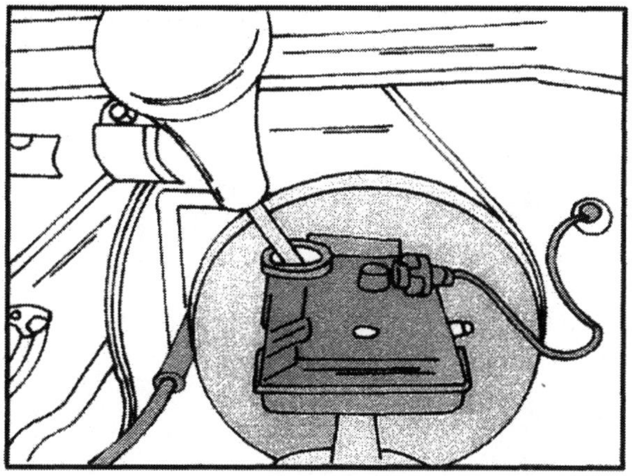

Fig. 2.8 – Brake fluid can be removed from the fluid reservoir in the manner shown.

- Disconnect the fluid hose (3) from the combination fluid reservoir (for brake and clutch). Close the end of the hose in suitable manner to prevent fluid from running out.
- Remove the connecting hose (5) to the slave cylinder and remove the cylinder from the vehicle.

The installation of the master cylinder is a reversal of the removal procedure. Make sure that the push in connector for the connecting hose is properly in position. Fit the push rod with the flange facing towards the clutch pedal and make sure the circlip engages into its groove. Finally bleed the clutch system (Section 2.5).
Start the engine. Depress the clutch pedal and engage the reverse gear. Grating noises will immediately indicate a faulty part of the clutch operation.
Fully depress the clutch pedal and check that it reaches its stop. Otherwise the clutch cannot fully disengage.

2.4.2. CLUTCH PEDAL – REMOVAL AND INSTALLATION

The removal of the clutch pedal is a fairly complicated operation, but can be carried out as follows:

- Move the drivers seat back to the full extent backwards.
- Remove the clutch pedal switch from the pedal.
- Remove the bolts securing the clutch master cylinder to the clutch pedal bracket (see Fig. 2.7).

The clutch pedal retracting spring must be removed as follows:

- Remove the pedal retracting spring from the side of the pedal. At the bottom of the spring you will find a retaining clip which must be removed. In the inside of the spring there is a spring guide. After removal of a bolt remove the spring guide with the retracting spring. The bolt must be coated with "Molycote" silicone grease. Obtain some from a Mercedes dealer.
- Detach the end piece from the spring guide and remove the spring.

The installation of the pedal retracting spring is a reversal of the removal operations. After installation check the clutch pedal for proper functioning.
After removal of the spring proceed as follows:

- Remove the securing clamp from the clutch pedal fulcrum pin. The clamp is located at the upper end of the clutch pedal (looking at the pedal on the inside). The fulcrum pin can now be pushed or driven out. The pin must be greased with the grease mentioned above during installation.
- Follow the clutch master cylinder to the connection on the pedal and remove a retaining clip. The clutch pedal can now be removed from the pedal bracket. The master cylinder push rod is now removed from its fulcrum pin. Again use the grease mentioned above before fitting the pin.

The installation of the clutch pedal is a reversal of the removal procedure, noting the points already given above.

2.4.3. HYDRAULIC LINE TO CLUTCH OPERATOR – REMOVAL AND INSTALLATION

Fig. 2.9 shows the installation of the pipe. To prevent loss of brake fluid a suitable clamp should be used to carefully clamp off the pipe at position (2) without damaging it. Removal and installation is carried out as follows. The vehicle must be resting on secure chassis stands, as the removal must take place from below the vehicle.

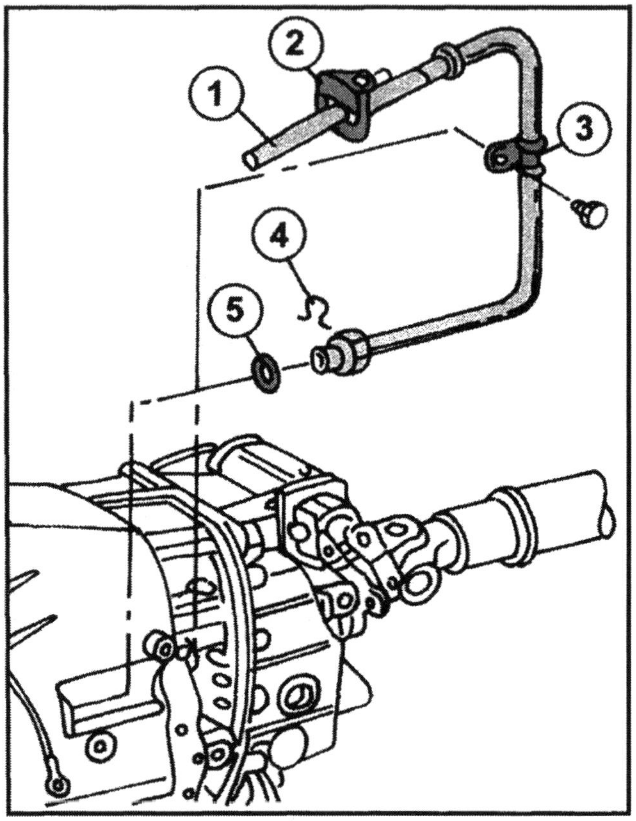

Fig. 2.9 – Details for the removal and installation of the clutch fluid line to the clutch operator inside the gearbox housing.

1 Fluid line
2 Retainer
3 Retainer
4 Spring clip
5 Sealing ring

- Clamp off the hydraulic hose (1) with the clamp (2) as explained.
- Unscrew the retaining bracket (3) for the pipe from the transmission.
- Remove the spring clip (4) and remove the pipe at the position shown out of the transmission. The sealing ring (5) must always be replaced. Take care of dripping brake fluid.

The installation is a reversal of the removal procedure. Bleed the clutch system as described below.

2.5. Bleeding the Clutch System

A pressure bleeder is used by Mercedes workshops. The following description involves the brake system and is therefore to be treated with caution. Make absolutely sure that the brakes have correct operating pressure after the clutch has been bled. A transparent hose of approx. 1 metre (3 ft.) in length is required. Proceed as follows:

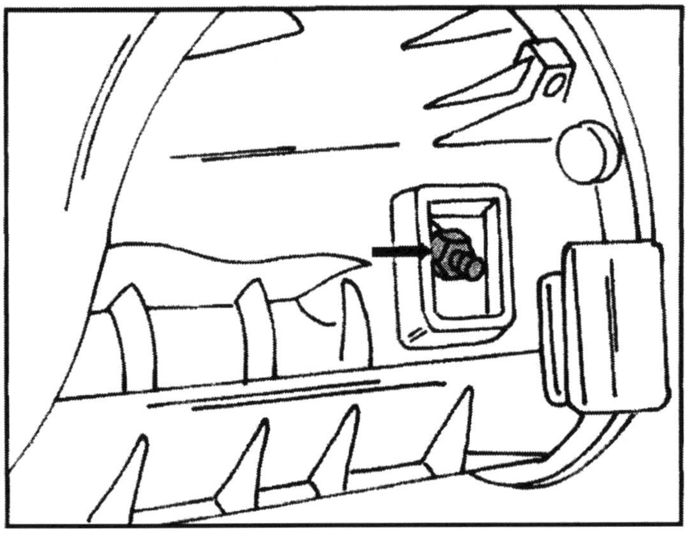

Fig. 2.10 – The location of the bleeding screw for the clutch system on the side of the transmission (arrow).

- Fill the brake/clutch fluid reservoir.
- Remove the dust cap of the bleeder screw in the side opening of the gearbox, shown in Fig. 2.10 and push the hose over the bleeder screw and place the hose end into a glass container filled with some brake fluid. Open the bleeder screw.
- Ask a second person to operate

the brake pedal until the hose is completely filled with brake fluid and no more air bubbles can be seen. Place a finger over the hose end to prevent fluid from running out.

- Tighten the bleed screw (0.6 kgm) and remove the hose from the end of the screw.
- Check the fluid level in the reservoir and, if necessary, top it up to the "Max." mark. Do not overfill. If necessary extract some fluid in the manner shown in Fig. 2.8. Start the engine, depress the clutch pedal and engage reverse. No grating noises should be heard.

3. MANUAL TRANSMISSION

The vehicles covered in this manual are either fitted with a six-speed transmission, with the technical designation type 716.637 or type 716.652. The type is important when it is intended to fit a second-hand transmission during the life-span of your vehicle.

Note: The type of oil used in the transmission has been changed during production. Transmissions are now filled with SAE 75 W 85 transmission oil. If you are familiar with earlier Sprinter models make sure that only the specified oil is used when topping up the oil level or changing the oil in the transmission.
The overhaul of the transmission is not described in this manual. The description in the overhaul section is limited to some minor repair operations, not involving the gear train or the gear shafts. If the transmission appears to be damaged or faulty, try to obtain an exchange unit. Transmission overhaul is now limited to specialised workshops which are equipped with the necessary special tools.

3.0. Technical Data

Fitted Transmission:
- Six-speed transmission:..716.637 or 716.652

Transmission Ratios (typical):

- First gear	5.010 : 1
- Second speed	2.830 : 1
- Third speed	1.790 : 1
- Fourth speed	1.260 : 1
- Fifth speed	1.000 : 1
- Sixth speed	0.800 : 1
- Reverse speed	4.570 : 1

Oil capacity: ....Refer to page 6, 1.2 or 1.5 litres, depending on gearbox
Lubrication oil:.. See note above

3.1 Removal and Installation

The following text describes the general removal and installation operations. The transmission is heavy and the necessary precautions must be taken when it is lifted out.

The workshop removes the transmission on a platform lift as the operations are carried out from below. The following instructions assume that the vehicle is lifted up and supported on secure chassis stands. The transmission must be lifted with a jack during the removal in order to remove the rear transmission crossmember from the transmission and the side members. The removal is carried out as follows.

- Disconnect the battery earth cable from the battery and change the gearbox into neutral.
- In the case of models 639.7/8 from VIN number 055347 onwards remove the rear engine compartment panelling at the bottom.
- Drain the transmission oil (see later on).
- Detach the hydraulic clutch fluid line from the central clutch operator as described in the clutch section. The retaining clip securing the hydraulic line must be removed and the line pulled out. Push the line to one side, out of the way. Brake fluid may run out and must be collected or absorbed.
- The transmission must now be supported from below. The workshop uses, of course, a special lifting bracket. The engine compartment panelling underneath must, of course be removed. Use a suitable jack and jack up the transmission until it is under tension.
- Remove the engine crossmember at the rear. In the case of all engines remove the bolt connecting the rear engine mount to the transmission carrier and the bolt securing the transmission carrier to the longitudinal member. The bolts are located on the outside of the crossmember (two) and in line with the propeller shaft (two). Different tightening torque values apply. The bolt securing the rear engine mount to the transmission carrier (centre bolts) are tightened to 2.0 kgm (14.5 ft.lb.), the bolt securing the transmission carrier to the longitudinal member (outside bolts) are tightened to 4.0 kgm (29 ft.lb.).
- Disconnect the two gearchange cables from the transmission. To do this, press off the two ball sockets from the ball heads on the transmission (screwdriver). Detach the cables from the bracket on the transmission and place the cables to one side. Use a piece of wire to tie them up securely. During installation grease the ball sockets and the ball heads with long-life grease. Check that the cable locking plates securing the cables to the bracket lock properly. They may have to be replaced.

Fig. 3.1 – The propeller shaft must be removed from the connecting flange of the gearbox.

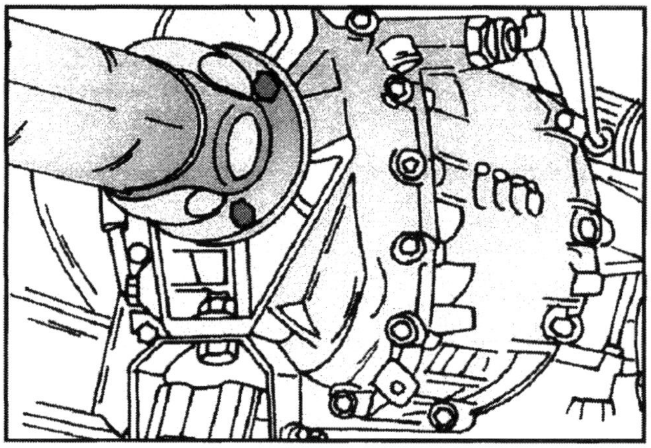

- At the rear of the transmission disconnect the propeller shaft. Hold the bolt heads with a ring spanner, unscrew the nuts on the other side with a socket. Rotate the shaft to reach all bolts. Fig. 3.1 shows the connection. Push the propeller shaft towards the right and tie it up with wire. The propeller shaft flange bolts are tightened in two stages. First tighten the bolts to 1.5 kgm (19 ft.lb.) and then by a further 90° (quarter of a turn).
- Locate the exhaust system bracket on the transmission and remove four nuts. The nuts are tightened to 2.3 kgm (16.5 ft.lb.).
- Locate the earth cable on the transmission and remove it. Place the cable to one side.

- Remove the starter motor without disconnecting the cables and place the starter motor to one side where it cannot get in the way during the subsequent removal of the gearbox.
- Place a mobile jack underneath the transmission, noting that it must be reaching high enough to support the transmission. Otherwise place a suitable wooden block onto the jack head to reach the gearbox. Operate the jack until the gearbox is under slight tension. Ensure that the transmission is well supported from below. Remove the gearbox-to-engine bolts. The bolts are tightened to 3.8 kgm (27.5 ft.lb.) during installation.
- Slowly withdraw the transmission towards the rear from the engine and lower it on the jack. Take care not to rest the weight on the clutch shaft, as this may bend the shaft or damage the clutch driven plate. You will need a helper to disengage and lower the transmission, as it must remain on the jack until it is free and can be withdrawn from underneath the vehicle. A helper is, of course, required – **Attention: Danger of accidents.**

When the transmission is removed for any reason, always check the condition of the clutch mechanism. Remove the clutch and check as described in the previous chapter. Also check the hydraulic clutch operator/release bearing.

The installation is a reversal of the removal procedure, but the following points must be noted:

- Lightly grease the centering lug and the splines of the clutch shaft with long-term grease. Place the transmission onto the jack and lift it into the approximate fitting position.
- Engage a gear (not reverse), align the transmission with the engine and push the transmission against the engine. Rotate the drive flange at the end of the transmission to and fro until the clutch shaft has engaged with the driven plate. Fully push the transmission against the engine. Note that the two dowel sleeves must engage in the locating holes in the transmission.
- Insert the bolts into the transmission, tighten them evenly until the gap between engine and transmission is nearly closed and then tighten the bolts to 3.8 kgm (27 ft.lb.). The same torque is also valid for the earth cable.
- Refit the rear crossmember from below, fit the transmission mounting in the centre and the bolts on the outside. Tighten the bolts as specified above.
- Connect the propeller shaft flange to the drive flange of the transmission, insert the bolts, fit the nuts and tighten the bolts as specified above. The bolt heads must be held against rotation, using a ring spanner.
- Refit the starter motor (4.5 kgm) and re-connect the battery earth cable. Remember to enter the radio security code and reset the radio stations and the electric clock.
- All other operations are carried out in reverse order. If the oil has been drained, fill the transmission with the correct quantity of the recommended oil (see Section 3.0.).

3.2 Transmission Repairs

Many repairs can be carried out without dismantling the transmission completely. The following sections describe some of the jobs which do not need special tools. Complete dismantling of the transmission requires special tools. A damaged transmission should be replaced by an exchange unit.

3.2.0. CLUTCH SHAFT OIL SEAL

The transmission must be removed from the vehicle to replace the oil seal inside the guide sleeve. You must also remove the hydraulic clutch operator from the inside of the transmission. Refer to Fig. 3.2 and remove the bolts (5) and lift off the clutch operator (4).

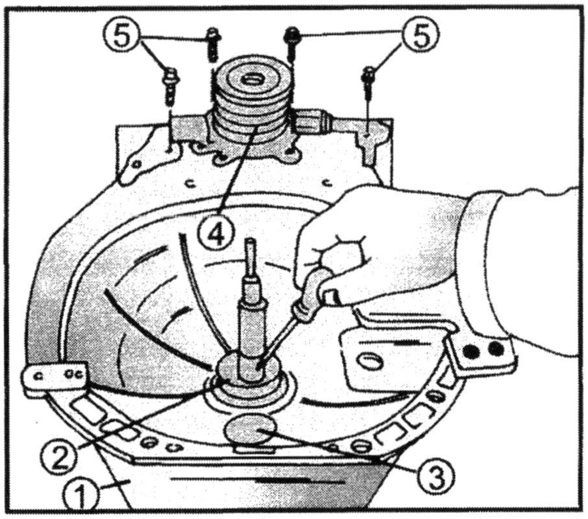

Fig. 3.2 – Details for the removal of the clutch shaft oil seal and the hydraulic clutch operator.

1 Gearbox
2 Radial shaft seal
3 Sealing cover
4 Clutch operator
5 Operator securing bolts, 0.9 kgm

Remove the oil seal, using a screwdriver as shown in the illustration. The easiest way is to punch two holes into the oil seal at opposite sides (carefully) and then insert the screwdriver blade into the holes to lever the seal out of position. Fit a new seal as shown in Fig. 3.3. The illustration shows the special drift used in the workshop.

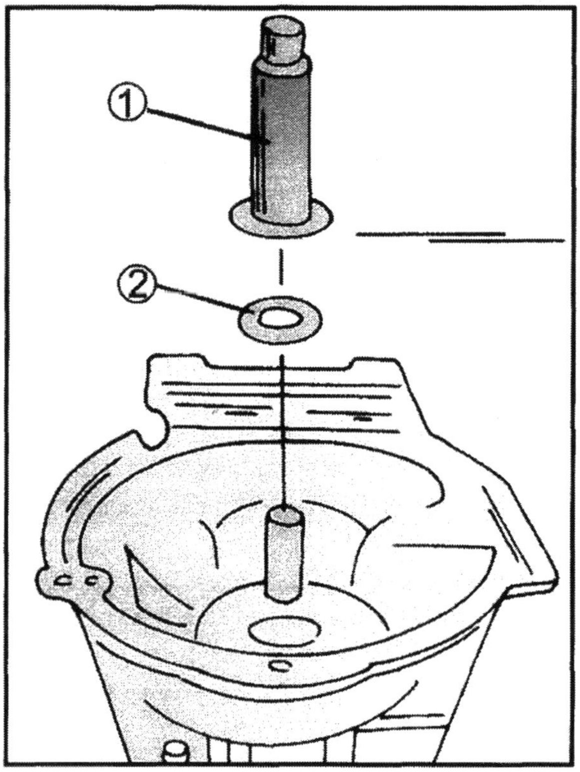

Fig. 3.3 – Installation of the clutch shaft oil seal. Use a drift (1) to drive the oil seal (2) in position until flush with the transmission housing.

A piece of tube or similar, covering the outside of the seal, can be used. Fill the space between the sealing lips with M.P. grease. A suitable drift must be available to drive in the oil seal, until the outer face is flush with the bore in the cover. Coat the splines of the drive shaft with long-term grease and refit the hydraulic clutch operator.

Check the oil level in the transmission if the oil has not been drained. To check, unscrew the oil filler plug. The oil must be 10 mm below the edge of the filler hole. Insert an angled forefinger to check. Tighten the oil filler plug to 3.5 kgm (25 ft.lb.).

3.2.1. TRANSMISSION HOUSING –Removal and Installation

The front and rear section of the transmission housing can be removed but the transmission housing must be dismantled to carry out the work. The transmission must be removed. A suitable mounting stand must be available to clamp the transmission in a horizontal position. Fig. 3.4 shows the partly dismantled transmission housing and should be referred to. Some knowledge is required to dismantle the transmission housing.

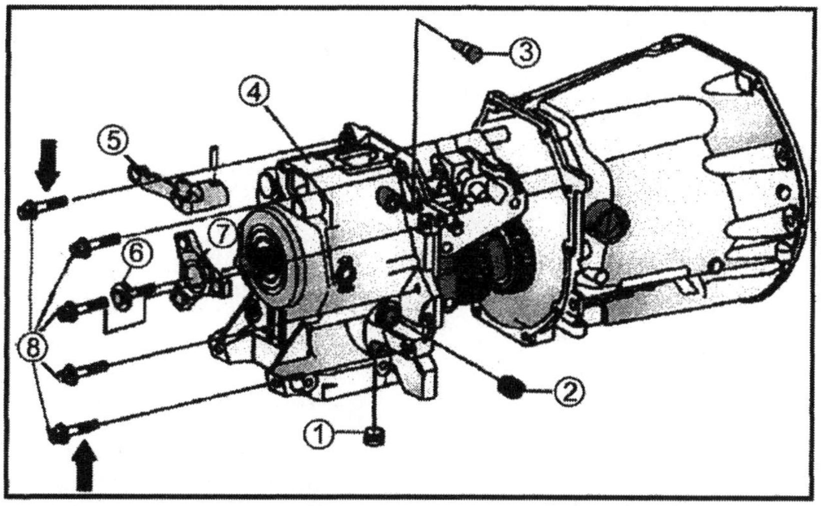

Fig. 3.4 – Details for the dismantling and assembling of the transmission housing. The numbers are referred to in the text.

- Attach the transmission to the mounting stand and turn it into horizontal position.
- Unscrew the drain plug (1) and drain the transmission oil. Collect the oil in a suitable container. Also unscrew the oil filler plug (2). The transmission can now be turned into vertical position.
- Remove the central clutch operator as described in section "Clutch" (see also Fig. 3.2).

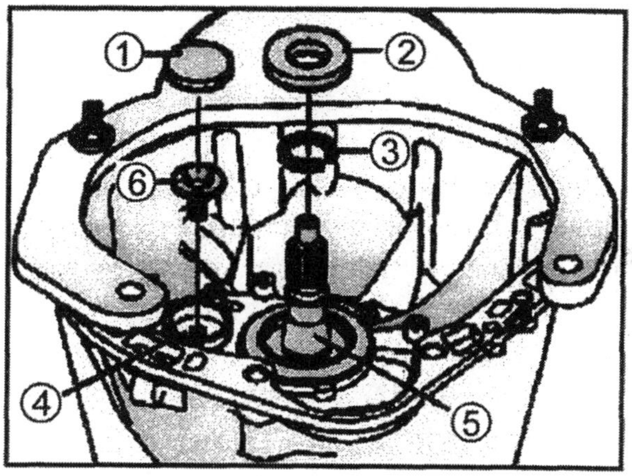

Fig. 3.5 – Details for the dismantling and assembling of the transmission housing. Transmission seen from the inside of the bell housing. The numbers are referred to in the text.

- Refer to Fig. 3.5 and remove the clutch shaft oil seal (2) as described earlier on. Also remove the sealing cover for the countershaft. In Fig. 3.2 the cover is identified with (3).
- Remove the circlip (3) from the input shaft (5), using a pair of circlip pliers.
- Unscrew the countershaft bolt (6). The output flange must be prevented from rotating when the bolt is slackened and removed.
- The reverse gear lock must be removed. The lock is located at the position shown in Fig. 3.6, but the removal requires an impact hammer and an extraction claw as shown in the illustration. The extractor claw, the shape of which is shown on the R.H. side (bottom) and is inserted as shown by (1). Attach the impact hammer (2) and move the weight in the direction of the arrow until the lock (3) is free.

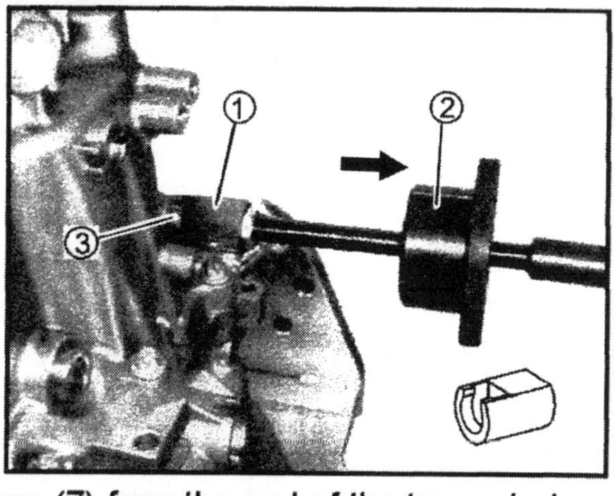

Fig. 3.6 – Removal of the reverse gear lock (3). The extractor claw (1) is fitted as shown. Attach the slide hammer (2) and hit the weight in the direction of the arrow to remove the lock.

- Remove the gear selector arm with the relay lever (5, Fig. 3.4).
- Insert a suitable rod into one of the drive flange holes, turn the flange until the rod rests against the transmission case and undo the bolt (6) in Fig. 3.4 a few turns without removing it completely.
- Use a two-arm puller to withdraw the flange (7) from the end of the transmission. The puller claws are placed underneath two of the flange ends, with the puller centre bolt resting against the bolt (6) which is still in position. Remove the flange,

fully unscrew the bolt (6) and the washer and remove the flange (7) from the shaft.

- Bring the transmission into a horizontal position and remove the bolts (8). The two bolts indicated by the arrows must be screwed back into place a few turns to retain the assembly in place. The transmission is now turned back into a vertical position.
- Drive out the countershaft using a plastic mallet.
- Turn the transmission into a horizontal position and fully remove the two bolts (8).
- The rear section and the front section of the transmission housing can now be removed as described below.

The assembly of the transmission housing is carried out as follows:

- Assemble the rear section and the front section of the transmission housing as described below and coat the sealing faces of the transmission housing with sealing compound.
- Insert the rear section of the transmission housing into the front section of the housing. Ensure that the countershaft and the shift rod move into the corresponding position. The teeth of the mainshaft must engage in the teeth of the input shaft. The input shaft can be rotated if necessary.
- Screw in the two bolts shown by the arrows in Fig. 3.4 to bring the housings together.
- Counterhold the output flange in suitable manner and fit the countershaft bolt (6) in Fig. 3.5. Tighten the bolt to 10.0 kgm (72 ft.lb.).
- Turn the transmission into a vertical position and fit the bolts (8). Tighten the bolts evenly to 3.0 kgm (22 ft.lb.).
- Refit the sealing ring for the output shaft as described earlier on.
- Refit the reverse gear lock. Use a plastic mallet and drive the lock in position until it is flush with the transmission housing.
- Install the shift arm with the relay lever (5) in Fig. 3.4.
- Knock the drive flange over the serrations of the transmission shaft until fully home and fit the bolt. Hold the flange against rotation and tighten the bolt to 11.0 kgm (80 ft.lb.).
- Install the central clutch operator as described in the clutch section and fit the oil drain plug (5.0 kgm/36 ft.lb.).
- Move the transmission into a horizontal position and fill the unit with oil. Note the differences. Transmission type 716.637 is filled with 1.2 litres, transmission type 716.652 with 1.5 litres. Screw in the filler plug and tighten it to 3.5 kgm (25 ft.lb.).
- The transmission can now be removed from the stand and refitted.

Rear Section of Transmission Housing – Dismantling and Assembling

Fig. 3.7 shows the rear section of the transmission housing after dismantling. The transmission housing must be dismantled as described above. Oil leaks may be the reason for the removal of the oil seal. A suitable drift is also required to fit the new oil seal. If a complete dismantling is necessary you will need a tow-arm puller to remove one of the bearings and a press to remove the other one.

- Remove the radial shaft sealing ring (12). An easy way to remove the seal is to screw two self-tapping screws at opposite points into the outside of the seal and then applying a screwdriver blade underneath the bolt heads to pry it out.
- Remove the shift arm with the relay lever (5, Fig. 3.4) and drive the securing roll pin out of the shift rod. Detach the connecting piece from the shift rod and remove

the radial sealing ring, using a suitable tool. The new oil seal is driven in position with a suitable piece of tube.

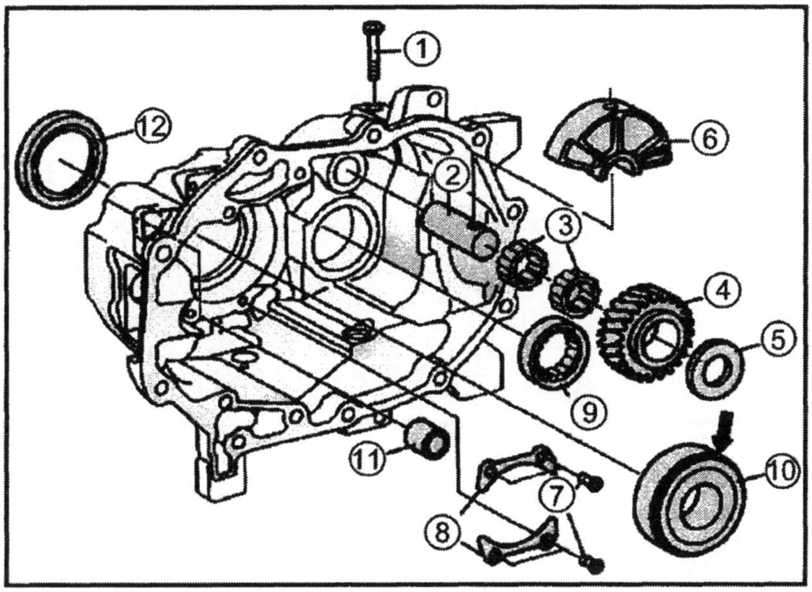

Fig. 3.7 – Dismantling and assembling the rear section of the transmission housing.

1 Bolts
2 Reverse idler shaft
3 Cylinder roller bearings
4 Reverse gearwheel
5 Washer
6 Bearing shell
7 Bolts
8 Retaining plates
9 Countershaft bearing
10 Mainshaft bearing
11 Shift shaft bearing
12 Radial shaft oil seal

- Remove the reverse idler shaft the reverse gear. Details of this operation are shown in Fig. 3.8. Remove the bolt (1) and pull the bearing shell (6) upwards. A wooden wedge can be inserted underneath the bearing shell to press it off. Remove the reverse idler shaft (2), the cylindrical roller bearings (3), the reverse gearwheel (4) and the washer (5). During installation: Lubricate the roller bearings (3), the idler shaft (2) and the shim (5) with oil and place the bearings, the reverse idler gearwheel and the shim reverse idler shaft. The arrangement of the individual parts are shown in Fig. 3.9 on the L.H. side. The threaded bore on the idler shaft, marked with the arrow, must be arranged as shown. Refer to the R.H. view of Fig. 3.9 and fit the bearing shell (6) together with the idler shaft (2) and the reverse gearwheel (4). Position the bearing shell correctly, insert (1) and tighten the bolt to 2.0 kgm (14.5 ft.lb.).

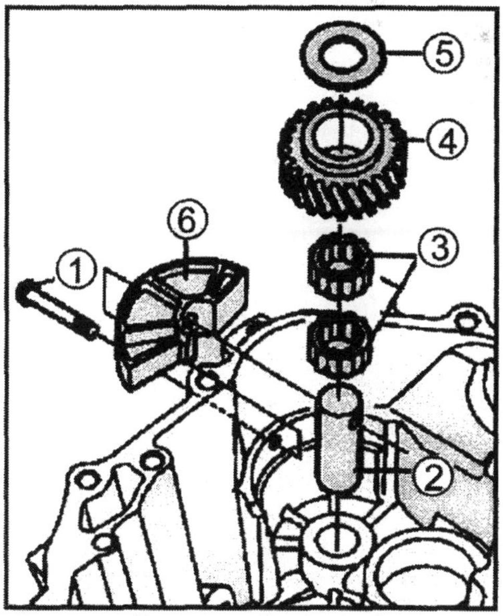

Fig. 3.8 – Removal and installation of the reverse idler shaft.

1 Bolt
2 Reverse idler shaft
3 Cylinder roller bearings
4 Reverse gearwheel
5 Shim
6 Bearing shell

- Unscrew the bolts (7) in Fig. 3.7 and take off the retaining plates (8). The bolts are tightened to 1.0 kgm (7.2 ft.lb.) during installation.
- Press out the mainshaft bearing (10) if necessary from the outside towards the inside, using a press mandrel, large enough to rest on the bearing outer race.
- Remove the countershaft bearing (9). The two-arm puller mentioned above is now required. Two opposite cut-outs are machined into the transmission housing to apply the claws of the puller underneath the bearing race.
- The shift shaft bearing (11) can also be removed.

The assembly is carried out in reverse order, noting the following points:

- If removed, fit the shift shaft bearing with a suitable drift.

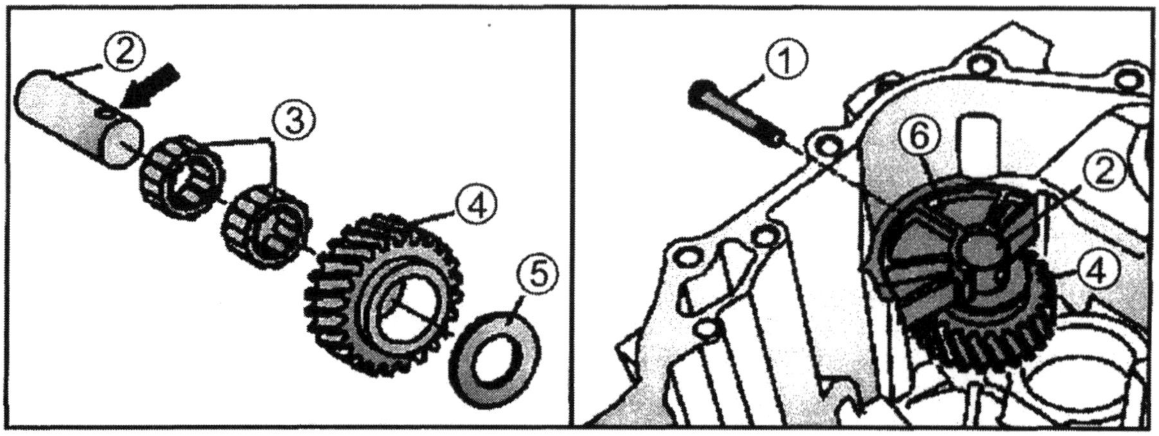

Fig. 3.9 – Assemble the reverse idler shaft as shown on the left, observing the threaded bore (arrow). The R.H. view shows the installation. The numbers refer to Fig. 3.8.

- Press the countershaft bearing in position, using a suitable drift. The marking on the bearing must point upwards in the direction of the front section.
- Press in the mainshaft bearing (10) with the groove of the bearing located as shown by the arrow in the illustration.
- Fit the locking plates (8) and attach them with new bolts (7). Tighten the bolts to 1.0 kgm (7.2 ft.lb.).
- Install the reverse idler shaft with the reverse gearwheel as described above.

Front Section of Transmission Housing – Dismantling and Assembling

Fig. 3.10 shows the parts that can be removed, if dismantling is required. The transmission housing must be dismantled as described above before the front section can be removed. Various drifts are required to remove the shaft bearings.

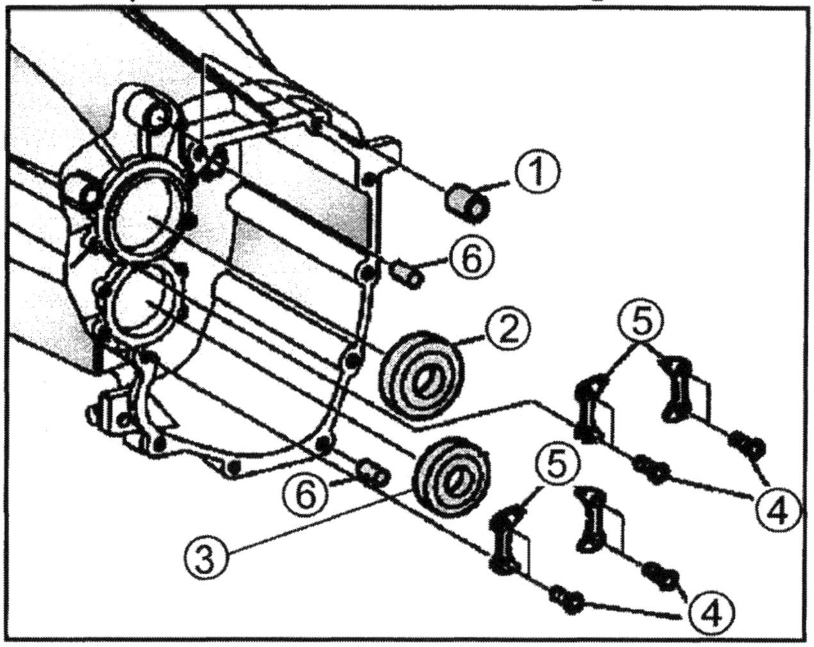

Fig. 3.10 – Details for the dismantling and assembling of the front section of the transmission housing.

1 Central shift shaft bearing
2 Input shaft bearing
3 Countershaft bearing
4 Bolts
5 Retaining plates
6 Guide sleeves

- Remove the input shaft with a pair of pliers.
- Remove the bolts (4) and take off the retaining plates (5).
- Drive out the input shaft bearing (2) and the countershaft bearing (3). Suitable drifts are required to drive out the bearings.
- Remove the guide sleeves (6) if required.

The assembly is a reversal of the dismantling procedure, but note the following points:

- If the guide sleeves (6) have been removed use new guide sleeves.
- Drive in the removed bearings with a suitable mandrel, placed over the outside of the outer faces.

- Fit the retaining plates (5) and fit the bolts (4). Tighten the bolts to 1.0 kgm (7.2 ft.lb.).
- Assemble the transmission housing as described above.

3.3 Gearbox Oil Level

The gearbox is filled with approx. 1.2 litres (type 716.637) or 1.5 litres (type 716.652) of the oil given in Section 3.0. Fluid or oil can only be lost if a leak has developed. The intervals between oil changes are sometimes changed and we recommend to check with your maintenance booklet when oil changes are to be carried out.

To check the oil level, remove the oil filler plug (1) in Fig. 3.11 from the side of the transmission (Allen head key) and check that the fluid/oil is up to the lower edge of the plug bore. If the oil cannot be seen, insert your forefinger to check whether you can reach it. If necessary top-up with the specified oil. A grease gun (thoroughly cleaned) can be used to fill in the required fluid/oil. Refit and tighten the filler plug (3.5 kgm/25 ft.lb.).

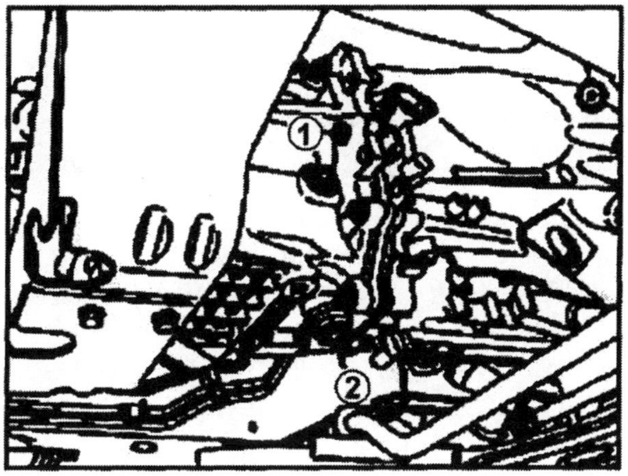

Fig. 3.11 – Location of the oil filler/oil level plug (1) and the oil drain plug (2).

Drain the transmission fluid/oil in the usual manner, i.e. remove the plug (2) in Fig. 3.11. The vehicle should be driven a short distance in order to warm up the oil. Then place the vehicle on chassis stands and place a suitable container underneath the gearbox. The filler plug should be removed to speed up the draining of the transmission. Make sure all oil has been drained. Allow the vehicle to stand for a while until all old oil has been drained.

Fit the drain plug and tighten it to 5.0 kgm/36 ft.lb.). Then fill the transmission with the oil specified in Section 3.0. A grease gun (thoroughly cleaned) can be used to fill in the required oil. If you cannot measure the exact quantity of oil fill the transmission until the oil reaches the lower edge of the filler hole. The filler/oil level plug is tightened to 3.5 kgm (25 ft.lb).

3.4 Manual Transmission/Clutch – Tightening Torques

Gearbox oil drain plug: .... 5.0 kgm (36 ft.lb.)
Gearbox oil filler plug: .... 3.5 kgm (25 ft.lb.)
Torx-head bolts, bearing locking plates, front and rear sections
to transmission housing: .... 1.0 kgm (7.2 ft.lb.)
Bolt, reverse gear shaft ("1" Fig. 3.9): .... 2.0 kgm (14.5 ft.lb.)
Central clutch operator to transmission: .... 0.9 kgm (7 ft.lb.)
Transmission to engine: .... 3.8 kgm (27 ft.lb.)
Propeller shaft flange to transmission flange: .... 1.5 kgm + 90° (11 ft.lb. + 90°)
Nut, bracket to front exhaust system: .... 2.3 kgm (16.5 ft.lb.)
Bolt, transmission carrier to longitudinal member: .... 4.0 kgm (29 ft.lb.)
Bolt, rear engine mount to transmission: .... 2.0 kgm (14.5 ft.lb.)
Bolt, transmission carrier to longitudinal member: .... 4.0 kgm (29 ft.lb.)
Clutch master cylinder to pedal bracket: .... 0.7 kgm (5 ft.lb.)

4. PROPELLER SHAFT

All vehicles are fitted with a two-part propeller shaft. The diameter of the front and rear shaft is 60 mm in the case of most models. The front and rear shaft are marked to each other. The shaft is correctly assembled when the yoke with the two marks are in line with the single mark on the rear shaft. The alignment is shown in Fig. 4.1.

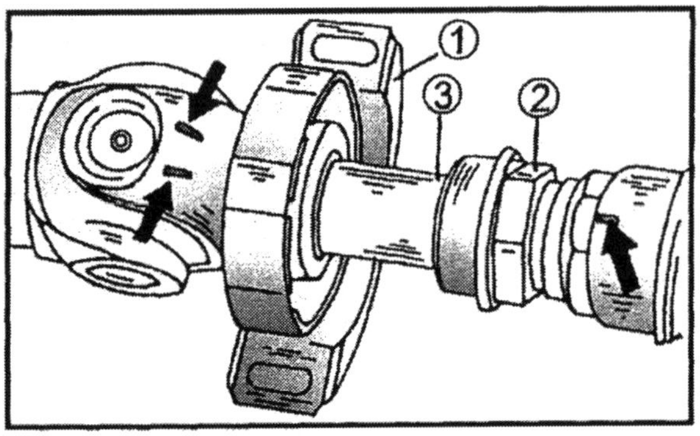

Fig. 4.1 – Marking of the propeller shaft. The markings of the front shaft (arrows) must be aligned with the mark on the rear shaft (R.H. arrow). Always assemble the shaft as shown.
1 Intermediate bearing
2 Locknut
3 Rubber sleeve

An intermediate bearing connects the two shafts in the centre. A so-called check strap retains the rear propeller shaft section to the vehicle underbody.

4.1. Propeller Shaft – Removal and Installation

- Place the vehicle on secure chassis stands to gain access to the propeller shaft mounting.
- Detach the front propeller shaft flange from the transmission flange. Move the shaft to one side, as disconnected, and tie it to the underbody, using a piece of cord or wire.

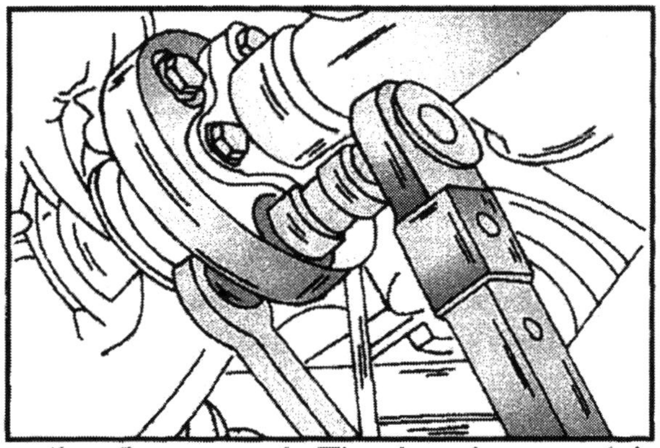

Fig. 4.2 – Removal of the propeller shaft from the rear axle flange. A socket and an open-ended spanner are required.

- Detach the rear propeller shaft flange from the rear axle flange (Fig. 4.2). Move the shaft to one side, as disconnected, and tie it to the underbody, using a piece of cord or wire.
- Remove the bolts securing the propeller shaft intermediate bearing from the floor panel. The bearing must be tightened after the shaft has been re-connected to the transmission and gearbox flanges.

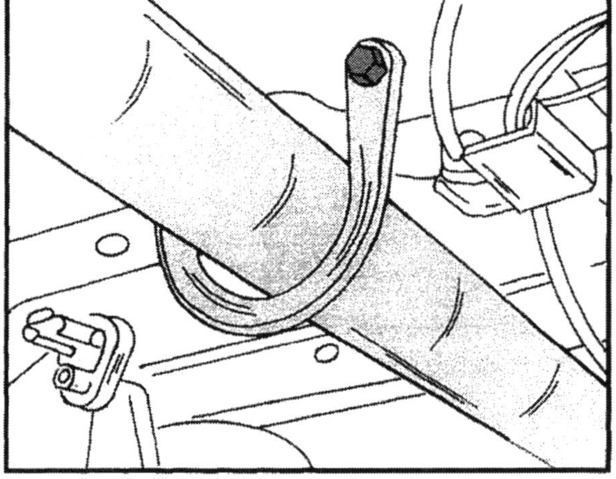

Fig. 4.3 – Fitted position of the "U" shaped bracket around the propeller shaft.

- A helper is required to hold the propeller shaft during the next operation. Refer to Fig. 4.3 and detach the propeller shaft check strap and take out the shaft. The shaft will drop if not held in position when the check strap is removed. Make a note of the fitted position of the check strap to ensure correct installation.
- Check all parts for damage or other faults before the shaft or shafts are refitted. If there is wear in the universal

joints, pull the propeller shafts apart and replace the damaged piece of propeller shaft. The intermediate bearing or the grooved ball bearing can be replaced as described below.

The installation is a reversal of the removal procedure.

- Refit the propeller shaft intermediate bearing to the vehicle floor, without tightening the bolts fully.
- Connect the front and rear ends of the propeller shaft to the drive flanges. Counterhold the bolts and tighten them to 1.5 kgm (11 ft.lb.) and from the final position a further 90° (quarter of a turn).
- Finally tighten the two intermediate bearing bolts. The remaining operations are carried out in reverse order.

Note the following tightening torques during the installation of the shafts:

Propeller shaft flange to transmission drive flange: .... 1.5 kgm (11 ft.lb.) + 90°
Propeller shaft flange to rear axle drive flange:......................... 1.5 kgm (11 ft.lb.) + 90°
Intermediate bearing to underbody:.. 5.8 kgm (42 ft.lb.)
Check strap to body:.. 5.8 kgm (42 ft.lb.)
Nut, transmission output flange (manual transmission):.................. 11.0 kgm (79 ft.lb.)
Nut, transmission output flange (automatic transmission):............ 20.0 kgm (145 ft.lb.)

4.2. Intermediate Bearing – Removal and Installation

Fig. 4.4 shows the intermediate bearing and its connection to the propeller shaft. The bearing can be replaced as follows:

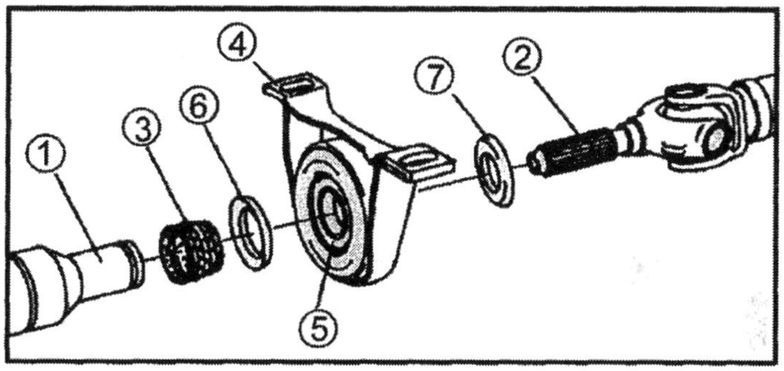

Fig. 4.4 – The propeller shaft intermediate bearing and its attachment to the propeller shafts. The numbers are referred to in the text.

The relationship of the front propeller shaft is marked as shown in Fig. 4.1.

- Separate the propeller shaft with the splined shaft support (1) from the propeller shaft with the splined shaft journal (2).
- Detach the rubber boot (3).
- Detach the rubber mount (intermediate bearing), shown with (4), together with the fitted ball bearing (5), the rear protective cap (6) and the front protective cap (7). The shafts must be clamped into a vice with soft-metal jaws to prevent damage. A two-arm puller is required to pull-off the assembly.

The installation is carried out as follows:

- Check the rubber boot (3) for damage and the rubber mount (4) for wear or damage and replace as necessary. The same applies to the ball bearing (5).
- Push the front protective cap (7) onto the splined shaft journal (2). Note that the protective caps (6) and (7) are not the same. The front cap (7) must be pushed onto its seat without any resistance. Check with the illustration to fit the caps correctly.
- Slowly press a new ball bearing into the bearing seat until it rests against the contact surface.
- Press the rubber mount (4) with the fitted grooved bearing (5) onto the shaft journal (2) and fit the rear protective cap (6) into the rubber mount. The open side of the rubber mount must face into the direction of travel.

- Fit the rubber boot (3), making sure it is fitted in the direction shown in the direction shown in the illustration.
- Join up the propeller shafts, observing the correct position of the shafts relative to each other, as shown in Fig. 4.1. Coat the splined shaft journal (2) with a little long-life grease.

5. FRONT AXLE AND FRONT SUSPENSION

5.0. Technical Data

Wheel bearing play:	0.01 – 0.02 mm
Max. permissible play:	0.05 mm
Wheel bearing grease:	Heat resistant grease No. 00 989 49 51
Amount of grease necessary:	
- In hub grease cap:	15 grams
- In wheel bearings and hub:	70 grams

5.1. Front Suspension Strut – Removal and Installation

Fig. 5.1 shows the assembled front suspension on one side. The front axle half consists of the steering knuckle together with the shock absorber strut and the suspension arm (semi-trailing arm). The whole assembly can be removed as described below, but we must point out that the panelling on the centre section of the windscreen and the wiper arm must be removed to gain access to all parts to be removed. The suspension strut must be removed from its upper attachment:

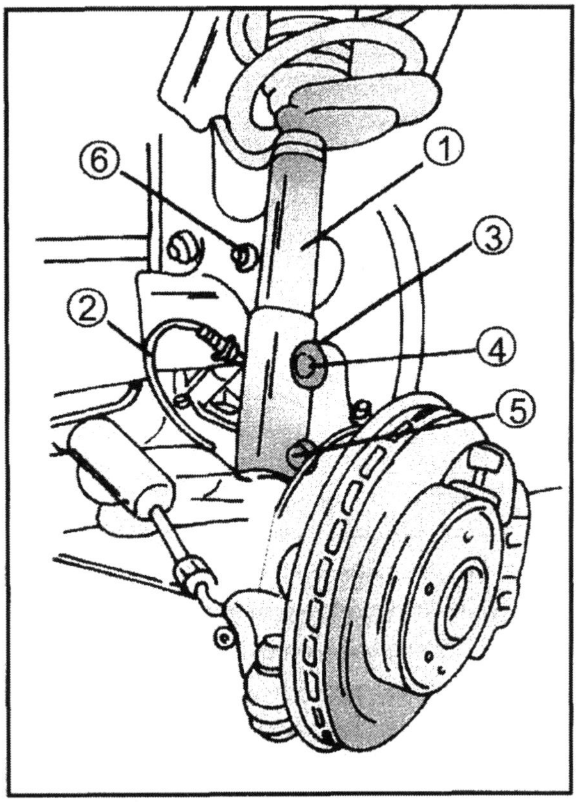

Fig. 5.1 – Fitted suspension strut on one side of the vehicle.

1 Suspension strut
2 Brake hose
3 Electrical wiring
4 Eccentric bolt
5 Collar bolt
6 Nut, stabiliser linkage

- Jack up the front end of the vehicle and place chassis stands in position. Remove the front wheels. From underneath the vehicle remove the noise dampening panel. Make sure that the vehicle is adequately supported.
- First remove the wiper arms. Remove the protective caps fitted over the wiper arm spindles. The wipers must be in the parking position.
- Disconnect the fluid hoses from the windscreen washer system (fitted around the wiper arm spindles) and then unscrew the nuts securing the wiper arms to the spindles. Mark The parked position of the wiper blades on the windscreen and pry

off the wiper arms. The installation is a reversal of the removal procedure, but make sure that the wiper blades take up their original position (wipers in park position). Fit and tighten the wiper spindle nuts to 1.8 kgm (13 ft.lb.).

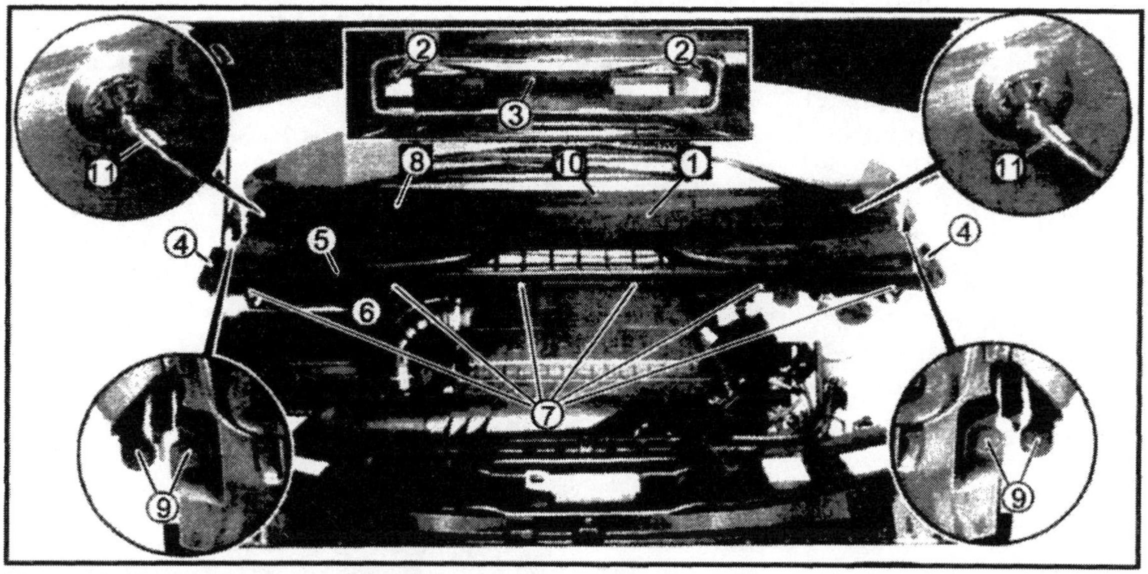

Fig. 5.2 – Details for the removal and installation of the panelling on the centre section of the windscreen. The numbers are referred to in the text.

- The panelling in the centre section of the windscreen must now be removed. This is a fairly complicated operation, but can be carried out. The operations must be carried out by referring to Fig. 5.2. Remove the large cover (1) with a suitable tool. Remove the two screws (2) and take off the clamping plate (3). On the L.H. and R.H. side remove the foam material (4). During installation first insert the clip into the foam material, then insert the foam material into the specified holes. Ensure that the clip is seated properly in the lip (5).
- Remove the clips on the sides of the centering panel (8) with a suitable tool. The clips may have to be replaced.
- Unscrew the bolts (7). During installation pay attention to the bracket (6).
- Lift the panelling from the centre section (8). Lift the panel until the L.H. and R.H. water hoses (11) are accessible and remove the two water hoses.
- The panelling can now be removed from the centre section. Slide the panelling downwards and free it from the window frame seal (10). Then remove the panelling upwards between the windscreen and the engine bonnet.

Install the panelling in reverse order.

- Remove the rubber plug from the radiator tank.
- Remove the nut securing the suspension strut from the suspension strut tower. To remove the R.H. suspension strut (1) in Fig. 5.1, turn the steering into the L.H. lock, to remove the R.H. strut turn the steering into the L.H. lock. A suitable socket wrench is required. Tighten the nut during installation to 5.6 kgm (40 ft.lb.).
- Remove the cover from the fuse box when removing the R.H. strut. In this case also remove four clips from the fire wall panelling and remove the nuts securing the firewall panelling from the radiator tank on the R.H. side. The firewall panelling can now be removed, but just enough to provide access to the stop plate under the suspension strut nut. The stop plate can now be removed. The stop plate must be replaced during installation.
- Remove the front wheels.
- Refer to Fig. 5.1 and detach the brake hose (2) and the electrical wiring (3) from the suspension strut (1). Check the brake hose and the electrical wiring for damage before installation.

- Detach the stabiliser bar connecting link from the strut after removal of the nut (6). The nut is tightened to 8.5 kgm (61 ft.lb.) during installation.
- Remove the bolt (4) securing the suspension strut to the steering knuckle. This is an eccentric bolt and its position must be marked before removal. As the fitted position of the bolt will influence the wheel alignment it is important that the bolt is refitted in its original position. The bolt should be replaced during installation, i.e. the marks must be transferred to the new bolt. The workshop tightens the bolt during the alignment check. The bolt is tightened in four stages. First tighten the bolt to 15.0 kgm (108 ft.lb.), then slacken it by half a turn (180°), tighten it to 12.0 kgm (86.5 ft.lb.) and from the final position a further 75°. The angle must be estimated, i.e. just a little less then a quarter of a turn. If not absolutely sure, have the wheel alignment checked after installation.
- Remove the bolt (5) and take out the suspension strut. If a new strut is fitted, take the old one to your parts supplier where it can be disposed.

The installation of the strut is a reversal of the removal procedure, noting the points already given above.

5.2. Front Axle Half – Removal and Installation

The front axle half is removed together with the suspension strut and the transverse control arm.

- Remove the suspension strut at the upper end as described in section 5.1.
- Remove the bolt securing the ABS sensor to the steering knuckle on both sides of the suspension. Tighten the bolts to 0.8 kgm (6 ft.lb.) during installation.
- Detach the brake hose from the suspension strut. Depending on the operations foreseen, it is also possible to disconnect the brake hose from the brake pipe. In this case counterhold the hexagon of the brake hose, unscrew the union nut on the other side of the bracket and remove the spring plate until the hose is free. Close the open ends of hose and pipe in suitable manner to prevent entry of dirt. **Note the following during installation.** The union nut is tightened to 1.5 kgm (10 ft.lb.) without allowing the brake hose to twist. After installation have the steering wheel turn from left to right and the car bounced up and down whilst checking the brake hose to make sure it cannot come near other parts of the front suspension.
- In the area of the brake hose connection free the cable from the ABS speed sensor (both sides) and the cable connected to the brake pad wear indicator. Both are attached to the suspension strut. The connectors must be located, depending which side is to be removed. Follow the cables and disconnect. Pull the cable harness through the rubber grommets and free them from their retaining clamps.
- Remove the front brake caliper as described in section "Brake System". If the brake hose remains connected there is no need to disconnect the brake line, but tie the caliper to the front suspension to prevent it from hanging down under its own weight.
- Unscrew the stabiliser bar connecting link at the upper end. To do this, remove the nut (6) in Fig. 5.1. Tighten the nut to 8.5 kgm (61 ft.lb.) during installation.
- Remove the nut securing the track rod ball joint to the steering lever and separate the ball joint with a suitable puller. Immediately check the rubber dust cap and the ball joint for damage or excessive clearance. Fit new nuts during installation and tighten the nut to 5.0 kgm (36 ft.lb.) and from the final position a further 60° (graduated disc or considering that 90° equals a quarter of a turn).
- Place a trolley jack underneath the suspension arm (transverse control arm) and lift the arm until just under tension. In this position remove the nuts securing the

suspension arms to the subframe. Fig. 5.3 shows where the nuts and the bolts on the other side can be found. The nuts and bolts are tightened to 15.5 kgm (112 ft.lb.) during installation.

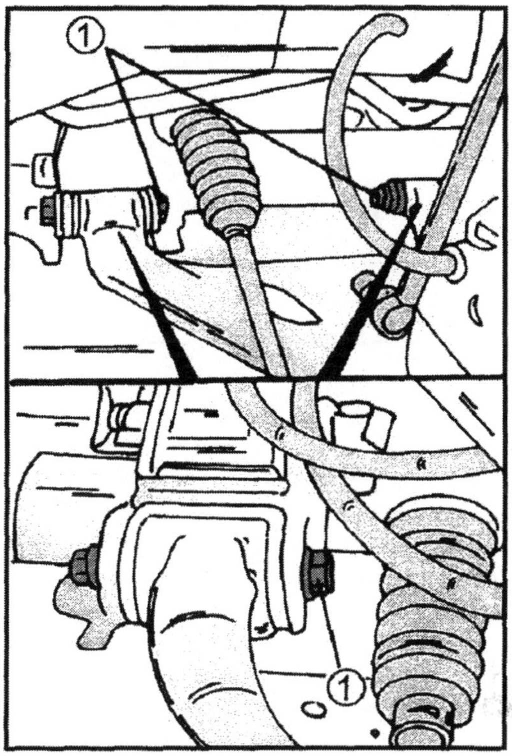

Fig. 5.3 – The bolts (1) secure the suspension arm (transverse control arm) to the subframe. One bolt and nut secure the arm on both sides.

- The complete front axle half assembly, i.e. spring strut with suspension arm (1) can now be lifted out as one unit as shown in Fig. 5.4.
- The front and rear bushes of the lower suspension arm can be replaced, but we recommend to take the suspension arm to a workshop to have the bushes replaced, if you find them worn beyond use. The spring strut can be dismantled as described below.

The installation of the front axle half is a reversal of the removal procedure, but the following points should be noted:

- The tightening torques specified above must be observed.
- Place the front axle half onto a mobile jack and slowly lift it in position to guide the piston rod through the opening in the mounting bearing in the body and refit the suspension strut as previously described. As the eccentric bolts are used to adjust the camber angle and the castor you will have to have the wheel alignment checked after completed installation.

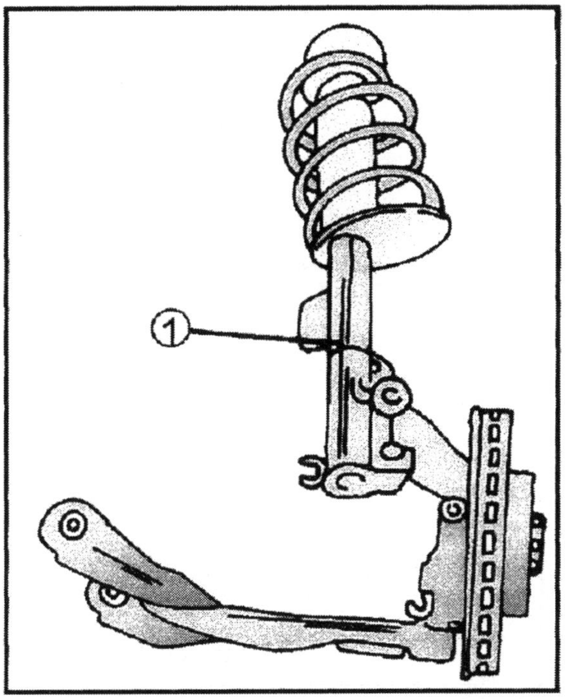

Fig. 5.4 – The front axle half (1) is removed as one assembly as shown.

- Fit the suspension arm to the frame crossmember.
- Fit the stabiliser bar connecting link to the suspension strut.
- Re-connect the brake hose. Note the note already given earlier on.
- Re-connect all electrical leads. Push the rubber sleeve into place. Sharp metal edges could damage the cables.
- Re-connect the track rod ball joint to the steering lever and tighten the nut. If the ball joint pin rotates it will be possible to insert an Allen key into the end of the pin to prevent it from rotating.
- Fit the wheels and lower the vehicle to the ground. Check once more that the eccentric bolts in the springs struts are fitted in accordance with the marks made before removal and tighten the nut on the other side to 15.0 kgm (108 ft.lb.).
- Have the front wheel alignment and the headlamp adjustment checked in a workshop.

5.3. Front Suspension Strut – Dismantling and Assembly

The parts shown in Fig. 5.5 can be replaced individually, bearing in mind that the piston rod (8) is part of the spring strut. (9). A spring compressor is required to compress the coil spring when the strut is dismantled.

Only the coil spring, the complete spring strut, the spring strut bearing and the sundry parts shown in Fig. 5.5 can be replaced. The spring strut cannot be dismantled, i.e. a faulty shock absorber must be rectified by fitting a new spring strut. Never try to dismantle a spring strut, without compressing the spring first. Fig. 5.6 shows a compressor as used in the workshop. A similar tool, available from tool hire companies for example, is suitable as long as the hooks can be placed over four or five spring coils, as shown in the R.H. view.

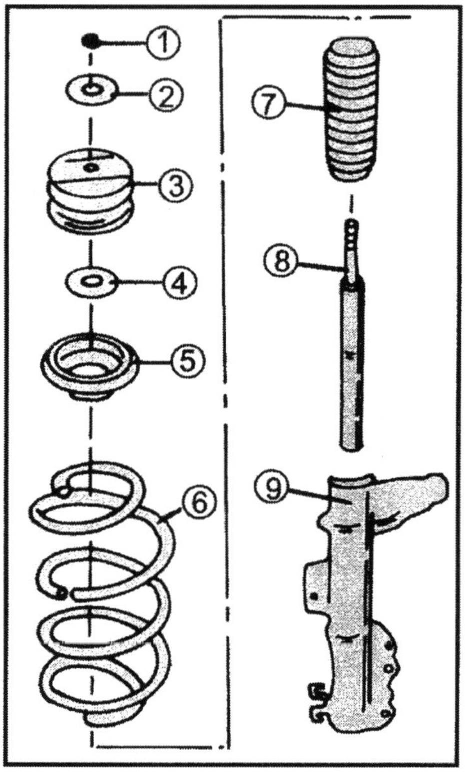

Fig. 5.5 – Dismantled spring strut.

1 Piston rod nut, 6.6 kgm
2 Washer
3 Rubber mount
4 Shim
5 Upper spring seat
6 Coil spring
7 Rubber gaiter
8 Piston rod
9 Spring strut

Dismantling

- Clamp the spring strut into a vice and fit the spring compressor. Fig. 5.6 shows the use of the spring compressor on the R.H. side. Make absolutely sure that the coils are securely retained in the compressor. Tighten the spring compressor until the spring or the spring cup (6) can be rotated freely between the spring seats.
- Slacken the piston rod nut, using a flat or ring spanner of suitable size. Fully unscrew the nut (1) and remove the shim (2).

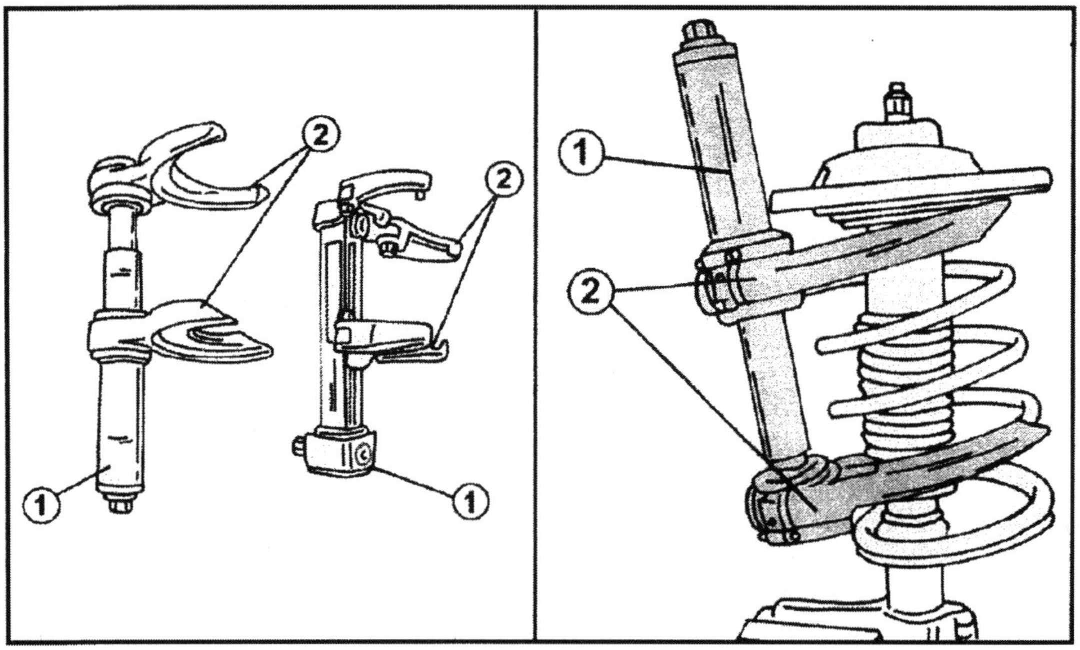

Fig. 5.6 – Special spring compressor (1) for the dismantling of a front spring strut. The R.H. view shows how the compressor (1) is placed around the coil spring, with the claws (2) placed around the coil spring windings.

- Remove the rubber mount (3) from the end of the piston rod and remove the shim (4) and the spring cup (5), i.e. the spring seat from the upper end of the coil spring.
- Remove the rubber protector (7) and take off the suspension strut (9). Take care not to damage the smooth surface of the piston rod when the spring strut is stored for later assembly. If a new spring strut is required take the old one to your parts supplier where it can be disposed of in a suitable manner.

Always quote the chassis number, the vehicle model and the model year when a coil spring is replaced, as not all vehicles may not have the same coil springs.

The shock absorber can be checked for correct operation by clamping the spring strut in upright position into a vice. Screw the nut back onto the piston rod and grip the nut with a pair of pliers. Move the piston rod up and down. The piston rod should move with even resistance. Any "dead play" indicates a faulty shock absorber.

Assembling

- Assemble the parts by referring to Fig. 5.5. Place the rubber gaiter over the piston rod and attach it to the spring strut. Clamp the spring strut into a vice in vertical position.
- Fit the compressed spring over the spring strut, engaging the lower coil with its stop. Place the upper spring seat (5) over the spring.
- Fit the shim (4) over the piston rod, followed by the rubber mount (3) and the clamp plate (2).
- Fit a new self-locking nut to the end of the piston rod and tighten the nut to 6.6 kgm (47.5 ft.lb.). The piston rod must be held against rotation until it is possible to apply a socket and torque wrench to the nut. The latter should be tight enough to keep the piston rod in position.
- Slowly release the spring compressor, at the same time checking that the spring coils enter the locations at top and bottom. The ends of the coils must be resting against the stop on the spring strut and a similar stop on the upper end.

5.4. Steering Knuckles and Wheel Hubs

5.4.0. REMOVAL

With the front end of the vehicle resting on secure chassis stands, proceed as follows:

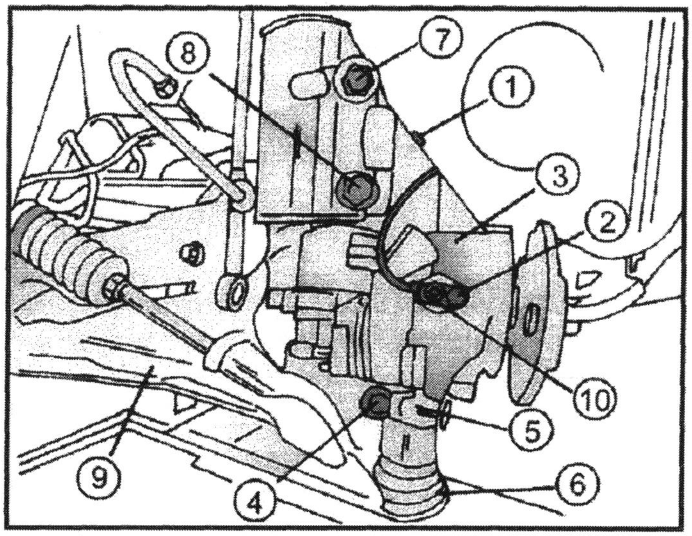

Fig. 5.7 – Details for the removal and installation of the steering knuckle and the wheel hub. The numbers are referred to during the removal and installation instructions.

Fig. 5.7 shows a view of one side of the suspension with the parts to be removed or disconnected. All operations referring to the brake system are described in section "Brake System". The front suspension arm/steering knuckle must be suitably supported from below as the various parts are removed.

- Remove the brake disc. If the steering knuckle (3) is to be replaced also remove the brake anchor plate.
- Remove the screw (1) securing the cable holder, take off the holder and place it to one side where it is out of the way.
- Remove the bolt (2) and withdraw the ABS wheel speed sensor (10) on the side in question.
- Remove the bolt (4) at the bottom of the steering knuckle. The bolts secures the steering knuckle to the suspension arm. Note the tightening torque sequence during installation.
- Remove the nut (5) from the track rod ball joint pin and using a suitable puller separate the track rod ball joint from the steering lever on the steering knuckle. Before installation check the track rod ball joint end and replace it if wear or other damage can be seen. Section "Steering" gives information for the replacement of a track rod ball joint.
- Mark the position of the steering knuckle (3) in relation to the suspension strut. The marking is necessary to obtain the correct camber setting. Although we recommend to have the camber checked in a workshop, you may be able to obtain the correct setting straight away.
- Remove the eccentric bolt (7), referred to by Mercedes-Benz as the cam bolt. This bolt is responsible for the adjustment of the camber setting. Mark the position of the bolt head as it is fitted with a spot of paint or similar and remove the nut on the other side. A new bolt must always be fitted, i.e. you will have to transfer the marking to the new bolt. The workshop does not fully tighten the bolt and nut during installation, as the final tightening takes place after the camber has been checked/adjusted.
- Remove the bolt and nut (8) securing the steering knuckle to the suspension strut. Again the bolt must be replaced and is treated in the same manner as described for bolt (7).
- Press the steering knuckle (5) out of the suspension arm (transverse arm) and remove it.

5.4.1. INSTALLATION

The installation is a reversal of the removal procedure, noting the following points:

- Refit the steering knuckle to the suspension strut and the suspension arm. Tighten the bolt (4) securing the knuckle to the suspension arm to 16.2 kgm (117 ft.lb.).
- Insert the bolts (7) and (8) and fit the nuts on the other side. Refer to the notes during the removal instructions regards the tightening of the bolts. Both bolts are tightened in the same manner after the camber has been adjusted. Insert the eccentric bolt (7) after the bolt head has been rotated until the marks made before removal are in line. Tighten both bolts to 15.0 kgm (108 ft.lb.) and from the final position slacken the bolts by one half of a turn. Now tighten the bolts to 12.0 kgm (86.5 ft.lb.) and then by a further 75°. This is, of course, best obtained by using a graduated disc with the socket. Otherwise estimate the angle, considering that 90° represents a quarter of a turn.
- Refit the track rod ball joint end to the steering lever. The nut is tightened in two stages. First to 5.0 kgm (36 ft.lb.) and then angle-tighten by 60°. Again estimate the angle if a graduated disc is not used.
- Refit the brake anchor plate if a new steering knuckle is fitted and fit the brake disc as described in section "Brake System".
- Refit the ABS wheel speed sensor (2). During installation the sensor(s) must be greased with long-life MP grease. The bolt (2) is tightened to 0.8 kgm (6 ft.lb.).

- Refit the cable holder screw (1) and tighten it to 0.8 kgm (6 ft.lb.).
- Finally lower the vehicle to the ground and tighten the wheel bolts. As recommended above, have the wheel camber checked at a workshop dealing with wheel alignments.

5.5. Suspension Arm – Removal and Installation

Fig. 5.8 shows a view of the area where the suspension arm is fitted. Fig. 5.3 shows the bolts securing the suspension arm to the subframe. Removal is carried out as follows, with the front of the vehicle resting on secure chassis stands.

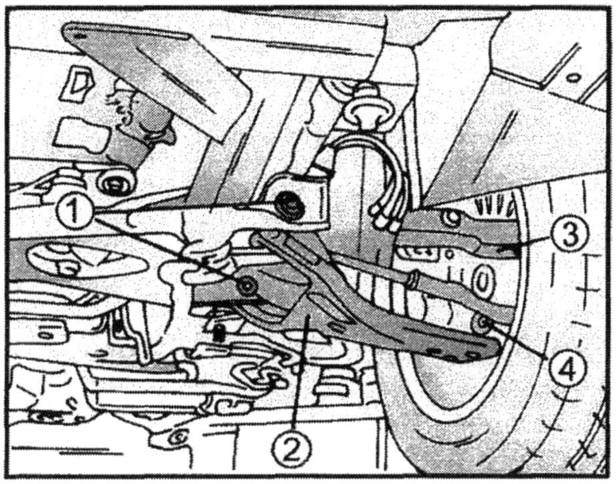

Fig. 5.8 – Details for the removal and installation of the suspension control arm (transverse control arm).
1 Control arm
2 Control arm mounting bolts
3 Steering knuckle
4 Ball joint mounting bolt

- Remove the front wheel on the side in question.
- Detach the control arm (1) from the subframe. These are the bolts and nuts (2), shown already in Fig. 5.3.
- Remove the bolt (4) from the steering knuckle (3) and take out the control arm (1).

The installation is a reversal of the removal procedure. Check the condition of the control arm before installation. In the illustration some arrows are shown. Rivets are inserted which must have a tight fit. A ball joint with excessive wear or a damaged rubber boot requires the replacement of the complete control arm. The rubber mounts of the arm can be replaced but we recommend to take the arm to a dealer to have the mounts replaced, if necessary.

Tighten the control arm mounting bolts and nuts, but note the difference. The bolt and nut securing the control arm to the subframe is tightened to 15.5 kgm (112 ft.lb.), the bolt and nuts securing the control arm to the steering knuckle is tightened to 16.2 kgm (117 ft.lb.).

5.6. Front Wheel Hubs and Wheel Bearings

The wheel bearings will be destroyed if they are removed from the inside of the steering knuckle. We strongly recommend to remove the steering knuckle as described earlier on and take the knuckle to a dealer to have the wheel bearing replaced, as special tools are required to remove and install the bearing. The same applies if a new steering knuckle is to be fitted. You will receive the knuckle with the wheel hub fitted, ready for installation.

5.7. Stabiliser Bar

5.7.0. REMOVAL AND INSTALLATION OF STABILISER BAR

The attachment of the stabiliser bar is shown in Fig. 5.9. The front end of the vehicle must be resting on secure chassis stands to remove the bar. The bolts securing the two engine mounts must be removed to take out the bar. This means that the engine must be lifted out of its mountings in a suitable manner, using a hoist or small handcrane to have the mountings free of tension. Remove the stabiliser bar as follows:

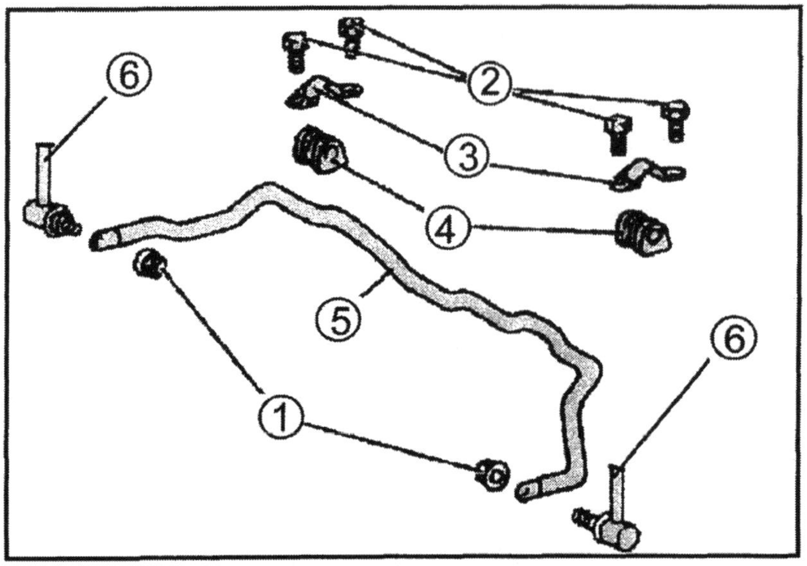

Fig. 5.9 – Attachment of the stabiliser bar.

1 Nut, links to stabiliser bar
2 Mounting clamp bolts
3 Mounting clamps
4 Rubber bushes
5 Stabiliser bar
6 Connecting links

- Remove the L.H. front wheel.
- On both sides of the engine remove the bolts securing the engine mounts from above. The engine must now be lifted slightly to release any tension on the engine mounts. The bolts are tightened to 5.8 kgm (42 ft.lb.) after the engine has been lowered back in position.
- In the case of a model 639.7/8 after VIN number 055347 remove the rear engine compartment panelling at the bottom – Check your VIN number.
- Refer to Fig. 5.9 and remove the nuts (1) securing the stabiliser connecting links to the ends of the stabiliser bar (5) and separate the links on both sides from the bar. The bolts are tightened to 8.5 kgm (61 ft.lb.). There is no need to disconnect the upper ends of the connecting links.
- If dealing with a petrol engine you will find some electrical cables secured by a bracket (looking from underneath, against the panelling). Unhook the electrical connectors and unscrew the bolts from the brackets. Tighten the bolts to 3.8 kgm (27.5 ft.lb.) during installation.
- Remove the bolts (2) from the mounting clamps (3). The bolts are tightened in two stages during installation. First tighten the bolts to 2.0 kgm (14.5 ft.lb.) and then by a further 60°. A graduated disc is required to tighten the bolts. Otherwise estimate the angle (90° = quarter of a turn).
- Remove the stabiliser bar (5) towards the L.H. side of the vehicle.

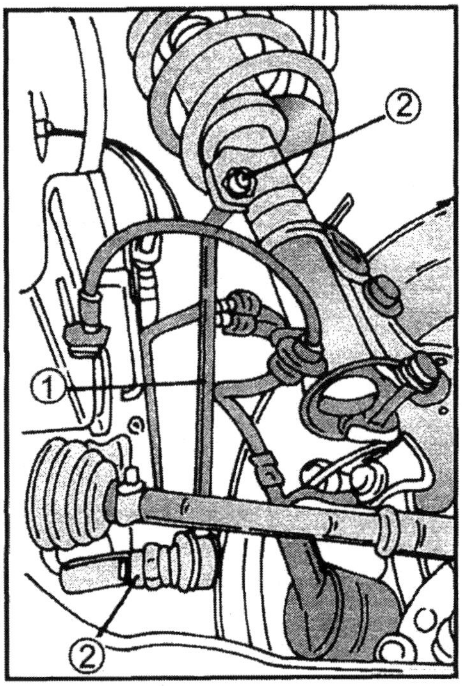

Fig. 5.10 – View of the front suspension with the fitted stabiliser bar connecting links (1) and the mounting bolts (2).

The installation is a reversal of the removal procedure. Check the rubber mountings (4) before installation and replace if necessary. The tightening torque values given above must be followed.

5.7.1. REMOVAL AND INSTALLATION OF LINKS

There is no need to remove the stabiliser bar to replace one of the links. Fig. 5.10 shows the links in fitted position on the front suspension. The links are connected by means of ball joints, which may need replacement some time during the life of the vehicle. The front end of the vehicle must be resting on secure chassis stands.

The removal and installation is straight forward. Remove the nuts at the upper and lower ends of the connecting links with an open-ended spanner. The ball joint pin could rotate and must be held with a suitable tool if this is the case. The workshop uses a special open-ended spanner to apply a torque wrench during installation. Try to use a socket or estimate the torque. Both nuts are tightened to 8.5 kgm (61 ft.lb.).
Before installation check the rubber boots around the ball joints. A new link must be fitted if damage, cuts or similar can be seen.

5.8. Front Wheel Alignment

The front wheel alignment cannot be checked or adjusted by ordinary means as a so-called electronic "inclinometer" is required to measure the vehicle level at the front and rear axle. The instrument will show the level of the front axle in degrees and the camber is adjusted in accordance with a table to the applicable value. The camber angle can be adjusted by the upper mounting bolt securing the suspension strut to the steering knuckle. The bolt is an eccentric bolt (cam bolt), shown in Fig. 5.7 with (7), which is rotated to adjust the camber setting. Only your Mercedes-Benz dealer will therefore be able to adjust the camber setting.
If the bolt is removed, for example during the removal of the spring strut or the steering knuckle, mark the bolt head and its position of the spring strut at opposite points with a centre punch and refit the bolt during installation with the bolt head positioned as marked. As a new bolt must be used, transfer the mark on the bolt head exactly to the new bolt.

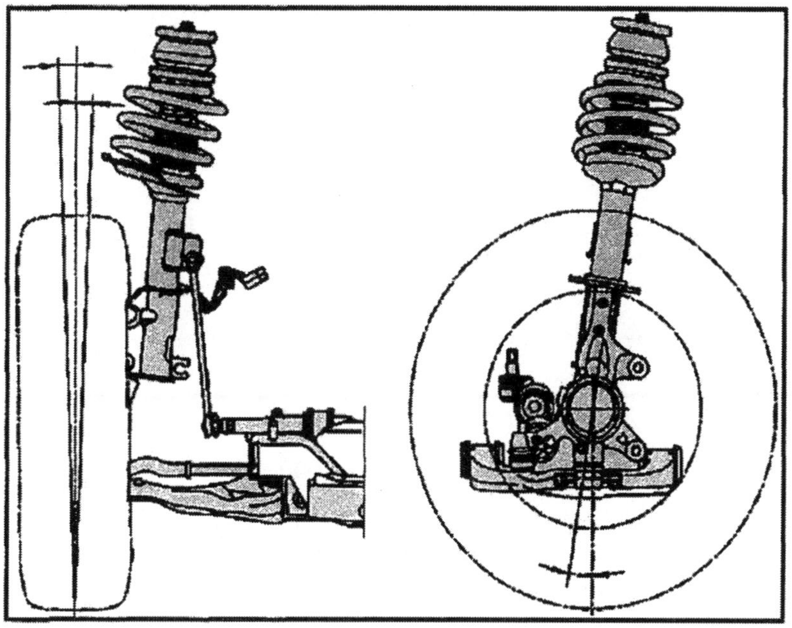

Fig. 5.11 – The camber angle on the left and the castor angle on the right.

During installation tighten both bolts to 15.0 kgm (108 ft.lb.) and from the final position slacken the bolts by one half of a turn. Now tighten the bolt to 12.0 kgm (86.5 ft.lb.) and then by a further 75°. This is, of course, best obtained by using a graduated disc with the socket. Otherwise estimate the angle, considering that 90° represents a quarter of a turn. The lower bolt suspension strut-to-steering knuckle is tightened in the same manner.
Fig. 5.11 shows on the L.H. side the camber angle. As you can see, the angle can either have a plus inclination (towards the left from the vertical line) or a minus inclination (towards the right from the vertical line). Two examples for the technically minded:
If the vehicle level has a value of **plus** 11.2°, the camber would be adjusted to **plus** 0° 30' (30 minutes, half a degree), if on the other hand the vehicle level is read as **minus**

0° 8°, the camber angle would be adjusted to **minus** 0° 32'. In both cases a difference of 20' is permissible.

The castor angle, shown on the right-hand side of Fig. 5.11 cannot be adjusted. The workshop will, however, measure the castor angle during an alignment check to verify the correct chassis geometry.

Toe-in Setting

As in the case of the camber and castor setting the vehicle level must be established by the electronic inclinometer mentioned above. Apart from the instrument a wheel spreader is required to press the wheels apart in order to remove any tension produced by the lower transverse control arm bushes and the steering linkage.

The toe-in itself is adjusted on each track rod, with the steering in the centre position, after slackening of the track rod locknut. The locknut is tightened to 5.0 kgm (35 ft.lb.).

5.9 Front Axle/Suspension – Tightening Torques

Steel wheel bolts: .... 18.0 kgm (130 ft.lb.)
Light alloy wheel bolts: .... 15.0 kgm (108 ft.lb.)
Suspension strut to steering knuckle (upper and lower bolts):
- 1st stage: .... 15.0 kgm (108 ft.lb.)
- 2nd stage: .... Slacken by half a turn (180°)
- 3rd stage: .... 12.0 kgm (86.5 ft.lb.)
- 4th stage: .... Angle-tighten a further 75°

Front engine mounting to engine support: .... 5.8 kgm (42 ft.lb.)
Stabiliser bar to subframe (mounting clamps): .... 2.0 kgm (14.5 ft.lb.) + 60°
Stabiliser bar connecting links, upper and lower ball joint: .... 8.5 kgm (62.5 ft.lb.)
ABS sensor and wear indicator cable brackets: .... 0.8 kgm (6 ft.lb.)
Steering knuckle to transverse control arm: .... 16.2 kgm (116.5 ft.lb.)
Transverse control arm to subframe: .... 15.5 kgm (112 ft.lb.)
Track rod ball joint to steering lever: .... 5.0 kgm (35 ft.lb.) + 60°
Track rod locknut: .... 5.0 kgm (35 ft.lb.) + 60°
Piston rod nut (spring strut, upper end): .... 6.6 kgm (47.5 ft.lb.)
Nut, windscreen wiper arm to wiper spindle: .... 1.8 kgm (13 ft.lb.)

6. STEERING

6.0. Technical Data

Type:	Rack and pinion steering with servo assistance, fitted behind the engine. Progressive ratio.
Type of fluid used:	As in automatic transmissions
Fluid used:	MB steering fluid
Fluid capacity:	0.8 to 1.0 litre

6.1. Steering Unit – Removal and Installation

The following precautions must be taken when work is carried out on the steering system:

- Before removing any connections clean the immediate surrounding and the connection.

- All removed parts must be placed onto a clean workbench and covered up in suitable manner if not refitted immediately.
- Never use fluffy rags to clean component parts. New parts must be unwrapped from their packing immediately before they are fitted. Only use original spare parts.
- Never re-use fluid which has been drained from the system.

Remove the steering as described below. The engine must be lifted out of its mountings as the engine mountings must be removed during the removal of the steering. The workshop is using an engine lifting device as shown in Fig. 6.1. With the help of the illustration you may be able to make up a construction of suitable bars. One other method is to use a hand crane or other lifting device and attach the engine with chains or ropes to the engine lifting brackets and lift the assembly until the engine mountings are free of tension.

The subframe underneath the steering must be removed from the longitudinal frame member to take out the unit. We would therefore like to point out that the removal and installation of the steering gear is not an easy operation.

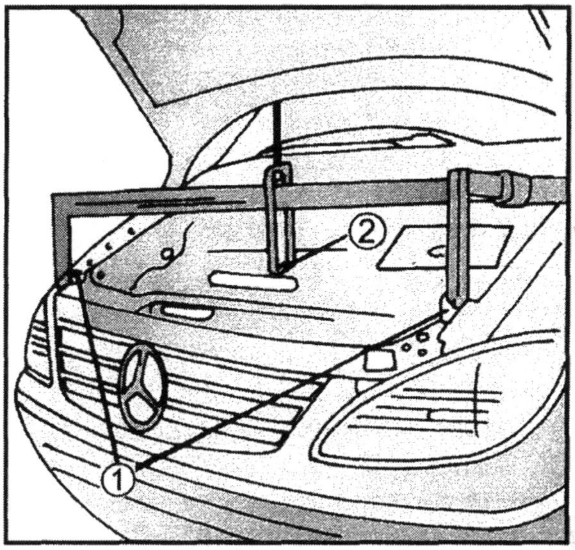

Fig. 6.1 – The engine lifting device used in the workshop is attached with its legs at (1). The bar (2) is attached to the engine. By tightening the spindle the engine will be lifted up.

It is assumed that the front end of the vehicle is resting on chassis stands:

- Remove both front wheels.
- Remove the nut (1) in Fig. 6.2 and separate the track rod ball joint with a suitable puller from the lever on the steering knuckle. The illustration shows a suitable puller. Both track rod ball joints must be removed. Remove the ball joint end of the L.H. track rod (4) after slackening the locknut (3). Unscrew the ball joint end, at the same time counting the turns (and half turns). Write it onto a piece of paper. During installation screw the track rod end by the same number of rotations onto the track rod. This will bring the toe-in to the approximate value. The locknut is tightened to 5.0 kgm (35 ft.lb.). The nut securing the ball joint to the steering lever is tightened in two stages. First tighten to 5.0 kgm (35 ft.lb.) and from the final position a further 60° (estimate the angle, 90° is a quarter of a turn). An Allen key can be inserted into the end of the track rod ball pin to prevent it from rotating during the initial stages of tightening.

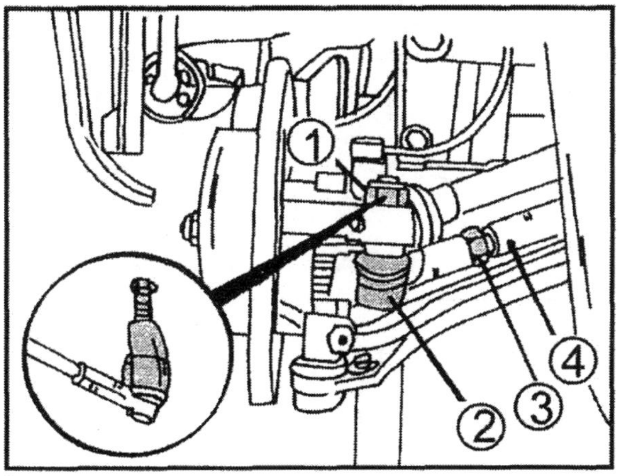

Fig. 6.2 – View of the track rod connection on the steering lever. The ball joint is separated (inset) after removal of the nut (1).

1 Nut
2 Track rod ball joint
3 Locknut
4 Track rod

- Turn the steering into the centre position, i.e. both front wheels must be in the straight-ahead position. Also check at the entry of the steering pinion that the two marks shown in Fig. 6.3 are opposite each other. Make sure that the steering wheel cannot move during subsequent operations out of the centre position. If necessary tie the wheel in position.

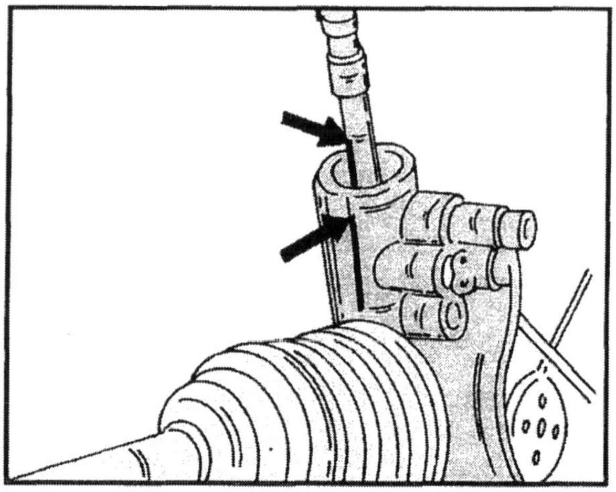

Fig. 6.3 – The two marks shown by the arrows are opposite each other when the steering is in the centre position.

- Remove the air cleaner housing as described for the engine in question.
- Remove the fluid out of the fluid reservoir, using a syringe.
- Locate the high-pressure steering hose and disconnect it from the power steering pump. Some fluid could run out. Collect or absorb it. Also disconnect the fluid return line below it. The pipe is secured by a clamp which must be removed. Again some fluid will run out. Fig. 6.5 shows in the L.H. illustration a view of the steering with the location of the high-pressure hose and the fluid return pipe.
- The engine must now be lifted out of its mountings as explained above. From above, remove the bolt securing the engine mount to the engine support on both sides of the engine. These are the bolts shown in Fig. 6.4. Tighten the bolts to 5.8 kgm (42 ft.lb.) during installation. The engine can now be lifted, but take care not to damage the viscous fan.

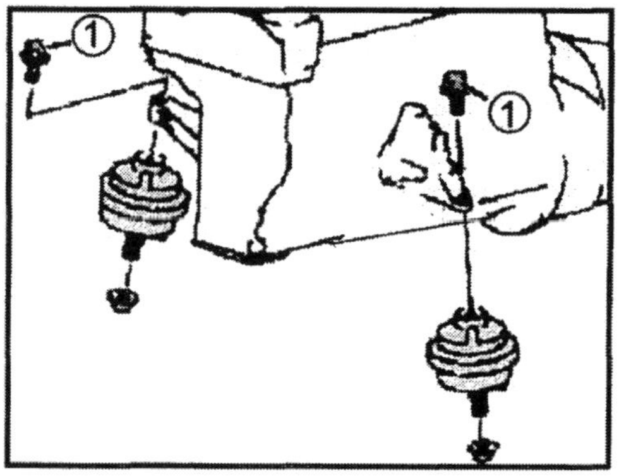

Fig. 6.4 – The two bolts (1) must be removed to detach the engine mounting blocks in order to lift the engine.

- Refer to the L.H. view of the 6.5 and remove the nut (1). In the same area underneath the subframe remove two clamps securing a pipe. Remove the bolt (2) (tighten to 1.8 kgm/13 ft.lb.) and remove the high-pressure hose (3) and the fluid return line (4). An "O" sealing ring is fitted under the connections which should always be replaced.

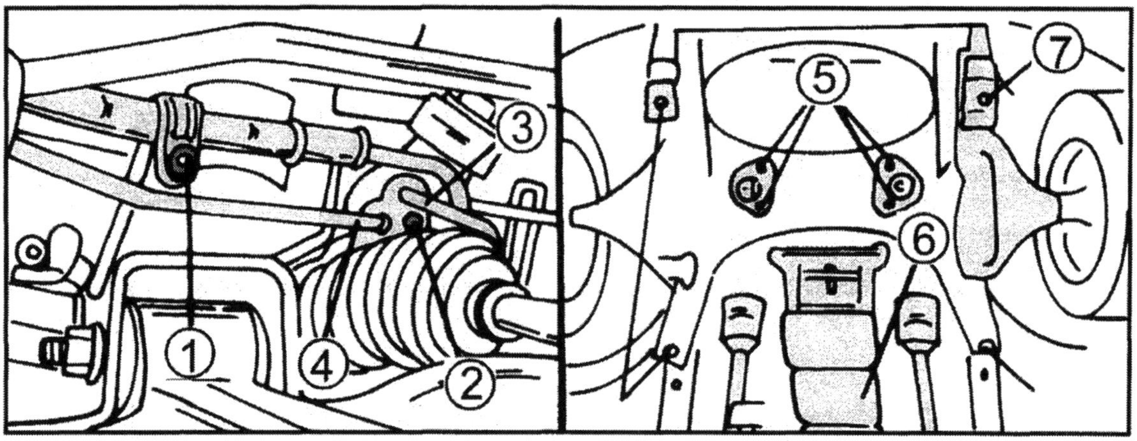

Fig. 6.5 – Details for the removal and installation of the steering gear. The operations referred to are carried out from underneath the vehicle.

- Remove the bolt securing the universal joint to the steering pinion and withdraw the pinion from the shaft. Fig. 6.6 shows where the universal joint is connected. The bolt is tightened to 2.8 kgm (20 ft.lb.) during installation.

- Refer to the R.H. view of Fig. 6.5 and remove the bolts (5) securing the steering gear to the axle/engine subframe (6). Note that the bolts have a different length and must be marked accordingly before removal. The bolts are tightened to 12.0 kgm (86.5 ft.lb.) during installation and from the final position a further quarter of a turn (90°).
- The subframe (6) must now be supported from below using a suitable jack, hydraulic type or otherwise. Remove the bolts (7) in Fig. 6.5 to separate the subframe from the longitudinal frame member. The bolts are tightened to 15.0 kgm (108 ft.lb.). Make sure the bolts are fitted to their original position.
- Remove the steering gear towards the L.H. side. At the top of the pinion shaft a washer is fitted for the plastic stop (2 in Fig. 6.6.) which must not be adjusted.

The installation of the steering gear is a reversal of the removal procedure, following the tightening torques already given above and any other notes. Also note the following:

- Before the steering is fitted, turn the steering pinion to one stop and from this position turn the pinion to the other lock, counting the numbers of turns. Now turn the steering pinion the other way by half the number of turns. Check that the alignment marks in Fig. 6.3 are opposite each other. This is the centre position.
- Slide the universal joint over the shaft, align the bore in the joint with the groove in the steering pinion and insert the bolt. Tighten the bolt to 2.8 kgm (20 ft.lb.).
- Fit the track rod ball joint end over the track rod (in the original position) and connect the ball joint stud to the lever on the steering knuckle.
- Fill the system with the recommended fluid and bleed it as described later on.
- Lower the vehicle to the ground and tighten the wheels.
- Check that the steering is in the centre position (wheels in straight-ahead position, steering wheel in correct position). If this is not the case, remove the universal joint once more and turn it by one tooth.
- Finally check the steering system for leaks and re-check the fluid level in the steering fluid reservoir.

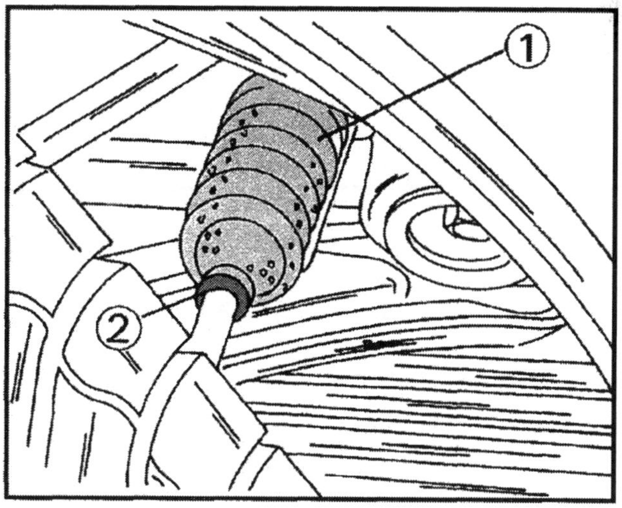

Fig. 6.6 – Slacken the gaiter clamp (2) on the outside and remove the gaiter (1) from the steering box. The large clamp on the inside is already removed.

6.2. Track rod Ball Joints – Replacement

The replacement of a ball joint follows the description during the removal and installation of the steering until the track rod ball joint has been separated from the steering lever on the steering knuckle, with the wheel removed on the side in question.

Remove the ball joint end of the track rod (4) in Fig. 6.2 after slackening the locknut (3). Unscrew the ball joint end, at the same time counting the turns (and half turns). Write it onto a piece of paper. During installation screw the track rod end by the same number of rotations onto the track rod. This will bring the toe-in to the approximate value. The locknut is tightened to 5.0 kgm (35 ft.lb.). The nut securing the ball joint to the steering lever is tightened in two stages. First tighten to 5.0 kgm (35 ft.lb.) and from the final position a further 60° (estimate the angle, 90° is a quarter of a turn). An Allen key can be inserted into the end of the track rod ball pin to prevent it from rotating during the initial stages of tightening.

The steering gaiter can be replaced if required. Slacken the gaiter clamp on the inside and the small one on the outside, as the position shown in Fig. 6.6 and remove the gaiter. Make sure that the new boot is seated correctly before the clamps are fitted.

6.3. Checking the Steering Mechanism

The following information will help you to check some parts of the steering system. With the results of the check you will be able to decide to take the necessary measures to correct the fault(s).

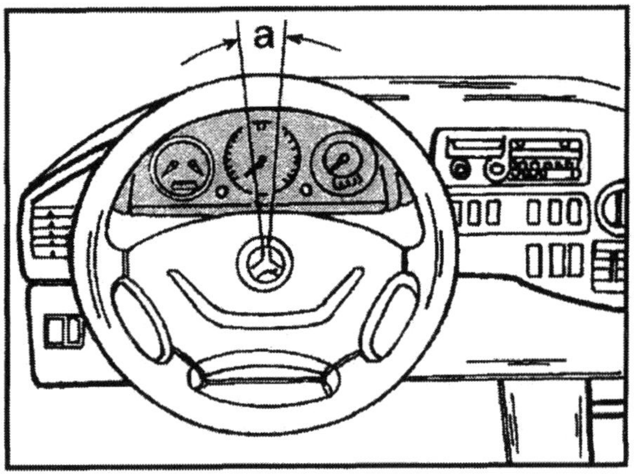

Fig. 6.7 – Play in the steering can be checked by moving the steering wheel as shown. If the dimension "a" exceeds 30 mm, you can assume that the steering gear is worn.

Start the engine and set the front wheels to the straight-ahead position. Grip the steering wheel on the rim and move it to and fro as shown in Fig. 6.7. If the free play is more than 30 mm (approx. 1 inch), you can check the moving parts of the steering for wear or excessive play. If the play is in the actual steering, you will have to fit a new steering mechanism, as they are no longer repaired.

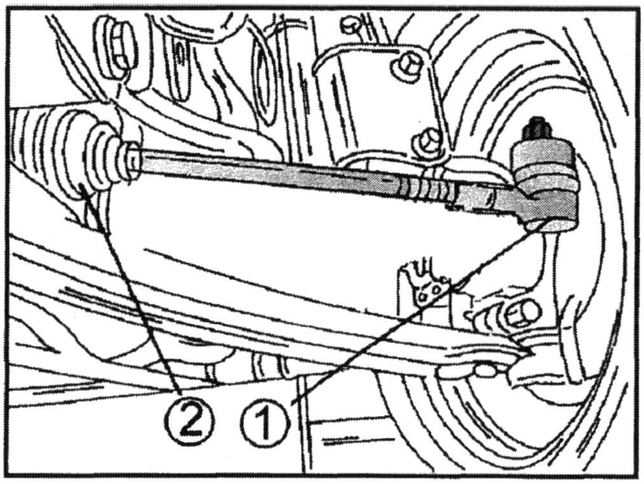

Fig. 6.8 – View of a fitted track rod. The track rod ball joint (1) and the steering gaiter (2) must be checked.

Checking the track rod ball joints and the steering rubber gaiters: With the wheel fitted and from underneath the vehicle grip the ball joint (1) in Fig. 6.8 and move it up and down. A worn ball joint can be detected by an up and down movement of more than 2.0 mm. The rubber dust caps must have no cuts. Dust caps cannot be replaced individually, i.e. a new ball joint is due.

Check the steering gaiters (2) in Fig. 6.8 over their entire length and circumference for cuts or other damage. Mainly the underside of the gaiters can get damaged. Make sure that the gaiter is securely attached to the track rod ball joint and the steering box.
Gaiters can be replaced as already described during the replacement of the ball joints.

6.4. Operations on the hydraulic System

6.4.0. FILLING THE HYDRAULIC SYSTEM

After the installation of the steering or after draining the system, it must be filled and bled in accordance with the instructions below.

Fluid drained during removal operations must not be re-used. Note that the capacity is given in Section 6.0. The fluid level is checked after unscrewing the filler cap as shown in Fig. 6.9. Check if the fluid level is between the "Min" and "Max" mark on the dipstick. The fluid level will be higher or lower, depending on the fluid temperature. To make sure you obtained the correct reading, clean the dipstick and re-insert it, screwing the cap fully down.

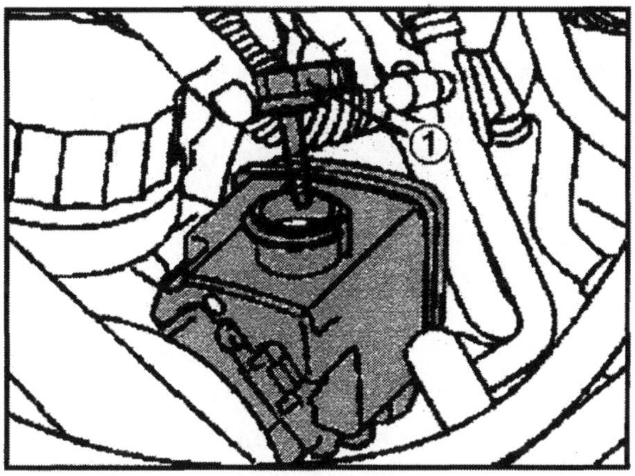

Fig. 6.9 – Checking the steering fluid level. The cap (1) must be removed to check the fluid level. The dipstick is at the bottom of the cap.

Check the filler cap seal before refitting it. The fluid level in the reservoir should be checked during each engine oil change. Take care that no foreign matter can enter the fluid reservoir.

6.4.1. BLEEDING THE HYDRAULIC SYSTEM

After the fluid level in the reservoir remains steady after the engine has been started, turn the steering wheel several times from one lock to the other to expel the air. Ask a helper to observe the fluid level during this operation. If necessary correct the fluid level to keep it on the "Max" mark. No air bubbles must be visible when the steering wheel is turned as described above.

6.5. Steering Pump Drive Belt

The steering pump is driven by the same belt as the other units at the front of the engine. To replace the belt, refer to Section "Cooling System".

6.6. Steering Pump – Removal and Installation

The attachment of the steering pump is not the same on all models. During the following instructions you will have to refer to the engine in question.

With Petrol Engine

Fig. 6.10 shows details for the removal and installation of the power steering pump. The fluid reservoir must be removed in the case of a petrol engine in order to take out the steering pump.

- Remove the air cleaner housing.
- Slacken the Poly V-belt until it can be lifted off the belt pulley (5) of the power steering pump (3) in Fig. 6.10.
- Remove the steering fluid reservoir as the reservoir is fitted immediately above the pump. The operations are carried out with the help of Fig. 6.11. Remove the reservoir cap in accordance with Fig. 6.9 and empty the reservoir (4). A suction tool with a rubber ball can be used to suck out the fluid. Slacken the bolt securing the hose clamp (5) and detach the return hose (6) from the reservoir. Close the open end of the hose in suitable manner. The hose can remain on the pump. Remove the bolt (3) securing the reservoir. The bolt is tightened to 1.4 kgm (10 ft.lb.) to the timing case. Use a suitable tool and remove the retaining clamp (7) from the reservoir at the position shown. The reservoir can now be withdrawn from the power steering pump. The sealing ring (1) must be replaced during installation.

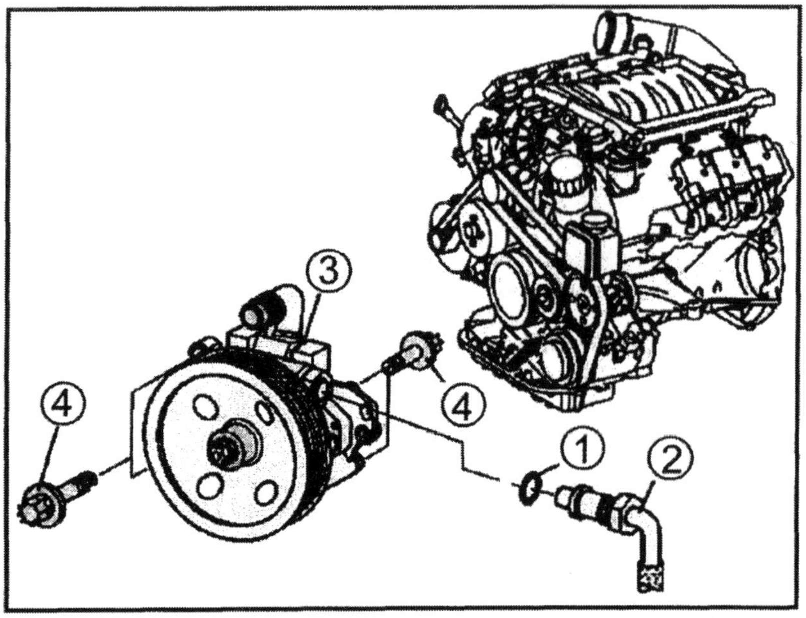

Fig. 6.10 – Details for the removal and installation of the power steering pump as fitted to a petrol engine.

1 Sealing ring
2 High-pressure hose
3 Power steering pump
4 Securing bolts

- Refer to Fig. 6.10 and disconnect the high-pressure hose (2) from the steering pump (3), using an open-ended spanner. Close the open connections of hose and pump in suitable manner. The sealing ring (1) must be replaced during installation. The union nut is tightened to 3.8 kgm (27.5 ft.lb.) during installation.

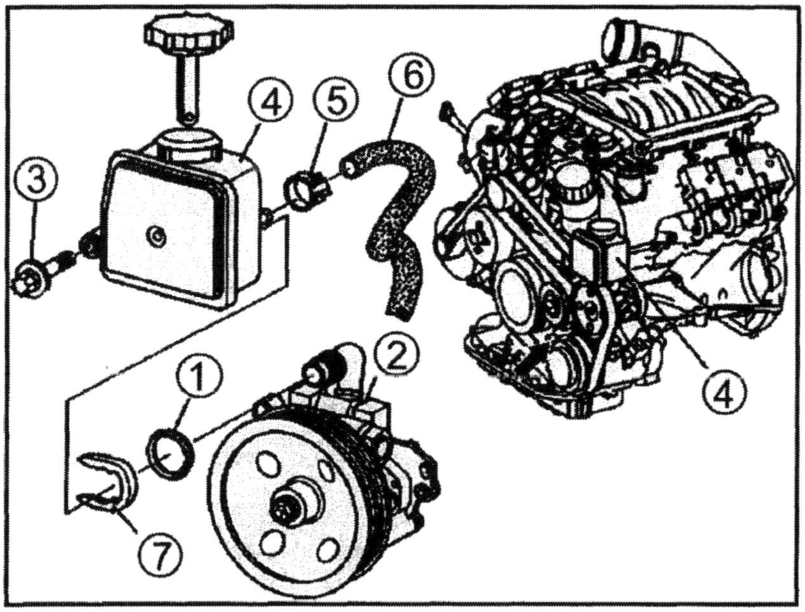

Fig. 6.11 – Details for the removal and installation of the steering fluid reservoir. The numbers are referred to in the text. The pulley (8) can be replaced after the pump has been removed.

- Remove the bolts (4) securing the power steering pump to the timing case cover and remove the pump (3). The bolts are tightened in accordance with the marking in the bolt heads. Bolt head marked with "8.8" tighten to 2.0 kgm (14.5 ft.lb.), bolt head marked with "10.9" tightened to 3.5 kgm (25 ft.lb.). Check the marks before tightening the two bolts.

The installation is a reversal of the removal procedure, noting the torque values given above. Refit the Poly V-belt as described in section "Cooling System". Finally fill and bleed the steering system as described earlier on. Start the engine and check the steering system for leaks and correct functioning.

With Diesel Engine

- Remove the reservoir cap in accordance with Fig. 6.9 and empty the reservoir. A suction tool with a rubber ball can be used to suck out the fluid. Refit the cap after the reservoir has been emptied.
- Slacken the Poly V-belt until it can be lifted off the belt pulley of the power steering pump. The arrangement is similar as shown in Fig. 6.10.
- At the bottom of the fluid reservoir disconnect the fluid return line and the fluid pressure hose. The latter is attached as shown in Fig. 6.10. Close the open connections of hose, pipe and pump in suitable manner. The sealing ring (1) must

be replaced during installation. The union nut is tightened to 3.8 kgm (27.5 ft.lb.) during installation

- If a new steering pump is to be fitted remove the belt pulley. Three bolts must be removed. The bolts are tightened to 2.5 kgm (18 ft.lb.) during installation.

  Remove the bolts securing the power steering pump top to the timing cover. You can see the three bolts around the reservoir filler cap looking at the cap from above. The pump can now be withdrawn towards the front. The bolts are tightened in accordance with the marking on the bolt heads. Bolt head marked with "8.8" tighten to 2.0 kgm (14.5 ft.lb.), bolt head marked with "10.9" tightened to 3.5 kgm (25 ft.lb.). Check the marks before tightening the three bolts.

The installation is a reversal of the removal procedure, noting the torque values given above. Refit the Poly V-belt as described in section "Cooling System". Finally fill and bleed the steering system as described earlier on. Start the engine and check the steering system for leaks and correct functioning.

6.7. Steering – Tightening Torques

Torx-headed bolt for universal joint: .... 2.8 kgm (20 ft.lb.)
High pressure hose to steering pump: .... 3.8 kgm (27.5 ft.lb.)
High pressure hose to steering gear: .... 1.8 kgm (13 ft.lb.)
Return flow line to steering gear: .... 1.8 kgm (13 ft.lb.)
Belt pulley to steering pump (petrol): .... 2.2 kgm (16 ft.lb.)
Belt pulley to steering pump (diesel): .... 2.5 kgm (18 ft.lb.)
Steering gear to axle/engine subframe: .... 12.0 kgm (86.6 ft.lb.) + 90°
Subframe to longitudinal chassis member: .... 15.0 kgm (115 ft.lb.)
Steering pump to timing cover:
- Bolt heads marked "8.8": .... 2.0 kgm (14.5 ft.lb.)
- Bolt heads marked "10.9": .... 3.5 kgm (18 ft.lb.)
Fluid reservoir to timing case (petrol): .... 1.4 kgm (10 ft.lb.)
Fluid reservoir to steering pump (diesel): .... 1.8 kgm (13 ft.lb.)
Steering pump/A/C compressor carrier to crankcase: .... 2.0 kgm (14.5 ft.lb.)
Track rod ball joints to steering lever, nut: .... 5.0 kgm (36 ft.lb.)+ 60°
Track rod locknuts: .... 5.0 kgm (36 ft.lb.)
Wheel bolts, steel rims: .... 18 kgm (130 ft.lb.)
Wheel bolts, light alloy: .... 15 kgm (108 ft.lb.)

7. REAR SUSPENSION

Depending on the model the rear suspension is either fitted with a coil spring, semi trailing arm and a hydraulic shock absorber with anti-roll bar (stabiliser bar) or the vehicle rear axle is equipped with pneumatic suspension and electronic level control, referred to as "ENR" (electronic niveau regulation). If this system is fitted it maintains the vehicle ride height irrespective of the vehicle load.
If the vehicle level is too high or too low while you are driving, the chassis could be damaged and unstable handling characteristics cannot be ruled out. The indicator lamp in the control panel must go out before driving is safe.

Although described in the Operation Instructions booklet, experience has shown that in many cases the booklet is not available, mainly if the vehicle has been obtained second-hand we therefore shall briefly refer to the main features.
The following "ENR" modes can be set:

- **Standard Mode:** Automatic control to maintain the standard ride height during a journey and when stationary.
- **Manual Mode:** You can raise or lower the vehicle level.
- **Passive Mode:** The vehicle ride height is not controlled. This is useful when work on the vehicle level control system is carried out, also to change a wheel.

Changing to a different Mode

Manual Mode:

- Open the tail gate or rear door, depending in the version.
- Press the R.H. button on the right-hand side panel in the rear/load compartment in the direction desired and hold it down for 2 seconds. The "ENR" system is now in manual mode and the vehicle can be raised or lowered as required. When the ignition is switched on, the indicator lamp in the instrument cluster will flash. The indicator lamp in the "ENR" on/off button will also flash.
- ***To lower the vehicle,*** press the lower section of the R.H. switch in the panel until the vehicle has reached the required position, ***to raise the vehicle***, press the upper section of the switch until the vehicle has reached the required position.
- ***Raising the vehicle to normal level automatically.*** Briefly press the upper section of the R.H. button in the panel if the vehicle is below the normal level. The "ENR" system will raise the vehicle to the normal level. If you want to keep the vehicle at a certain level, briefly press the button again or press the L.H. "ENR" on/off button in any direction.
- ***Changing to Standard Mode:*** Press the L.H. button in the panel for 2 seconds or drive faster than 2 mph. The indicator lamps in the instrument cluster and the button stops flashing.
- ***Passive Mode:*** Press the L.H. button in the panel for 2 seconds. The system is in the passive mode and no level control takes place. The indicator lamps in the instrument cluster and in the on/off button light up. You can change to standard mode as described above at any time.
- ***The "ENR" system is deactivated.*** Various reasons. The battery charge is low and the engine is not running, the vehicle has been raised and lowered repeatedly, thereby overheating the compressor, the vehicle is parked (ignition switched off, key removed). The system can be reactivated by opening a door or switching on the ignition.

7.1. Rear Coil Springs – Removal and Installation

Attention: Rear springs are under high tension. The spring must be compressed with strong spring compressors before the shock absorbers are removed. In addition a mobile jack must be placed underneath the suspension arm to support the assembly during removal. Never attempt to remove a rear spring without the equipment mentioned.

A coil spring can be removed as follows:

- Slacken the wheel bolts, jack up the rear end of the vehicle and place chassis stands in position. Remove the rear wheel.
- Place the jack underneath the bottom of the suspension arm and lift the arm until you can see that the shock absorbers are compressed. The rear spring can now be seen as shown in Fig. 7.1
- Place a spring compressor around three of the spring coils. A suitable compressor is shown in Fig. 7.2, the claws of which are placed on the left and right over the coils. Tighten the nuts of the adjusting mechanism until you can see that the upper and lower end of the spring has lifted from the spring seats. Make absolutely sure that the spring compressor is strong enough for the spring.

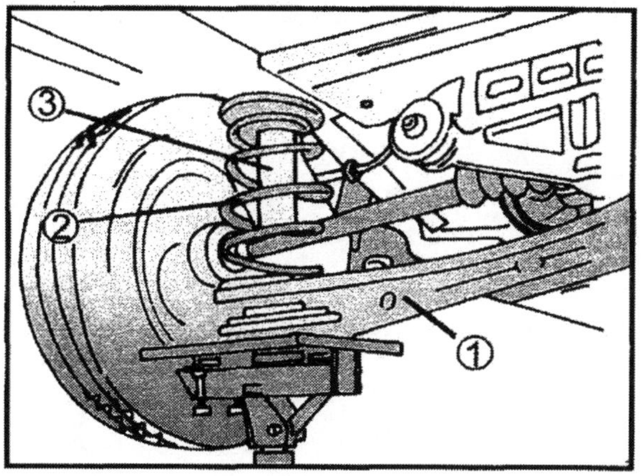

Fig. 7.1 – View of the rear suspension on one side.

1 Semi-trailing arm
2 Coil spring
3 Upper stop rubber

- Remove the bolt securing the shock absorber to the suspension arm and withdraw the bolt.
- Slowly lower the jack, at the same time observing the spring. As soon as the spring is free of tension you can remove it.
- Remove the upper spring seat (3) from the spring.

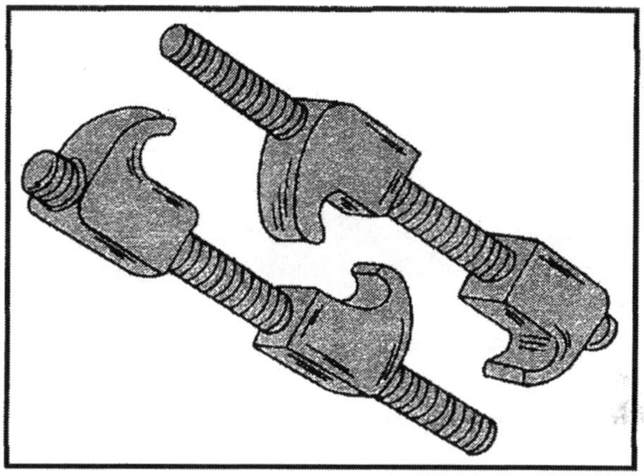

Fig. 7.2 – A suitable spring compressor.

Before refitting the old spring check it thoroughly. The protective coating of the spring must not be damaged, as rust will develop later on. Also check the condition of the upper and lower spring seats before you re-use them.
The installation of the spring is carried out as follows:

- Place the upper spring seats in position and insert the compressed coil spring between the seats. The spring must be fitted correctly.
- Lift the suspension arm with the jack until the shock absorber can be connected with its lower mounting. Insert the shock absorber bolt and slowly lower the jack, at the same time observing that the spring is in its correct position. After the suspension arm is in its normal position, tighten the shock absorber bolt to 13.0 kgm (94 ft.lb.).
- Remove the spring compressor. Again observe the spring to make sure it engages correctly into the rubber bearing on the floor panel on one side and with the guide in the suspension arm on the other side.
- Fit the wheel, lower the vehicle to the ground and tighten the wheel bolts.

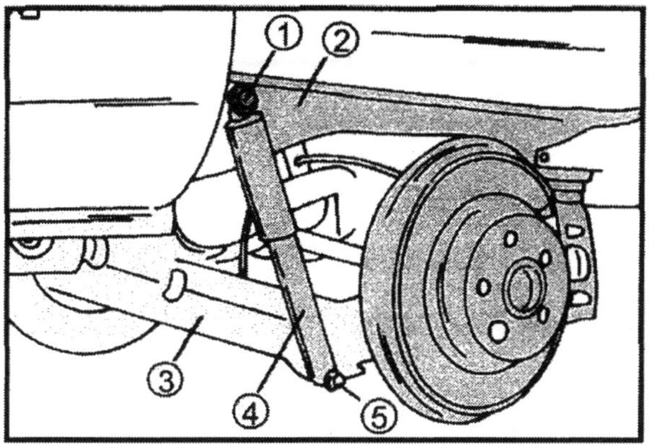

Fig. 7.3 – The shock absorber is attached to the suspension as shown.

1 Bolt, damper to body
2 Body panel
3 Semi-trailing arm
4 Shock absorber
5 Bolt, damper to trailing arm

7.2. Shock Absorbers – Removal and Installation

The attachment of the shock absorber at the lower and upper ends can be seen in Fig. 7.3. As the coil spring is "clamped" between the suspension arm and the body floor without any attachment, it will be necessary to compress the spring as described above before a shock absorber can be removed. Remove as follows:

- Slacken the wheel bolts, jack up the rear end of the vehicle and remove the wheel.
- Compress the coil spring as described during the removal of the spring.
- Place a trolley jack underneath the suspension arm, as shown in Fig. 7.1, and lift up the arm until you can see that the shock absorber is compressed. Then remove the upper mounting bolt.
- Remove the lower mounting bolt and take out the damper.

The shock absorber can be checked for correct operation by clamping the lower mounting eye in upright position into a vice. Grip the upper end and move the shock absorber up and down. The damper should move with even resistance. Any "dead play" indicates a faulty shock absorber. The shock absorber must also be replaced if the upper or lower mounting bush is worn.

The installation of the shock absorber is carried out in reverse order. Tighten the upper and the lower bolt to 13 kgm (94 ft.lb.). A tolerance of 1.0 kgm is permissible in both cases.

7.3. Rear Semi-trailing Arms – Removal and Installation

The operations described cover the different models with and without air suspension. Note that some operations must only be carried out if it is known that a new trailing arm must be fitted. It is obvious that the work is best carried out on a lifting platform or over a pit. Otherwise lift the vehicle and place secure chassis stands in position when required. Fig. 7.4 shows a view from underneath the rear suspension with the location of the various parts on a vehicle with coil springs. Proceed as follows with the removal, but note the operations which are only necessary when a new trailing arm is to be fitted.

- On a vehicle with air suspension lower the vehicle on the rear axle as described above. Switch the electronic level control to the passive mode, again as described earlier on.

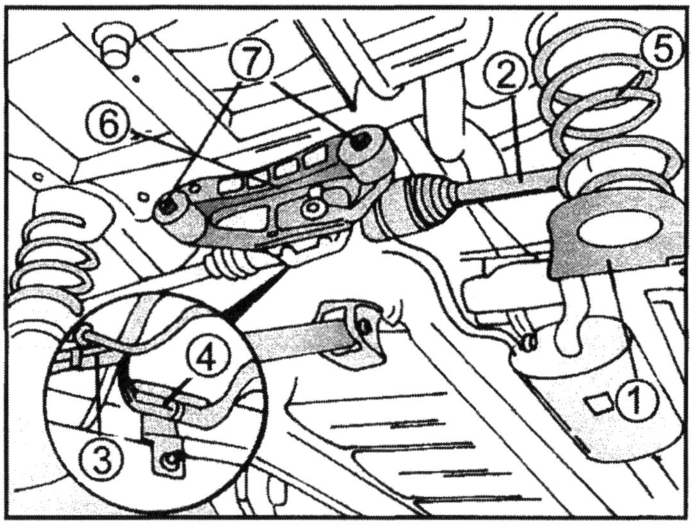

Fig. 7.4 – View from underneath the rear suspension with parts to be removed.
1 Semi-trailing arm
2 Rear axle shaft
3 Electrical wiring
4 Speed sensor
5 Coil spring
6 End cover on rear axle centre
7 Mounting bolts, 5.0 kgm + 90°

- If the trailing arm (1) in Fig. 7.4 is to be replaced, operate the parking brake (handbrake). Also if the arm is to be replaced remove the hub cap, but note that this does not apply if light alloy rims are fitted.
- If light alloy rims are fitted and the trailing arm is to be replaced, remove the rear wheel and remove the plastic cover in the centre of the light alloy rim. The rear wheel must now be refitted, but two of the bolts are sufficient to attach and tighten the wheel. With the wheel remounted lower the vehicle until the tyre touches the ground.
- Undo the hexagon collar nut securing the rear axle drive shaft (2) to the wheel hub if a new arm is to be fitted. Do not remove the nut completely to prevent the

rear axle shaft from dropping down during removal. The vehicle must not be moved with a slackened hexagon nut (wheel bearing damage).

- Release the parking brake (if arm is to be replaced), place the rear end of the vehicle on chassis stands and remove the wheel.
- For the next operation you will have to know the vehicle identification number (VIN). In all cases disconnect the electrical connector for the left and right speed sensors and brake pad wear sensors (4) in Fig. 7.4 at the position shown, but note that from vehicle identification number 030230 onwards the connector must be detached from the bracket by unlocking the catches. On vehicles up to VIN 030229 push the wire clip upwards to unlock the connector.
- Break the cable tie on the underbody and expose the electrical wiring (3). Electrical wiring will be found on the L.H. rear speed sensor, the R.H. rear sensor and the two brake pad wear sensors. The cable ties must be replaced during installation.
- Remove the rear brake cable through the passage in the substructure. Section "Brake System" will give further information on the removal of the cable. The parking brake may have to be adjusted.
- On a vehicle with air suspension disconnect the connecting link for the stabiliser bar from the trailing arm. Also on these models the operating lever for the level sensor on the trailing arm must be disconnected.
- Remove the bolts securing the brake caliper mounting bracket and remove the complete bracket together with the caliper. Use a piece of wire and tie up the mounting bracket to some part of the rear suspension. From the caliper remove the bolt securing the brake pad wear indicator. The caliper bracket mounting bolts are tightened to 10.6 kgm (76 ft.lb.) during installation, the bolt for the pad wear indicator to 0.9 kgm (7 ft.lb.). Check the brake hose before installation and replace if necessary.
- Detach the electrical connector for the brake pad wear contact sensor on the L.H. and R.H. side.
- Support the trailing arm from below as shown in Fig.7.1 and unscrew the shock absorber mounting bolts from the trailing arm. The shock absorber bolt is tightened to 13.0 kgm (94 ft.lb.) during installation. Also refer to "Removal and Installation of Shock Absorber" described earlier on.
- Remove the rear coil spring (if fitted) as already described.

Fig. 7.5 – The two bolts and nuts (1) and (2) secure the semi-trailing arm to the chassis. The differences between the two bolts are explained below.

- On a vehicle with air suspension remove the spring bellows as described later on. Release the residual pressure by undoing the union nut for the L.H. and R.H. air lines. The air spring bellows are unhooked downwards from the semi-trailing arm.
- The trailing arm is now removed, but note the following. Fig. 7.5 shows the two bolts securing the trailing arm to the chassis. Bolt (1) is an eccentric bolt and is used to adjust the rear wheel alignment. Before the bolt and the nut on the other side are slackened, mark the position of the bolt head on the mounting bracket, using a centre punch or paint. The bolt and nut must be replaced during installation, i.e. you will have to transfer the marking precisely to the new bolt head. The second bolt (2) is a standard

collar bolt with a nut on the other side. Again a new nut and bolt must be used during installation. The bolts are tightened to 16.0 kgm (115 ft.lb.) and from the final position a further 120° (consider that 90° represent a quarter of a turn). After both bolts have been removed withdraw the trailing arm from its chassis mounting.

Note: After installation you will have to decide. If the eccentric bolt has been fitted correctly, the rear wheel alignment should be correct. If in doubt, have the alignment checked at a dealer.

- Remove the rear axle shaft out of the rear axle centre assembly and completely remove the arm.

If a new trailing arm is to be fitted, the following operations must be carried out:

- Check the rubber mountings in the trailing arm and have them replaced if necessary.
- Remove the rear axle shaft out of the wheel hub as described later on.
- Remove the rear brake disc.
- Remove the rear brake cable from the trailing arm.
- Remove the speed sensor from the trailing arm.

The installation of the trailing arm is a reversal of the removal procedure, noting the tightening torques and other assembly notes already given during the removal. Finally check the oil level in the rear axle centre assembly and if necessary top it up with SAE 85 W-90 hypoid gear oil. Tighten the filler plug to 9.0 kgm (65 ft.lb.). As already mentioned, have the rear wheel alignment checked at a dealer, if necessary.

7.4. Rear Axle Shafts – Removal and Installation

Some differences must be noted, depending on the type of suspension fitted which will be specified when applicable. Differences within the various wheel types and sizes must be followed. Read carefully:

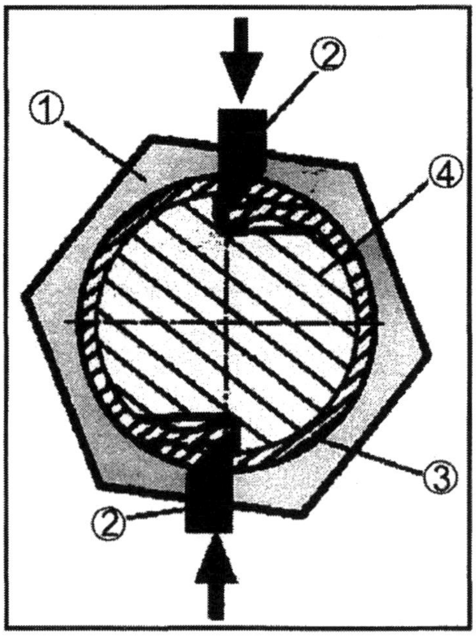

Fig. 7.6 – The collared nut at the outside of the rear axle drive shafts. The numbers are referred to below.

- Operate the parking brake.
- On models with steel rims 6.5J x 16 and 6J x 16 remove the hub cap, on models with light alloy rims 6.5 J x 16 and 7J x 17 remove the wheel.
- On models with light alloy 6.5J x 16 and 7J x 17 remove the wheel hub covers in the centre of the light alloy rim and then refit the wheel. Two wheel bolts will be enough.
- Study the collared nut at the outside of the axle shaft and bend up the two locking tabs, using a chisel or similar. Slacken the nut. After the nut has been slackened the vehicle must no longer be moved on its wheels (wheel bearing damage). A new nut must always be fitted. During installation fit and tighten the nut to 16.5 kgm (119 ft.lb.) and from the final position a further 60° (90° = quarter of a turn). The nut must now be locked as shown in Fig. 7.6. Press the collar (3) into the notches in the rear axle shaft (4) as shown, using a suitable tool (2). Make sure the nut (1) is locked correctly. No gap should remain between the notches and the locking tabs.
- Remove the wheel once more and release the parking brake.

- In the case of a vehicle with air suspension lower the vehicle at the rear axle and switch the electronic level control to the passive mode. Both operations are described at the beginning of this chapter.
- Remove the semi-trailing arm as described earlier on. **Note:** When removing both rear axle shafts only remove the L.H. trailing arm since the rear axle centre will also be removed from the vehicle. In this case support the rear axle centre from below using a suitable jack and the necessary blocks and remove the bolts (7) in Fig. 7.4. The bolts are tightened to 5.0 kgm (36 ft.lb.) and a further quarter of a turn during installation.

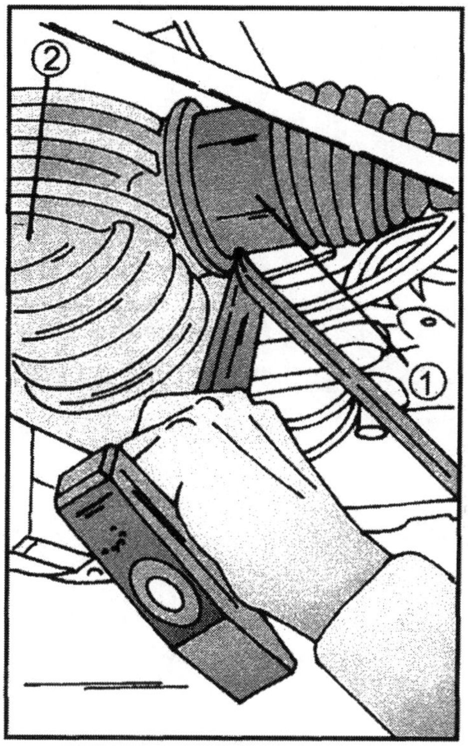

Fig. 7.7 – Removal of a rear drive shaft on the inside.

- Remove the rear axle centre assembly (6) in Fig. 7.4 from the underbody.
- Press the rear axle shaft out of the rear axle centre assembly. This is carried out as shown in Fig. 7.7, i.e. a tyre lever and a hammer handle is used to press out the shaft. At the end of the shaft you will find a circlip which must be replaced during installation. When fitting the shaft, push it into the rear axle centre with a short, sharp push. You should feel or hear the locking of the circlip. Try to pull out the shaft to ensure that the circlip has engaged.

Fig. 7.8 – Removal of a rear axle shaft from the wheel hub.

- Use a puller, of the type shown in Fig. 7.8 and press the rear axle shaft out of the wheel hub. Lower the front end of the shaft and withdraw it from the rear suspension.

Note that not all drive shafts are the same for all models. If a new shaft is required, quote the fitted engine and the VIN identification. Some shafts are fitted with an additional weight near the stub axle end.

The installation is a reversal of the removal procedure, following the tightening torques given above. After installation check the oil level in the rear axle and if necessary top-up with hypoid gear oil SAE 85 W 90. The filler plug is tightened to 9.0 kgm (65 ft.lb.).

7.5. Rear Axle Shafts – Repairs

7.5.0. REPLACING THE RUBBER BOOTS

Fig. 7.9 shows a dismantled drive shaft. Before a shaft is dismantled, you will have to obtain some grease that is used in the workshop for the constant velocity joints, but note that the inner CV joint has a different capacity, depending on the fitted engine. A special pair of pliers (made by Hazel) is used to remove the boot clamp on the inside of the outer boot. Remove the boot (2) or (3) to be replaced as follows:

- Clamp the axle shaft into a vice, using soft-metal jaws and remove the boot clamp (1) in Fig. 7.10 at the inner end and the boot clamp (2) at the outer end with a pair of side cutters.

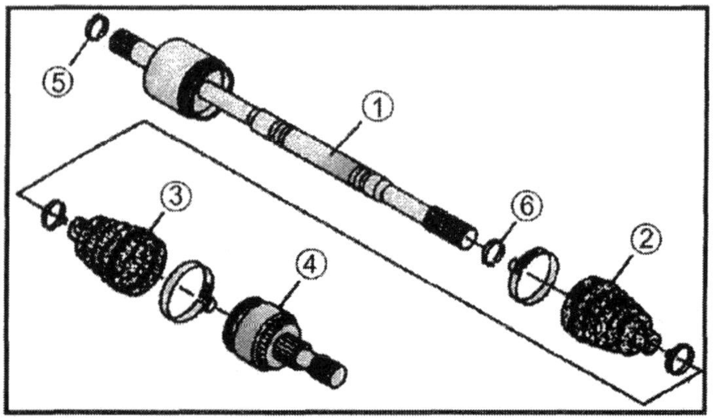

Fig. 7.9 – Dismantled rear axle drive shaft.
1 Rear axle shaft
2 Inner rubber boot
3 Outer rubber boot
4 Outer CV joint
5 Circlip
6 Circlip

- Remove the rubber boot and clean out the visible grease.
- Fill the inner boot, the outer boot, the inner joint and the outer joint with the recommended grease, but note the differences. The outside CV joint is filled with 145 grams of grease, irrespective of the fitted engine. The inner CV joint is filled with 160 grams of grease if a diesel engine of type 646.983 is fitted or with 185 grams of grease if a petrol engine or a diesel engine of type 646.982 is fitted. You will have to identify the engine.

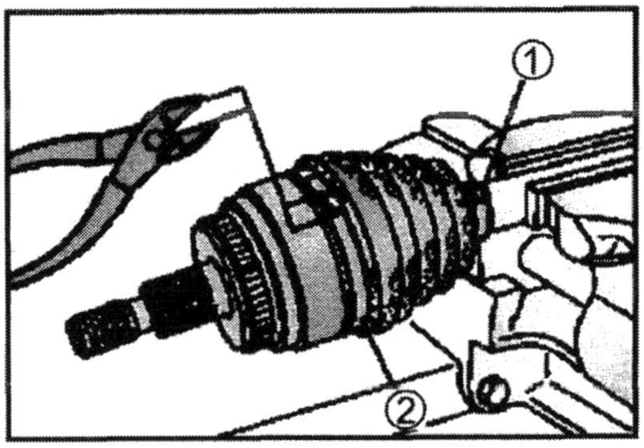

Fig. 7.10 – Removal of the rubber boot clamps (1) and (2).

- Fit the rubber boot over the CV joint at both ends and tighten the boot clamps with commercially available pliers. Note that the connections of the clamp (1) and the clamp (2) must be offset by 180° for the inner joint relative to the outer joint to ensure that the balance is not disturbed.

7.5.1. REPLACING A CV JOINT

We recommend to have the inner CV joint replaced in a workshop, as it must be extracted and pressed on. The outer joint can be replaced without special tools, but you need a 32 mm open-ended spanner. Clamp the drive shaft into a vice and apply the spanner as shown in Fig. 7.11. Use a hammer and hit against the spanner until the joint is free. Hold the joint from the other end to prevent it from dropping down. Remove the circlip from the end of the shaft.

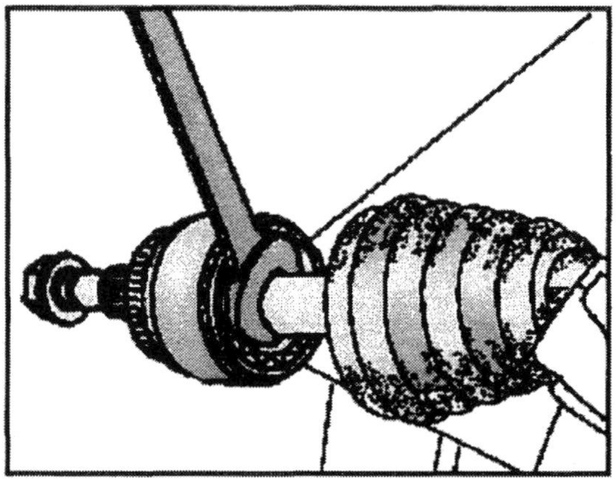

Fig. 7.11 – Removal of the outer CV joint with the open-ended spanner mentioned in the text.

- Fit a new circlip and press the outer CV joint over the shaft until you can hear the circlip snap into its groove.

7.6. Oil Seals in Rear Axle Centre - Replacement

The oil seals at the outlets of the rear axle shafts can be replaced after the axle shaft has been removed as described earlier on. Use a tyre lever or similar tool and lever out the old oil seal. Coat the sealing lip of the new oil seal with oil and drive the seal into the housing until the outside is flush. A new circlip is required for the rear axle shaft.

After installation check the oil level in the rear axle centre assembly. Fit the filler plug and tighten it to 9.0 kgm (65 ft.lb.).

7.7. Stabiliser Bar – With Air Suspension

Fig. 7.12 shows the rear suspension from underneath with the attachment of the stabiliser bar. Two mounting clamps (2) secure the stabiliser bar (1) to the underfloor. Each end of the stabiliser bar is connected with a link (4), one end to the bar and the other end to the trailing arm (3) which carries the air bellows.
With the vehicle securely supported on chassis stands, remove the bolts securing the connecting links to the semi-trailing arms and remove the mounting clamps from the underbody and remove the bar.

Fig. 7.12 – Details for the removal and installation of the stabiliser bar. The numbers are referred to in the text.

Check the rubber bushes in the connecting links and in the mounting clamps for wear and replace if necessary.
The installation is a reversal of the removal procedure. Tighten the mounting clamp bolts to 2.0 kgm (14.5 ft.lb.) and the link rod bolts to 6.0 kgm (43.5 ft.lb.).

7.8. Rear Hub Flange and Wheel Bearings - Removal and Installation

Certain extractors are required for this operation. Provided that these can be obtained, proceed as follows:

- Remove the rear axle drive shaft as described above and remove the brake caliper. Use a piece of wire to suspend the caliper to the rear suspension.
- Fit a puller, as shown in Fig. 7.8 to the hub flange and press the flange out of the wheel bearing.
- Use a pair of pliers and remove the circlip visible in the bearing bore. The wheel bearing must now be removed from the wheel carrier by means of the extractor 201 589 04 43 00. If this extractor cannot be obtained and the bearing must be replaced, remove the wheel bearing under a press. A bearing race remains on the wheel hub flange and must be removed with a suitable puller.
- Check all parts for wear or damage. Bearings once removed should always be replaced, mainly if the bearing race has been removed from the hub flange.

Fit the new bearing until the circlip can be inserted. Check that the circlip has entered its groove fully. Draw the wheel hub flange in position with the special tool or carefully try to drive it in position with a soft-metal drift. The small thrust washer must press against the inner bearing race. There is no need to adjust the wheel bearings.

7.9. Rear Axle Oil Level Check and Oil Change

An oil level check plug is fitted into the rear cover of the axle drive. A further plug is fitted to the bottom of the axle housing to drain the oil. Fig 7.13 shows where the two plugs can be found on different axles. An Allen head key is required to remove the plugs. The surrounding area of the plug must be cleaned before the plug is unscrewed.

The oil level in the rear axle is correct, when the oil can be seen on the lower edge of the filler hole. If necessary top-up with SAE 90 or 85W90 hypoid oil. The plugs are tightened to 9.0 kgm (65 ft.lb.).

To change the rear axle oil proceed as follows:

- Drive the vehicle for a few miles to warm up the oil in the axle.
- Place the rear end of the vehicle on chassis stands and unscrew the drain plug at the position shown in Fig. 7.13, using the Allen key mentioned above. Again the area around the plug must be thoroughly cleaned. Take care if the oil is hot, as it will shoot out immediately.
- Clean the cover of the axle drive and unscrew the level/filler plug (1) in Fig. 7.13.
- Clean the drain plug and fit it (maybe with a new sealing ring) into the axle. Tighten it to 9.0 kgm (65 ft.lb.).

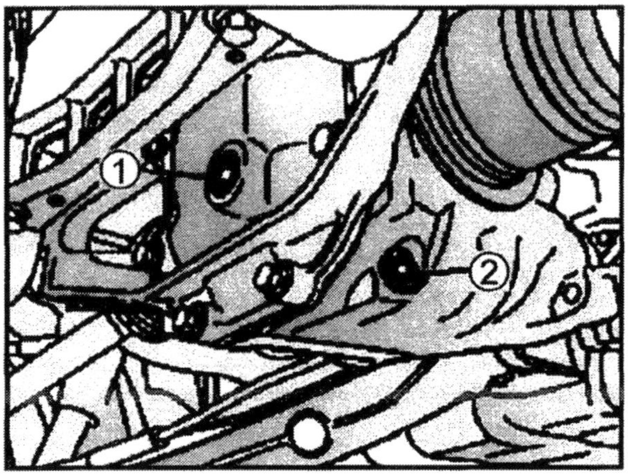

Fig. 7.13 – Location of the oil filler/oil level plug (1) and the drain plug (2).

- Fill the rear axle with the 1.5 litres of the specified oil. The oil must be level with the lower edge of the filler bore.
- Clean the filler plug and fit and tighten it to the torque given above.

7.10. Air Suspension Elements - Removal and Installation

Fig. 7.12 shows the location of one of the air suspension elements at the upper right. We must point out that the air pressure pipe must be treated in a certain manner and we advise you to read the instructions before you commence with the job.

- Press the knob in the dashboard for approx. 2 seconds to set the vehicle in the operating condition. This must be carried out before the rear end of the vehicle can be lifted up (passive mode). Now lift up the vehicle and remove the rear wheel. Check that the air suspension element is free on the bottom of the vehicle floor. Support the assembly from underneath.
- Release the air pressure in the element. To do this, disconnect the air inlet pipe located underneath the drive shaft connection after removal of the union nut. The nut is tightened to 1.0 kgm (7.2 ft.lb.).
- Remove the two bolts from the element at the upper end to detach the element from the substructure. The bolts are tightened to 1.9 kgm (14 ft.lb.) during installation. Make sure that the element is seated correctly before the bolts are tightened.
- At the lower end of the element disconnect the compressed air hose and then remove the bolt securing the element to the air spring element plate. Again the bolts are tightened to 1.9 kgm during installation.

The installation is a reversal of the removal procedure, noting the torque values given above. After installation set the electronic level control into operation and check that the air element is correctly seated.

7.11. Operations on the Air Suspension System

The removal and installation of the compressor (fitted to the L.H. side of the vehicle below the door on the bottom of the vehicle floor), the replacement of the electro-pneumatic valves (on the side of the compressor) and the removal and installation of the height sensor and the removal and installation of the regulating linkage should be entrusted to a workshop. In most cases the correct functioning of the air suspension system must be checked in a workshop in any case.

7.12. Rear Axle Centre Assembly – Removal and Installation

Fig. 7.14 shows a view from underneath the vehicle with the parts to be removed to take out the rear axle centre assembly. The vehicle should ideally be placed on a lifting platform. Otherwise place the vehicle on secure chassis stands. A jack must also be available to lower the axle assembly. Proceed as follows:

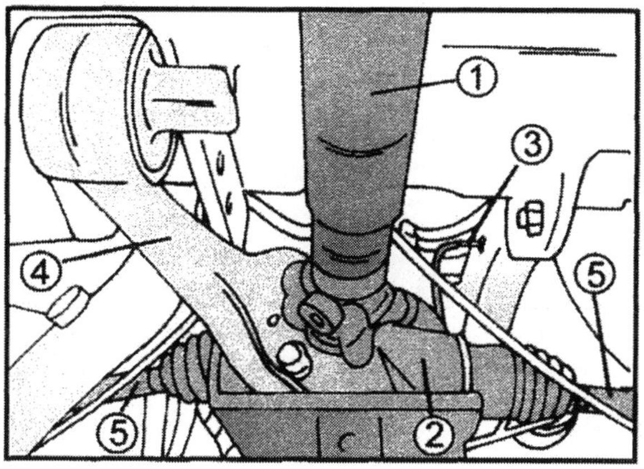

Fig. 7.14 – Details for the removal and installation of the rear axle centre assembly.

1 Propeller shaft
2 Rear axle centre assembly
3 Bleed (centre) hose
4 Front mounting
5 Drive shafts

- Remove the L.H. semi-trailing arm as described earlier on in this section and disconnect the propeller shaft (1) from the rear axle centre assembly. Use a piece of wire to tie the shaft to the underbody.
- Pull the vent hose (3) out of the underbody.
- Place a mobile jack securely with a suitable support underneath the axle assembly (2) and remove the two bolts securing the two bolts securing the axle assembly. The bolts are tightened to 5.0 kgm (36 ft.lb.) and a further 90° during installation.
- Unscrew the bolts securing the front mount of the rear axle centre assembly (4). Two bolts hold the mount in position. The bolt securing the mount to the rear axle centre assembly is tightened to 5.0 kgm (36 ft.lb.) and from the final position a further quarter of a turn. The bolt securing the mount to the body is tightened to 2.5 kgm (18 ft.lb.).
- Tie the R.H. drive shaft (5) to the underbody, using a suitable piece of wire.
- Press the R.H. shaft out of the axle centre assembly as described earlier on and lower the axle assembly on the jack

The installation is carried out as follows:

- Jack up the axle assembly into the installation position and undo the R.H. shaft from the wire sling.
- Install the front mount of the axle assembly and tighten the bolts to the torque values given above.
- Refit the two bolts securing the axle centre mount and tighten the bolts as specified above.
- Install the centre hose (3) in Fig. 7.14 into the underbody, refit the propeller shaft and refit the trailing arm.
- Finally check the oil level in the rear axle and correct if necessary. The toe-in of the rear wheels should be checked in a workshop if parts have been replaced.

7.13. Rear Wheel Alignment

The checking and adjusting of the rear wheel alignment requires the use of special equipment and special wrenches and we recommend to have the alignment checked in a workshop. If you have removed parts of the suspension this will alter the alignment and if you have followed the advice to mark the eccentric bolts and washers there will be no problems to drive the vehicle to the wheel alignment centre to have the settings checked.

7.14. Rear Axle and Suspension – Tightening Torques

Bolt, rear shock absorber, upper and lower mounting: .... 13.0 kgm (94 ft.lb.)
Bolt, semi-trailing arm to substructure:...16.0 kgm (115 ft.lb.) + 120°
Collar nut, wheel hub to rear axle shaft:..16.5 kgm (119 ft.lb.) + 60°
Stabiliser bar mounting to frame floor (underbody):... 2.0 kgm (14 ft.lb.)
Ball joints of linkage for stabiliser bar:... 6.0 kgm (43 ft.lb.)
Bolt, front suspension, rear axle centre piece to centre piece: ....5.0 kgm (36 ft.lb.) + 90°
Bolt, front suspension, rear axle centre piece to body: .... 2.5 kgm (18 ft.lb.)
Bolt, rear suspension (bearing bracket), rear axle centre piece to body: .... 7.2 kgm (52 ft.lb.)
Propeller shaft flange to rear axle drive flange:...1.5 kgm (11 ft.lb.) + 90°
Speed sensor to rear axle (ABS):... 0.9 kgm (7.0 ft.lb.)
Brake pad wear indicators to brake caliper:.. 0.9 kgm (7.0 ft.lb.)
Floating caliper to suspension arm: .... 10.6 kgm (76 ft.lb.)
Rear axle oil filler and drain plugs:.. 9.0 kgm (65 ft.lb.)
Air suspension elements (bellow), top and bottom: .... 1.9 kgm (14 ft.lb.)
Steel wheel bolts:... 18.0 kgm (130 ft.lb.)
Light alloy wheel bolts: .... 15.0 kgm (108 ft.lb.)

8. BRAKE SYSTEM

8.0. Technical Data

Type of system	See description below
Front Brakes	
Caliper piston diameter:	44.00 mm
Min. thickness of brake pad linings:	3.5 mm
Brake disc thickness:	28.0 mm
Brake disc diameter:	300.0 mm
Min. thickness of brake disc:	25.0 mm
Max. run-out of brake discs:	0.10 mm
Wear limit of brake discs, per side:	max. 0.05 mm
Rear Disc Brakes	
Brake Disc Diameter:	296.0 mm
Min. Brake Disc Thickness:	10.0 mm
Min. thickness of brake pads, incl. metal plate:	7.0 or 7.5 mm
Min. thickness of pad linings:	3.5 mm
Handbrake	
Operation:	Brake shoes inside brake disc

8.1. Short Description

All models covered in this manual are fitted with a hydraulic dual-circuit brake with vacuum-operated brake servo unit. The brake servo unit is supplied with vacuum from a separate exhauster pump (diesel) or from the inlet manifold (petrol). Sliding brake calipers with 4 pistons (two per caliper) are used for the front wheels. Disc brakes, working on the same principle as the front brakes are used on the rear wheels. The brake calipers are may not be the same on all models.

The brake system is diagonally split, i.e. one circuit serves one of the front brake calipers and the diagonally opposed rear brake caliper. The other circuit operates the other two brake assemblies accordingly. If one brake circuits fails, the brakes will operate as normal, but more brake force will be required.

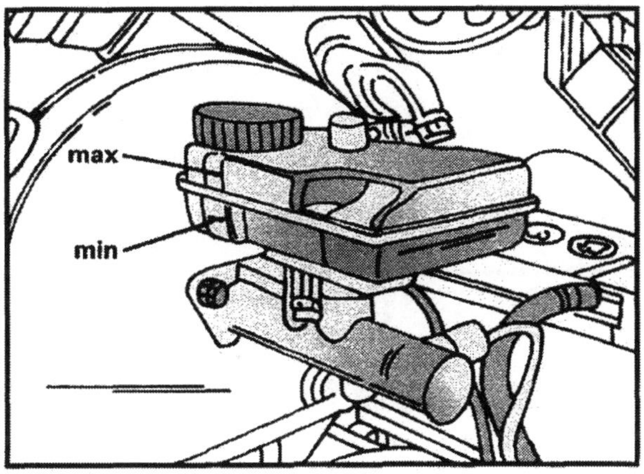

Fig. 8.1 – The brake fluid reservoir is fitted to the master brake cylinder. Note the "min" and "max" marks on the outside of the cylinder.

Checking the Brake Fluid Level

The brake fluid reservoir is fitted above the master brake cylinder in the position shown in Fig. 8.3. At all times make sure that the brake fluid is between the "Min" and "Max" marks on the outside of the cylinder. If the brake fluid level sinks below the "Min" mark there are other reasons which must be investigated.

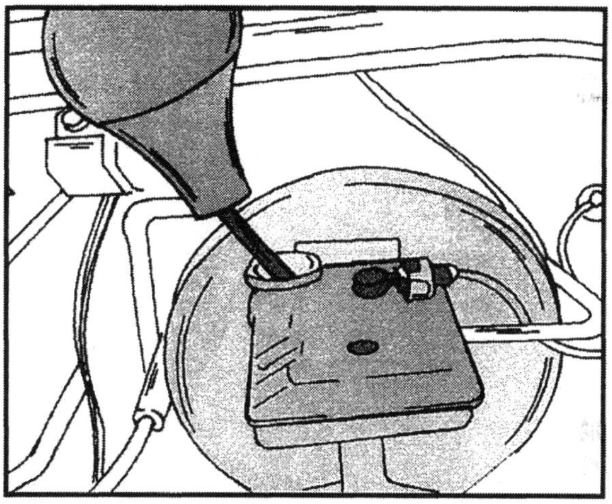

Fig. 8.2 – Brake fluid can be removed from the fluid reservoir as shown, when it is necessary to drain the brake system.

Brake fluid must sometimes be removed from the reservoir. If this is necessary we recommend the method shown in Fig. 8.2.

8.2 Front Disc Brakes

As already mentioned, sliding brake calipers with two pistons are fitted to the front wheels. The assemblies consist of a caliper mounting bracket bolted rigidly to the front axle steering knuckle and a separate caliper cylinder. When the brake is operated, the piston presses first with its brake pad against the brake disc. The caliper cylinder then slides on glide bolts and moves against the direction of the pressure, until the other brake pad is pressed against the brake disc on the other side. Only the caliper cylinder must be removed to replace the brake pads, the mounting bracket remains on the steering knuckle.

8.2.0. BRAKE PAD REPLACEMENT

If the remaining brake pad material thickness is less then 3.5 mm, a warning light will light up in the dashboard, indicating that the brake pads must be replaced. The thickness of the brake pad linings can also be checked through the openings in the brake caliper after the wheels have been removed. The brake pad thickness can also be checked through the fitted wheel. In this case shine a torch through the upper wheel rim opening to check the thickness of the outer brake pad. The inner brake pad must, however, be checked from the underside of the vehicle. We therefore recommend to remove the wheel to speed up the operation.

Proceed as follows. Remember that brake pads must be replaced on both sides of the front axle. If the old pads are to be refitted, mark their position during removal:

8.3 – The brake caliper in position after removal of the wheel.
1 Brake caliper mounting bracket
2 Brake caliper
3 Upper caliper slide bolt
4 Closing cap for caliper bolts
5 Brake hose
6 Bleeding valve
7 Lower caliper slide bolt

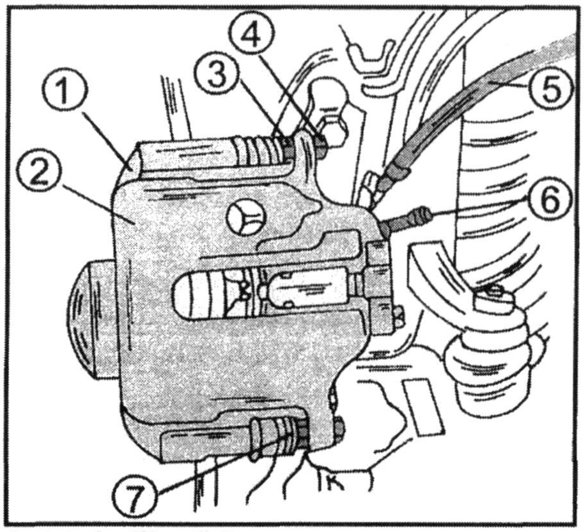

- Operate the brake pedal a few times to remove the vacuum from the brake system. Remove the brake fluid reservoir cap.
- Place the front end of the vehicle on chassis stands. The wheels must be removed. The caliper is then exposed as shown in Fig. 8.3.
- Disconnect the lead for the brake pad wear indicator from the centre of the caliper. To do this, remove the small cover and push off the connector, using a small screwdriver.
- Grip the brake caliper and with a short pull the top part of the caliper towards the outside. This will push the pistons approx. 3 mm into their bores. This clearance is needed so that when folded back the upper part of the brake caliper does not get caught on the rivets on the back of the brake pads.
- Unscrew the bolt in (2) Fig. 8.4, lift up the upper part of the brake caliper (1) and remove the brake pads (4) from the brake caliper mounting bracket (5).

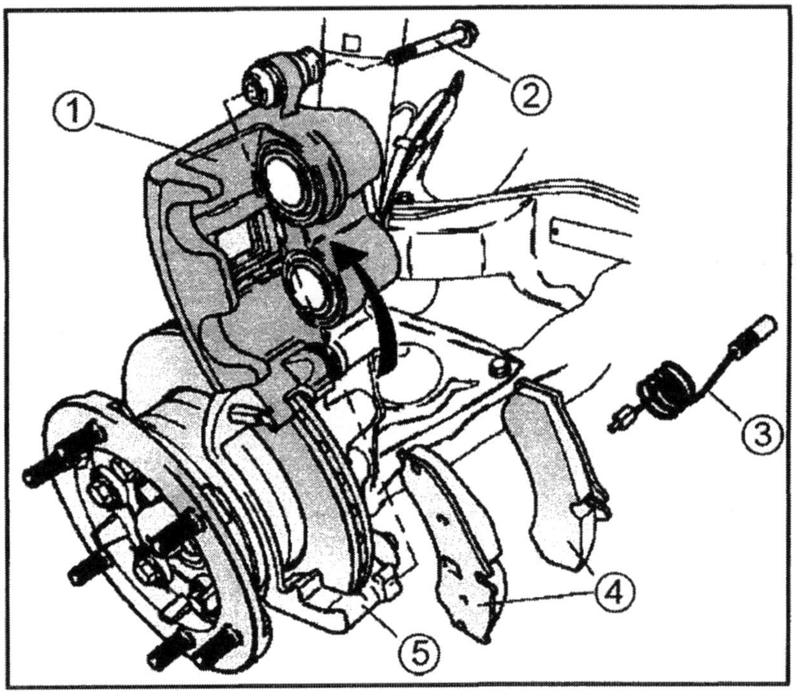

Fig. 8.4 – Details of the removal and installation of the brake pads. The numbers are referred to below. The pad wear sensor (3) must be disconnected.

If the brake pad material has worn to a thickness of 3.5 mm, fit a new set of pads. Although the min. permissible material thickness is considered to be 2.0 mm, you will find that the brake pad wear indicator warning light will light up, when the thickness has been reduced to 3.5 mm. Never replace one brake pad only, even though if the other one appears to be in good condition.

If brake pads have worn more than expected, there is a possibility that the piston(s) have seized inside the bores. In this case it may be necessary to have the brake calipers overhauled. The same applies if the rubber dust boots on the cylinder housing are damaged.

Check the brake disc faces and clean them thoroughly before new pads are fitted. If possible measure the brake disc thickness (as shown in Fig. 8.6) and compare the result with the dimensions given in Section 8.0. If the minimum thickness has been reached, replace the disc(s). Discs with uneven faces can perhaps be re-faced. Check with a MB workshop.

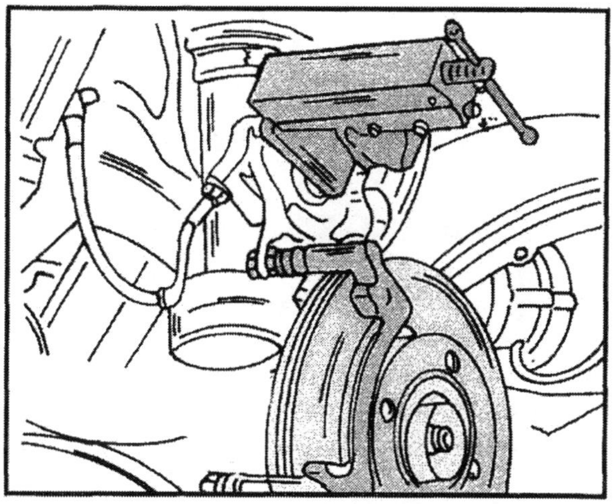

Fig. 8.5 – Piston reset pliers fitted to the caliper cylinder. Push the pistons to the bottom of the bore. A piece of wood can also be used.

The installation is carried out in reverse order to the removal procedure, noting the following points:

- Draw some fluid out of the fluid reservoir as already explained earlier on.
- Re-cut the threaded bores in the brake caliper mounting bracket (5) to clean out old thread locking compound. The bolts (2) must always be replaced.
- Push the pistons back into their bores as shown in Fig. 8.5 or in a similar manner, but note that both pistons must be pushed back. Take care not to damage the piston seals.
- Rub a thin film of brake pad paste on the side guides and piston contact surfaces of the carrier plate of the brake pads. Place the brake pads in position on the left and on the right.
- Carefully lower the brake caliper until the bolt (2) can be inserted. Tighten the bolt to 3.0 kgm (22 ft.lb.).
- All other operations are carried out in reverse order to the removal procedure, noting the instructions already given earlier on.

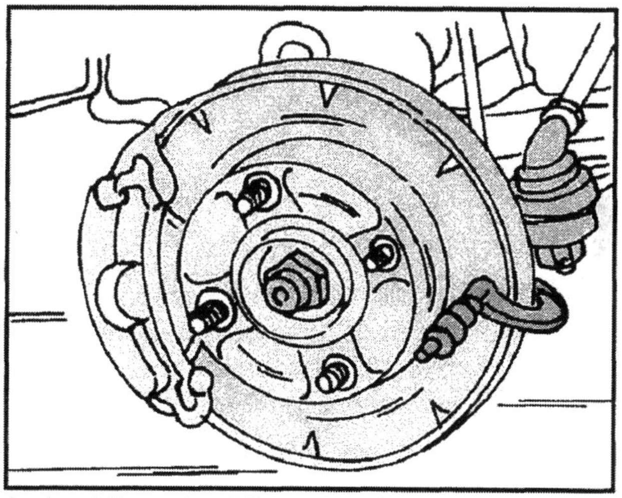

Fig. 8.6 – Brake discs can be measured with a micrometer as shown.

- Check the fluid level in the fluid reservoir and correct if necessary. Finally fit the wheels and tighten the wheel bolts to the correct torque setting.
- Treat the new brake pads with care during the first 100 miles or so, remember that the new material must "bed in" before they reach their full performance. To speed up the process, brake the vehicle several times from a speed of approx. 50 mph to 25 mph. Allow the brakes to cool between each braking operation.

8.2.1 Brake Caliper – Removal and Installation

Fig. 8.7 shows the attachment details of a front brake caliper. The brake pedal must be operated a few times to expel the vacuum in the system and the front of the vehicle must be on chassis stands, wheel(s) removed.

- Push a bleed hose over one of the bleed screws in the brake calipers and insert the other end of the hose into a jar. Open the bleed screw and operate the brake pedal until the brake fluid has been drained.
- Unscrew the union nut from the brake pipe connection (3) on the caliper. Close the open connections in a suitable manner to prevent entry of foreign matter.

Fig. 8.7 – Details for the removal and installation of a front brake caliper.

1 Electrical connector plug
2 Caliper securing bolt, 3.0 kgm
3 Brake hose, 1.8 kgm
4 Brake caliper
5 Brake caliper mounting bracket
6 Brake pad wear contact sensor

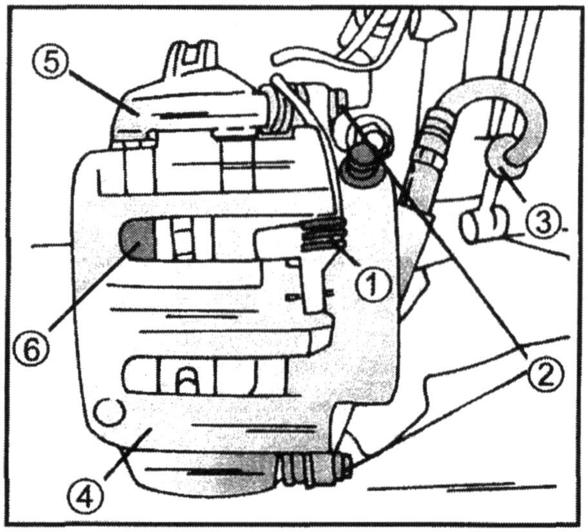

- Disconnect the electrical connector plug (1) from the brake caliper.
- Remove the caliper cylinder in the manner described above, but remove both bolts and lift off the caliper. The caliper can now be unscrewed from the brake hose. The brake pads can be removed if required.
- If necessary, unscrew the two bolts and remove the mounting bracket from the steering knuckle.

The installation is carried out in reverse order, noting the following points:

- Use new bolts for the mounting bracket. These are specially designed for the mounting bracket. Tighten the bolts to 17.2 kgm (124.0 ft.lb.).
- Screw the brake caliper onto the brake hose until nearly tight. The final tightening takes place when the cylinder is in position.
- Refit the brake pads and the cylinder as described above. The two bolts (2) are tightened to 3.0 kgm (22 ft.lb.).
- Tighten the brake hose connection to 1.8 kgm (13 ft.lb.). Make sure that the hose cannot twist during this operation. If this is the case, you can slacken the connection between hose and pipe and reset the hose accordingly. Move the steering from lock to lock and check that the hose cannot rub against other parts of the front suspension. The brake system must be bled of air after installation.

8.3. Rear Disc Brakes

8.3.1. REPLACING THE BRAKE PADS

The brake pads must be replaced if the thickness of the brake pad material is approx. 3.5 mm. The thickness can be checked with the calipers fitted to the rear axle, as has been described for the front calipers, i.e. with the wheels fitted or removed.

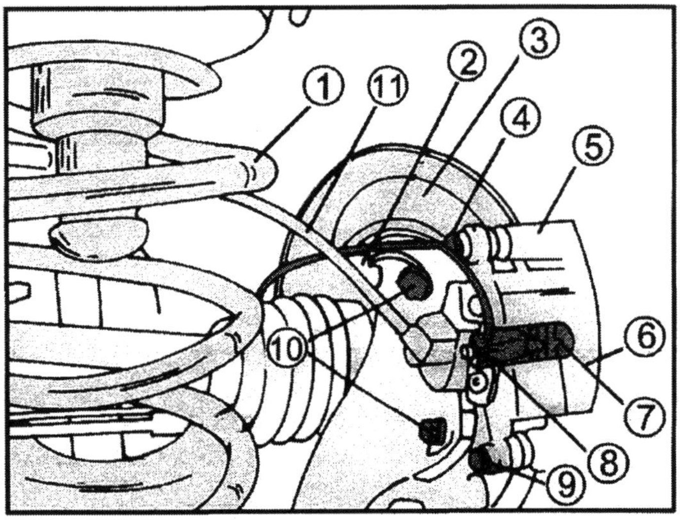

Fig. 8.8 – Details for the removal and installation of the rear disc brake pads.

1 Coil spring
2 Sensor, speed sensor
3 Rear wheel hub
4 Caliper mounting bolt
5 Brake caliper mounting bracket
6 Floating brake caliper
7 Brake pad wear indicator
8 Electrical connector
9 Caliper mounting bolt
10 Caliper mounting bracket bolts
11 Brake hose

Fig. 8.8 shows the component parts of a rear caliper and the brake pads on a vehicle with coil spring suspension and can be referred to during the following instructions. The operations refer to one caliper, but remember to remove the pads of the other caliper in the same manner if replacement is necessary. Always replace brake pads in sets.

- Open the brake fluid reservoir cap.
- Operate the brake pedal a few times to expel the vacuum from the system, jack up the rear end of the vehicle and place chassis stands underneath the rear end of the vehicle. Remove the rear wheels.
- On both sides of the vehicle disconnect the brake pad wear indicator cables (8) from the connector (7). Make a note of the cable routing to make sure it is reconnected correctly.
- Remove the two brake caliper mounting bolts (4) and (9) from the caliper mounting bracket (5) and lift off the caliper (6). Use a piece of wire and tie the caliper to the rear suspension. The two bolts must always be replaced and are tightened to 3.2 kgm (23 ft.lb.) during installation.
- The brake pads are now exposed and can be removed from the brake caliper mounting bracket. Immediately clean the caliper openings and check the piston seals for signs of leaks.

Refit the brake pads as follows:

- Push the piston back into the caliper bore, using a hammer handle.
- Clean the contact areas for the brake pads inside the caliper housing.
- Remove some brake fluid from the fluid reservoir, if it is fairly full, to prevent fluid overflow during the next operation. You will have the push the piston(s) into the caliper bore(s), to fit the new brake pads. Normally a special pair of pliers is used, which is employed in the manner shown for the front calipers earlier on.
- Fit the brake pads with the brake pad wear contact sensor on the outside. The side guide surfaces of the brake pads should be slightly coated with brake paste.
- Place the cylinder caliper carefully over the brake pads and insert the bolts (4) and (9) through the cylinder housing into the mounting bracket. Tighten the bolts to 3.2 kgm (23 ft.lb.)
- Roll the brake pad wear indicator cable to a spiral and connect to the connector spade terminal of the brake pad. Push the cover over the connector. Operate the brake pedal a few times to set the brake pads against the brake disc.
- Check the fluid level in the fluid reservoir and correct if necessary. Finally fit the wheels and tighten the wheel bolts to the correct torque setting.
- Treat the new brake pads with care during the first 100 miles or so, remember that the new material must "bed in" before they reach their full performance. To speed up the process, brake the vehicle several times from a speed of approx. 50 mph to 25 mph. Allow the brakes to cool between each braking operation.

8.3.2. BRAKE CALIPER – REMOVAL AND INSTALLATION

The removal and installation of the brake calipers is carried out by referring to Fig. 8.8
To remove:

- Open the brake fluid reservoir cap.
- Operate the brake pedal a few times to expel the vacuum from the system, jack up the rear end of the vehicle and place chassis stands underneath the axle tube. Remove the rear wheels.
- Push a bleed hose over the bleed valve and guide the other end into a container. Operate the brake pedal until the fluid has been drained.
- Slacken the brake hose (11) where it is connected to the brake caliper. It is also possible to disconnect the brake hose at the connection with the brake pipe. In

this case slacken and unscrew the union nut, knock out the spring plate and withdraw the hose. The hose connection at the caliper should be slackened as well to facilitate the installation later on. Open hose/pipe ends must be protected to prevent entry of dirt.

- Remove the bolt securing the pad wear indicator (7) and withdraw the cable (8). The bolt is tightened to 0.9 kgm (7 ft.lb.) during installation.
- Remove the two caliper mounting bolts (4) and (8) and take off the caliper. Unscrew the caliper from the brake hose and take it off.
- The brake caliper mounting bracket (5) can be removed from the semi-trailing arm (suspension arm), after removal of the two bolts (10). The M12 bolts can be re-used and are tightened to 10.6 kgm (76.5 ft.lb.).

The installation is a reversal of the removal procedure, noting the tightening torques given above. Operate the brake pedal a few times to set the brake pads against the brake disc. Check the fluid level in the fluid reservoir and correct if necessary. Finally fit the wheels and tighten the wheel bolts to the correct torque setting.

Treat the new brake pads with care during the first 100 miles or so, remember that the new material must "bed in" before they reach their full performance. To speed up the process, brake the vehicle several times from a speed of approx. 50 mph to 25 mph. Allow the brakes to cool between each braking operation.

8.2.2 Brake Calipers – OVERHAUL

We do not recommend the dismantling and/or overhaul of the brake calipers. Either fit a new caliper or have the cylinder overhauled in a workshop dealing with brake caliper cylinder units.

8.3 Brake Discs

8.3.1. FRONT DISCS – REMOVAL AND INSTALLATION

If new brake discs, or even one new disc is to be fitted, always replace the brake pads. The brake pads must be removed as described. Then remove the bolts securing the brake caliper to the caliper mounting bracket and lift off the mounting bracket. There is no need to remove or disconnect the brake hose, but tie the mounting bracket to some part of the front suspension. The brake disc is secured by a single bolt (must always be replaced) to the wheel hub. The wheel hub can be prevented from rotating as shown in Fig. 8.9, using a tyre lever or similar. Use a plastic or rubber mallet to remove a tight disc.

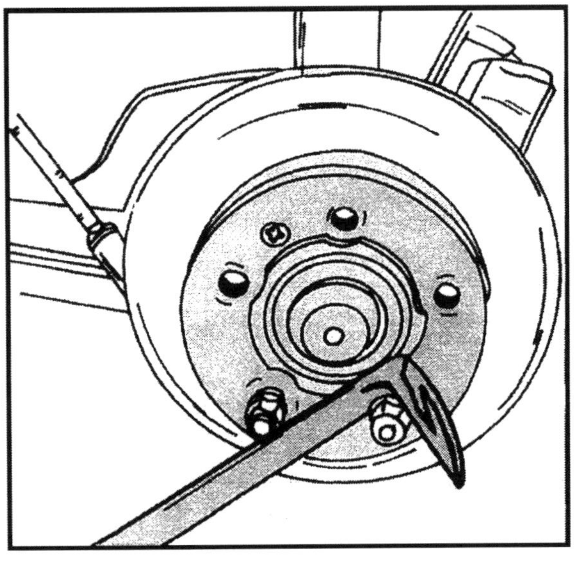

Fig. 8.9 – Counterhold the wheel hub by means of a tyre lever. Two wheel bolts must be screwed into the hub.

The installation is a reversal of the removal procedure. Tighten the disc securing bolt to 2.2 kgm (16 ft.lb.), the caliper mounting bracket bolts to 17.2 kgm (124 ft.lb.).

8.3.2. REAR DISCS – REMOVAL AND INSTALLATION

If new brake discs, or even one new disc is to be fitted, always replace the brake pads. The rear brake discs are attached to the wheel hub with a single bolt (2.2 kgm/16 ft.lb.). The bolt must always be replaced. To remove the bolt, apply the handbrake to lock the disc/hub. Release the handbrake after removal of the bolt. To remove a brake disc, remove the brake pads and caliper mounting bracket as described during the removal and installation of the rear brake pads/calipers. Suspend the caliper on a piece of wire without disconnecting the brake hose.

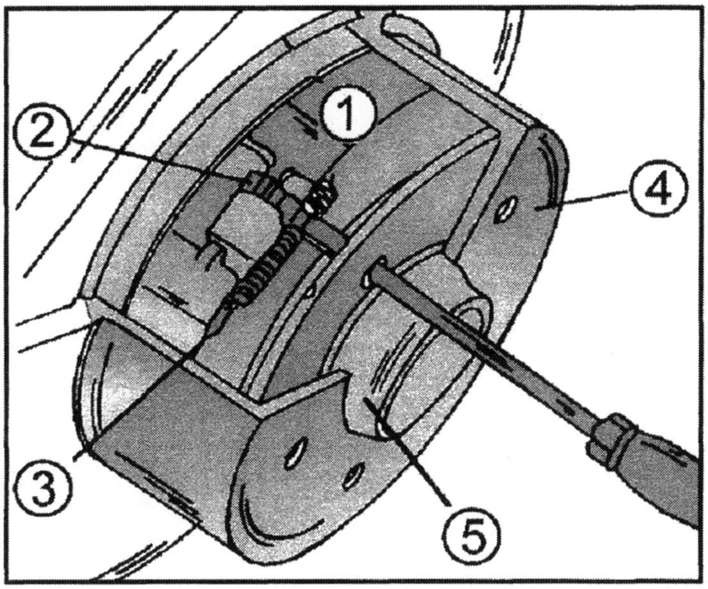

Fig. 8.10 – Insertion of the screwdriver to adjust the handbrake adjuster wheel inside the brake disc/drum.

1 Brake shoe
2 Adjuster wheel
3 Return spring
4 Brake drum/disc for handbrake
5 Wheel hub

The brake shoes in the inside of the handbrake must be reset in order to remove the disc. To do this, insert a screwdriver at the position shown in Fig. 8.10 and reset the adjuster wheel for the brake shoes. The brake disc may have to be rotated to clear the adjuster wheel. The brake shoes will now be away from the brake disc and the disc can be removed. A sticking disc can be freed with a rubber or plastic mallet.

The installation is a reversal of the removal procedure. Tighten the caliper mounting bracket to 10.6 kgm (76.5 ft.lb.).

The handbrake brake shoes must now be set to their basic position as described later on.

After installation check the operation of the handbrake and adjust if necessary.

8.3.3. RE-FACING THE BRAKE DISCS

Brake discs can be re-faced provided their minimum thickness has not been reached. The thickness of the front discs is the same on all models (28 mm), as is the min. permissible thickness (25 mm). Discs with a different thickness of 10 mm, with a different min. permissible thickness (8 mm) are used for the rear wheels. If you intend to have the discs re-faced, we advise you to have it done at an MB workshop, as all necessary data will be available.

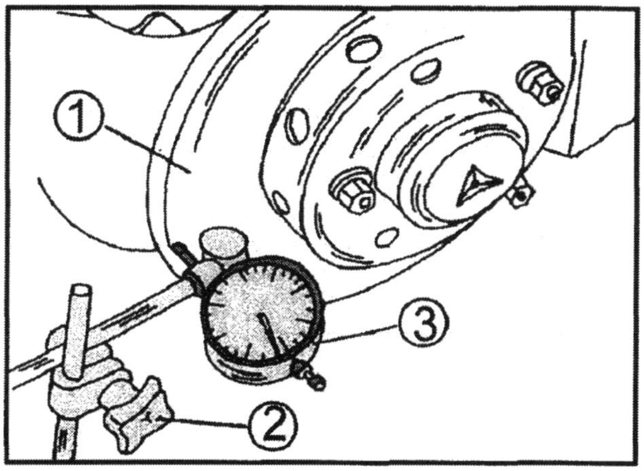

Fig. 8.11 – Checking a brake disc for run-out.

1 Brake disc
2 Dial gauge holder
3 Dial gauge

Discs should be checked for run-out after installation. To do this, attach a dial indicator with the stylus placed against the outer edge of the disc as shown in Fig. 8.11. Slowly turn the disc and read off the dial indicator reading. If the reading exceeds 0.12 mm remove the disc and check the disc and hub faces for foreign matter. Otherwise the disc may be out-of-true.

8.4. Master Brake Cylinder

All models are fitted with a tandem master cylinder with a double reservoir, supplying both circuits of the dual circuit with brake fluid. The brake lines are diagonally split.

The brake fluid level is indicated by means of a warning light in the dashboard, when it is below the "Min" mark.

8.4.0. REMOVAL AND INSTALLATION

The master cylinder is fitted to the front face of the brake servo unit and is removed from the engine compartment. Fig. 8.12 shows the parts of the cylinder together with the brake servo unit. The brake fluid reservoir must be removed to take out the cylinder. The cylinder can be removed as follows:

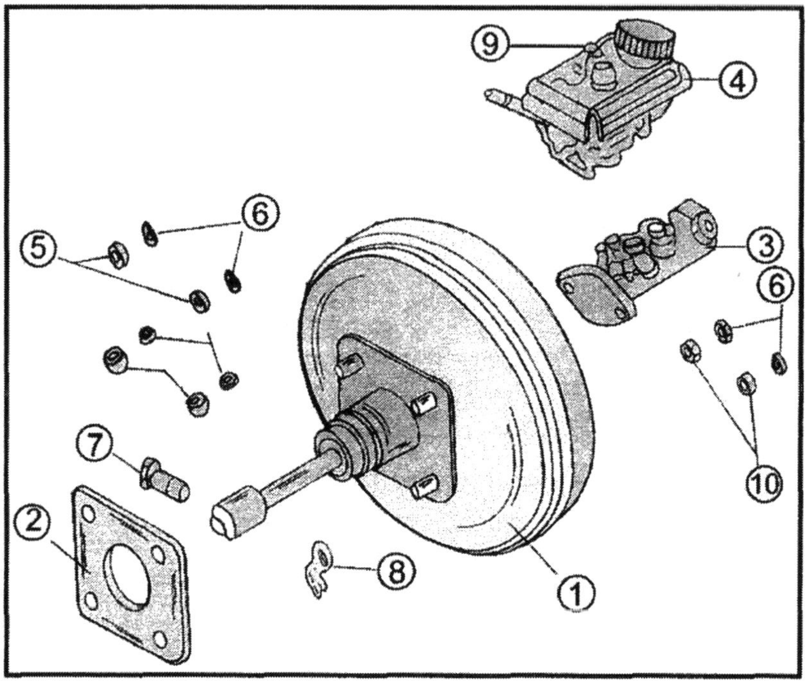

Fig. 8.12 – Details for the removal and installation of the brake master cylinder and brake servo unit.

1 Brake servo unit
2 Gasket
3 Master brake cylinder
4 Fluid reservoir
5 Securing nuts, brake servo
6 Washers
7 Clevis pin, push rod
8 Wire securing clip
9 Terminal for brake fluid level indicator
10 Securing nuts, cylinder

- Push a bleed hose over one of the bleed screws in a front brake caliper (remove the rubber dust cap first), suspend the other end of the hose into a container, open the screw and pump the brake pedal until the system has been drained.

The panelling on the centre piece of the windscreen must now be removed as follows:

- Disconnect the fluid hoses from the windscreen washer system (fitted around the wiper arm spindles) and then unscrew the nuts securing the wiper arms to the spindles. Mark the parked position of the wiper blades on the windscreen and pry off the wiper arms. The installation is a reversal of the removal procedure, but make sure that the wiper blades take up their original position (wipers in park position). Fit and tighten the wiper spindle nuts to 1.8 kgm (13 ft.lb.).
- The panelling in the centre section of the windscreen must now be removed. This is a fairly complicated operation, but can be carried out. The operations must be carried out by referring to Fig. 5.2. Remove the large cover (1) with a suitable tool. Remove the two screws (2) and take off the clamping plate (3). On the L.H. and R.H. side remove the foam material (4). During installation first insert the clip into the foam material, then insert the foam material into the specified holes. Ensure that the clip is seated properly in the lip (5).
- Remove the clips on the sides of the centering panel (8) with a suitable tool. The clips may have to be replaced.
- Unscrew the bolts (7). During installation pay attention to the bracket (6).
- Lift the panelling from the centre section (8). Lift the panel until the L.H. and R.H. water hoses (11) are accessible and remove the two water hoses.

- The panelling can now be removed from the centre section. Slide the panelling downwards and free it from the window frame seal (10). Then remove the panelling upwards between the windscreen and the engine bonnet.

Install the panelling in reverse order.

Remove the fluid reservoir as follows:

- Unlock the electrical connector plug on the front end of the master cylinder, remove the reservoir cap and empty the reservoir as already described earlier on.
- Remove the bracket for the reservoir from the body. Two nuts secure the bracket to the suspension strut tower and one bolt to the radiator tank.
- Remove a cotter pin on one side of the reservoir and pull out the retaining pin on the other side. The brake fluid reservoir can now be removed in an upward direction. Two sealing rings are fitted into the master cylinder opening which should be replaced if no longer in good condition.
- The installation is a reversal of the removal procedure.
- Withdraw the cable plug from the brake fluid level contact switch on the side of the reservoir cap.
- Disconnect the brake pipes from the cylinder (3) and push them carefully to one side. Close the pipe ends in a suitable manner to prevent entry of foreign matter. Take care not to drip brake fluid onto other parts in the engine compartment. Tighten the union nuts during installation to 1.2 – 1.4 kgm (9 – 11 ft.lb.).
- Unscrew the cylinder and lift it off in horizontal position. Take care not to tilt the cylinder, as this will bring the push rod out of alignment. A rubber sealing ring is inserted between cylinder and brake servo unit. Take care to collect it.

The installation is a reversal of the removal procedure. Always replace the sealing ring. Tighten the cylinder to 2.0 kgm (14 ft.lb.). There is no need to adjust the brake servo push rod. When fitting the reservoir coat the inside of the rubber grommets with brake fluid and press the reservoir in position until fully home.

Finally fill the reservoir and bleed the brake system as described later on.

8.4.1 MASTER CYLINDER – OVERHAUL

Master brake cylinders are no longer overhauled. A new cylinder must be fitted if the old one fails. Always quote the model and the chassis number when ordering a new cylinder, as cylinders are sometimes changed and a different cylinder may have been allocated to your particular vehicle.

8.5. The Handbrake

The handbrake is known as duo-servo-brake. "Duo" indicates that the braking effect is the same in both rotating directions of the brake disc.

"Servo" indicates that the operation of one brake shoe is transmitted to the other brake shoe. The brake shoes for the handbrake are located inside the rear brake discs. The remaining component parts of the handbrake system are a normal handbrake lever, a compensating lever assembly underneath the vehicle, the necessary handbrake cables and the operating levers on the two rear brake calipers.

8.5.0. BRAKE SHOES – REMOVAL AND INSTALLATION

A special tool is required to remove the brake shoes in order to remove the shoe return springs. Otherwise you will have to remove the rear wheel hub to expose the brake shoes. Fig. 8.13 shows the component parts of the handbrake brake shoes on one side. The following description assumes that the latter recommendation is followed, as it will not be possible to reach the springs. The operations are described for one side, but shoes must always be replaced as a set. The rear of the vehicle must be on chassis stands, with the wheels removed.

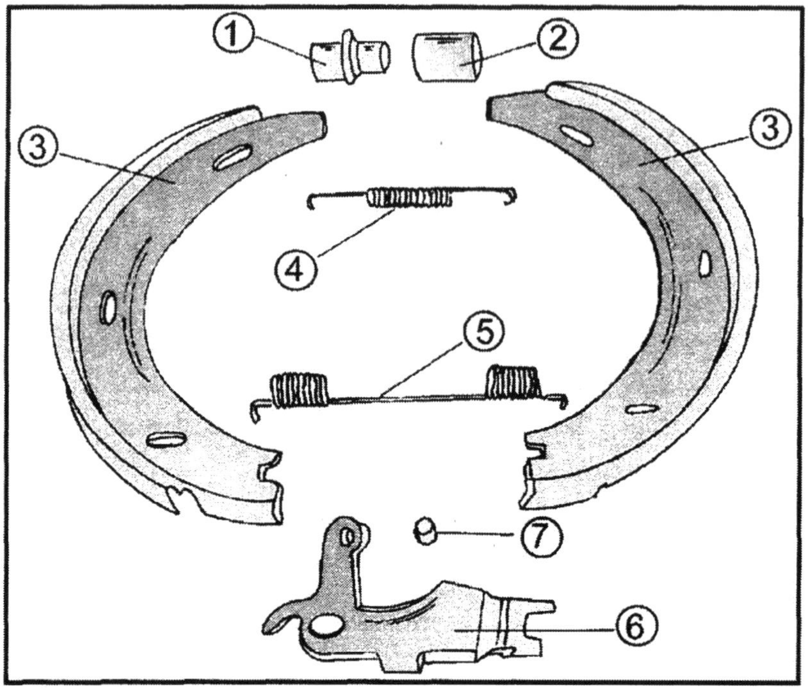

Fig. 8.13 – The brake shoes for the handbrake mechanism inside the brake disc.

1 Adjuster wheel
2 Threaded sleeve
3 Brake shoes
4 Upper shoe return spring
5 Lower shoe return spring
6 Shoe adjuster
7 Brake shoe operating roller

- Remove the rear brake disc as already described. If the disc cannot be removed immediately, slacken the nut at the end of the handbrake cable compensating lever.
- Remove the upper brake shoe return spring.
- Rotate the adjuster wheel (1) in Fig. 8.13 with a small screwdriver until the brake shoes are together as far as possible at the upper end.
- Disconnect the handbrake lever on the outside from its connection.
- Remove the brake shoe return springs. One of the springs must be compressed until the spring seat can be rotated by 90°. Remove the other springs with a pair of pliers.
- Expand the two brake shoes until they are free to be removed. The adjuster wheel at the lower end will come free during this operation.

The installation is carried out as follows:

- Connect the handbrake cable with the pin to the spreader lock and push the spreader lock into the brake back plate.
- Coat the thread of the adjuster push rod and the cylindrical part with long-term-grease and turn back the adjuster into the body.
- Fit the adjuster mechanism into the brake shoes with the adjuster wheel at the lower end. Fit the upper return spring from the front into the brake shoe.
- Expand the brake shoes until one shoe can be engaged with the spreader lock. Fit the spring from the side into the brake shoe.
- Compress the other spring from the front until the inner end can be engaged into the brake back plate and turn it by 90° to secure it.
- Fit the remaining spring between the brake shoes. A wire hook can be used to stretch and insert the spring.
- Refit the brake disc and the brake caliper assembly and adjust the handbrake as described below.

8.5.1. HANDBRAKE ADJUSTMENT

The handbrake must be adjusted if the handbrake foot pedal can be depressed with medium force by more than 3 to 4 "clicks" (from a total of 6) before the braking effect takes place. The handbrake is adjusted inside the brake discs, but it may also be necessary to adjust the handbrake cables (see following section). Proceed as follows to adjust the brake shoes:

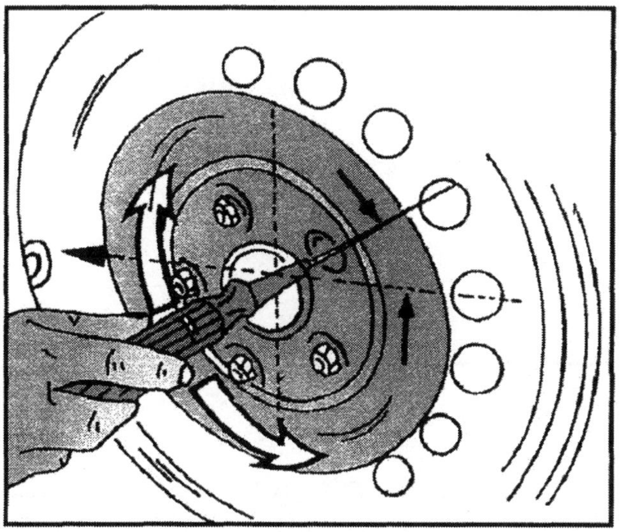

Fig. 8.14 – Turn the wheel into the position shown, remove the wheel bolt and insert a screwdriver into the hole to rotate the brake shoe adjuster inside the brake disc.

- Place the rear end of the vehicle on chassis stands and slacken the handbrake cables at the compensating lever. If steel wheels are fitted, you will have to remove the hub caps.
- Turn the wheel into the position shown in Fig. 8.14 and remove the wheel bolt. *If light alloy wheels are fitted the wheels must be removed.* Insert a screwdriver into the hole and engage it by "feeling" with the teeth of the adjuster wheel. Fig. 8.15 shows a cut-away view of the brake disc to show the brake adjuster with more details. Turn the adjuster wheel by 10 teeth backwards, until the wheel can be rotated freely (or the disc in the case of light alloy wheels). The screwdriver must be moved as follows during the adjustment:

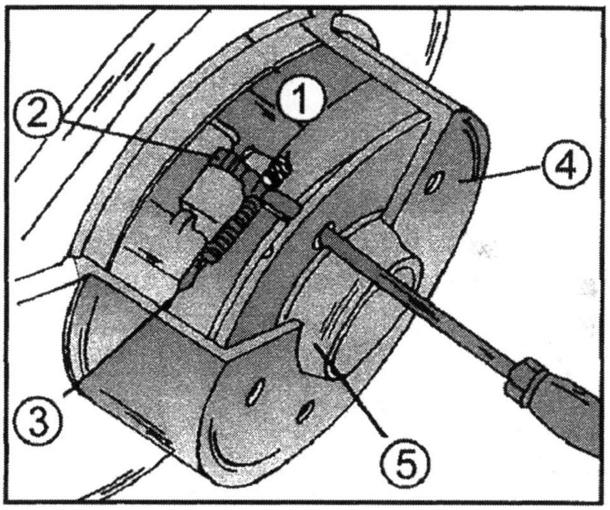

Fig. 8.15 – Insertion of the screwdriver to adjust the handbrake adjuster wheel inside the brake disc/drum.

1 Brake shoe
2 Adjuster wheel
3 Return spring
4 Brake drum/disc for handbrake
5 Wheel hub

- **On the L.H.** side from the bottom towards the top.
- **On the R.H.** side from the top towards the bottom.
- Depress the handbrake pedal a few times. After adjustment check that the brake shoes release the brake disc/drum. Otherwise re-adjust.
- Fit the removed wheel bolt (18 kgm) or refit the wheel, depending on the version, and tighten it to the correct torque.
- Adjust the handbrake cables as described below.

8.5.2. FRONT HANDBRAKE CABLE – REPLACEMENT

The attachment of the front handbrake cables is shown in Fig. 8.16. The cable is connected between the handbrake pedal and the handbrake cable compensating bracket on the underfloor of the vehicle.

The connection of the cable on the compensating lever can be seen in Fig. 8.17, where you will also find the adjusting nut. We would like to point out that the replacement of the front cable is a complicated operation as it involves the removal of the fuel tank, the removal of the front seat, the parking brake pedal and other time-consuming and difficult operations. Additionally a special tool is required to disconnect the end of the cable from the handbrake pedal. As the cable will, according to research, last a long time we refrain from describing the removal and installation.

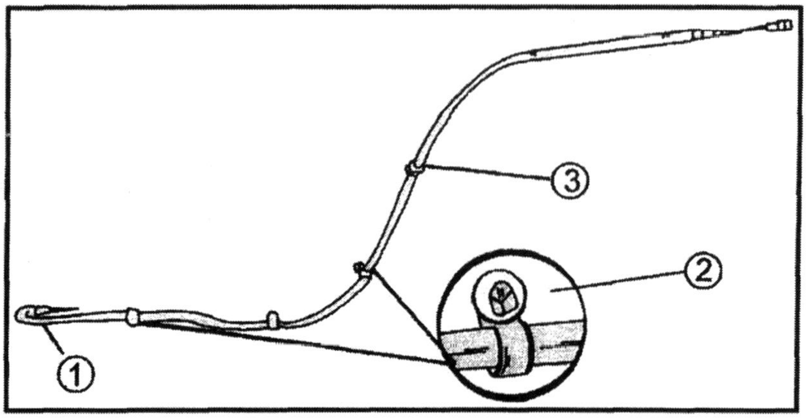

Fig. 8.16 – Attachment of the handbrake cable on the vehicle floor.
1 Cable end, connected to handbrake pedal
2 Cable securing clamps 0.8 kgm
3 Cable clamps

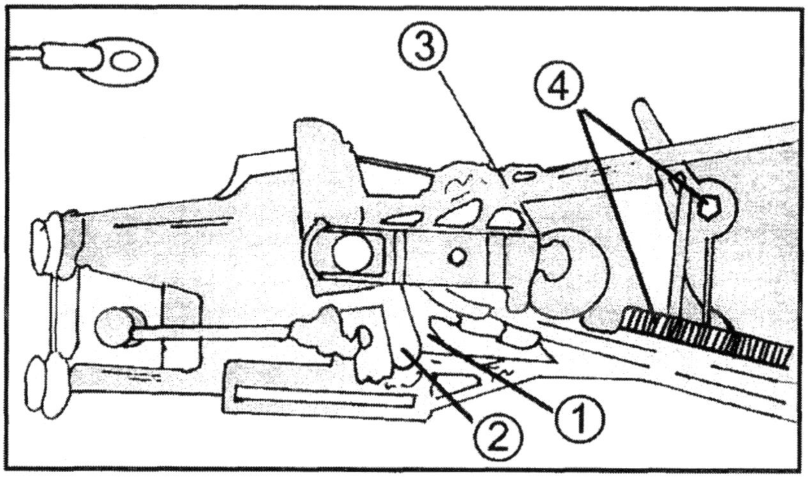

Fig. 8.17 – The attachment of the front handbrake cable on the handbrake cable compensating lever.
1 Compensating lever
2 Bolt, handbrake cable to compensating lever, 1.8 kgm
3 Adjusting bolt
4 Return spring

8.5.3. REAR HANDBRAKE CABLE – REPLACEMENT

Fig. 8.18 shows the routing of the rear handbrake cables to the brake assemblies. The rear cables, and also one of the cables, can be replaced without any difficulties. The operations are the same on both sides.

Fig. 8.18 – Layout of the rear handbrake cables.
1 Handbrake cables
2 Compensating lever
3 Return spring
4 Handbrake operating lever
5 Clevis pin to operating lever
6 Cable attachment to vehicle floor
7 Adjusting bolt

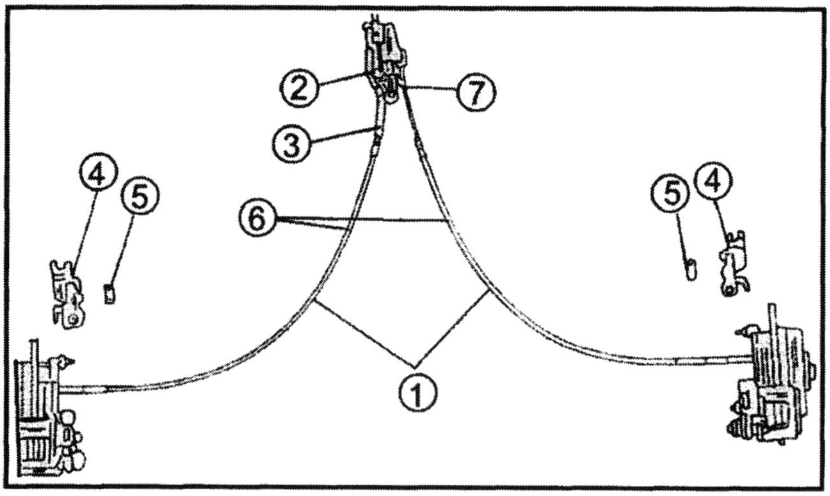

- Place the rear end of the vehicle on chassis stands and remove the wheel on the side in question.
- Unscrew the nut (7) in Fig. 8.18 until the handbrake cable is slack.
- Unhook the handbrake cable (1) from the handbrake compensating lever. To do this, use a 13 mm A/F open-ended spanner until the cable or both cables can be freed.
- If ABS is fitted, detach the electrical lead from the L.H. brake cable.

- Free the cable(s) from the cable retainers at position (6).
- Remove the rear brake disc as described above (remove the bolt) and disconnect the cable from the lever on the brake shoe. Withdraw the cable.

Always check the condition of the brake shoe linings after the disc has been removed.

The installation is a reversal of the removal procedure. The brake cables must be adjusted as follows after installation:

- Tighten the adjusting bolt (7) until all play has been eliminated.
- Operate the handbrake pedal two or three times.
- Tighten the adjusting bolt again until a handbrake pedal pressure of 190 – 260 kg is required to press the pedal to the first notch. This, of course, cannot be measured by normal means and must therefore be estimated.
- If the brake shoes have been replaced they can be "bedded in". To do this take the vehicle onto a clear road and at a speed of approx. 50 mph depress the handbrake pedal to the first notch, but not longer than 10 seconds. Then release the pedal. After stopping the vehicle operate the pedal 2 or 3 times fully and check the handbrake has fully released. The rear of the vehicle must, of course, be jacked up to carry out the check.

8.6. Brake Servo Unit

Brake servo units should not be dismantled, as special tools are required to dismantle, assemble and test the unit. Remember that a failure of the servo unit to act will not affect the efficiency of the brake system but, of course, additional effort will be required for the same braking distance to be maintained. The vacuum for the operation of the servo unit is supplied by a separate exhauster (vacuum) pump in the case of a diesel engine or from the inlet manifold in the case of a petrol engine.

Note: If you coast downhill without the engine running, for what ever reason, remember that the vacuum in the unit will be used up after a few applications of the brake pedal and the brake system will from then onwards operate without the power-assistance. Be prepared for his.

The removal of the brake servo unit is similar as the removal of the brake master cylinder, as the latter must be removed before the brake servo unit can be taken out. After removal of the cylinder take out the spring clip (8) in Fig. 8.12, push out the clevis pin (7) and remove the nuts (5) and (6) securing the servo unit to the pedal bracket. Remove the gasket on the rear of the servo unit.

The installation is a reversal of the removal procedure. Tighten the securing nuts to 2.8 kgm (20 ft.lb.).

8.7. Bleeding of the Brake System

Bleeding of the brake system should be carried out at any time that any part of the system has been disconnected, for whatever reason. If only one of the brake circuits (caliper) has been opened, either bleed the front left and rear right or front right and rear left. If both circuits have been opened, first bleed the primary circuit and then the secondary circuit. The procedure given below should be followed and it should be noted that an assistant will be required, unless a so-called "one-man" bleeding kit is available.

Always use clean fresh brake fluid of the recommended specification and never re-use fluid bled from the system. Be ready to top up the reservoir with fluid (a brake bleeding kit will do this automatically) as the operations proceed. If the level is allowed to fall below the minimum the operations will have to be re-started.

- Obtain a length of plastic tube, preferably clear, and a clean container. Put in an inch or two of brake fluid into the container and then go to the first bleed point. Take off the dust cap and attach the tube to the screw, immersing the other end of the tube into the fluid in the container.
- Open the bleed screw about three quarters of a turn and have your assistant depress the brake pedal firmly to its full extent while you keep the end of the tube well below the fluid level in the container. Watch the bubbles emerging from the tube and repeat the operation until no more are seen. Depress the brake pedal once more, hold it down and tighten the bleed screw firmly.
- Check the fluid level, go to the next point and repeat the operations in the same way. Install all dust caps, depress the brake pedal several times and finally top up the reservoirs.

Note: A pressure-bleeding apparatus is used in the workshop to bleed the brake system. If you are not 100% happy with the performance of the brakes, have the brakes bled at a dealership. Do not take chances. If you do the bleeding operations yourself, only use the brake fluids supplied by Mercedes-Benz. Although other fluids are suitable, we advise against it. Never use fluid which has been standing for a while, as fluid will deteriorate over the years. Also do not re-use fluid drained from the system.

8.8. Changing the Brake Fluid

The brake fluid should be changed every two years. In general this is normally not carried out, but we highly recommend it if heavy loads or passengers are carried. Empty the brake system by opening all bleeding screws and pumping the pedal. Old fluid should be collected into a jar as described above.

After filling the fluid reservoir, bleed the system in the manner described above. In this case it is more important to observe the fluid level in the reservoir, as it will sink rapidly. It is of advantage to have an additional helper, one to operate the brake pedal and another one to top-up the reservoir.

Make sure to fit the rubber dust caps over the bleed screws after the system has been bled. Operate the brake pedal a few times to check the pedal pressure. If the pedal "lifts" during the pumping operation, re-bleed the system.

8.9. Brakes – Tightening Torques

Front Brakes

Brake hose to brake caliper: .... 1.2 – 1.4 kgm (9 - 10 ft.lb.)
Brake caliper to caliper bracket:.. 3.0 kgm (22 ft.lb.)
Brake caliper bracket to steering knuckle:.. 17.2 kgm (124 ft.lb.)
Brake disc to wheel hub: .... 2.2 kgm (16 ft.lb.)
Brake hose to caliper:.. 1.8 kgm (13 ft.lb.)
Splash shield to steering knuckle: .... 0.9 kgm (7 ft.lb.)
Brake pad wear indictor bolt:.. 0.9 kgm (7 ft.lb.)
Bleeding screw to caliper: .... 1.0 kgm (7.2 ft.lb.)
Brake disc to wheel hub: .... 2.2 kgm (16 ft.lb.)
Steel wheels to hubs: .... 18.0 kgm (130 ft.lb.)
Light alloy wheels to hubs: .... 15.0 kgm (108 ft.lb.)

Rear Brakes

Brake caliper mounting bracket to semi-trailing arm: .... 10.6 kgm (76 ft.lb.)
Brake caliper to caliper mounting bracket: .... 3.2 kgm (23 ft.lb.)

Bleeding valve to caliper: .... 1.0 kgm (7.2 ft.lb.)
Brake pad wear indictor bolt: .... 0.9 kgm (7 ft.lb.)
Brake disc to wheel hub: .... 2.2 kgm (16 ft.lb.)
Brake hose to caliper and brake pipe: .... 1.8 kgm (13 ft.lb.)

Other Tightening Torques
Brake pipes to master cylinder: .... 1.2 - 1.4 kgm (9 - 10 ft.lb.)
Brake master cylinder to brake servo unit: .... 2.0 kgm (14 ft.lb.)
Brake servo unit to body: .... 2.0 kgm (14 ft.lb.)
Bolt for pedal stop: .... 0.8 kgm (6 ft.lb.)
Front handbrake cable to compensating lever: .... 1.8 kgm (13 ft.lb.)
Compensating lever to body: .... 2.7 kgm (19.5 ft.lb.)
Front handbrake cable to body: .... 0.8 kgm (6 ft.lb.)
Steel wheels to hubs: .... 18.0 kgm (130 ft.lb.)
Light alloy wheels to hubs: .... 15.0 kgm (108 ft.lb.)

9. ELECTRICAL SYSTEM

9.0 Battery

Voltage:	12 volts
Polarity:	Negative earth
Capacity:	Depending on equipment (66 or 88 Ah)

Condition of Charge:	
- Fully charged:	1.280
- Half charged:	1.200
- Discharged:	1.120

To check the voltage of the battery, use an ordinary voltmeter and apply between the two battery terminals. A voltage of 12.5 volts or more should be obtained.

If a hydrometer is available, the specific gravity of the electrolyte can be checked. The readings of all cells must be approximately the same. A cell with a low reading indicates a short circuit in that particular cell. Two adjacent cells with a low reading indicates a leak between these two cells.

A battery can be re-charged, but the charging rate must not exceed 10% of the battery capacity, for example 8.8 amps. The battery must be disconnected from the electrical system. Charge the battery until the specific gravity and the charging voltage are no longer increasing within 2 hours. Add distilled water only. Never add acid to the battery.

The level of the battery electrolyte should always be kept above the top of the plates.

9.1. Alternator

9.1.0. ROUTINE PRECAUTIONS

The vehicles covered in this manual employ an alternator and control unit. This equipment contains polarity-sensitive components and the precautions below **must** be observed to avoid damage:

- Check the battery polarity **before** connecting the terminals. Immediate damage will result to the silicon diodes from a wrong connection - even if only momentarily.

- Never disconnect the battery or alternator terminals **whilst the engine is running.**
- Never allow the alternator to be rotated by the engine unless **ALL** connections are made.
- Disconnect the alternator multi-pin connector **before** using electric welding equipment anywhere on the vehicle.
- Disconnect the battery leads if a rapid battery charger is to be used.
- If an auxiliary battery is used to start the engine, take care that the polarity is correct. **Do not** disconnect the cables from the vehicle battery.

9.1.1. REMOVAL AND INSTALLATION

The alternator is rigidly attached to the engine. The removal and installation of the alternator is a fairly comprehensive operation, as amongst other work the air cleaner housing (petrol engine only) and the Poly V-belt must be removed. The engine must be lifted out of its mountings to remove the engine mountings if a petrol engine is fitted. Remove as follows:

Petrol Engines

- Disconnect the battery.
- Remove the air cleaner housing.
- Slacken the poly drive belt as described in the section dealing with the cooling system and lift off the belt.
- Lift the engine out of its mountings as shown in Fig. 6.1 on page 113 or in a similar manner. Remove the bolts securing the engine supports to the engine mount on both sides of the engine as soon as the mountings are free of tension. The bolts are tightened to 5.8 kgm (42 ft.lb.) during installation.

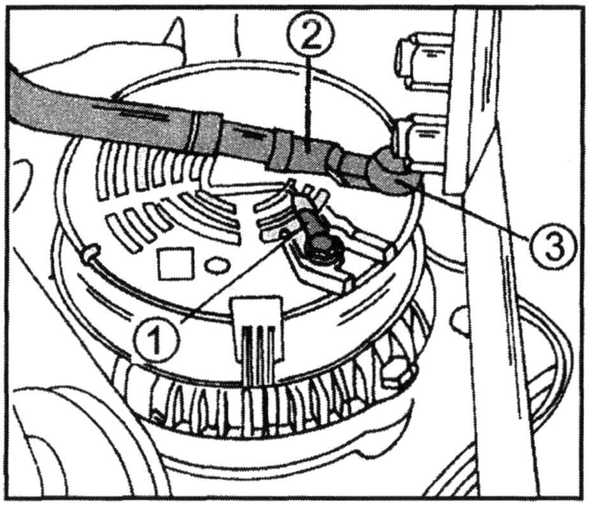

Fig. 9.1 – View of the rear of the alternator showing the connections.
1 Cable to terminal D+
2 Cable to terminal B+
3 Cable protector

- Remove the contact protector (3) in Fig. 9.1.
- Disconnect the two cables (1) and (2) at the rear of the alternator. One of the terminals is protected by the protector (3) mentioned above. The nut of the terminal (2) is tightened to 1.4 kgm (10 ft.lb.) during installation.
- Raise the engine as far as possible but make sure that the rear of the engine cannot contact the engine compartment closing panel, in order not to damage the mass air flow meter and the fuel line. A helper will have to observe the movement of the engine.
- Remove the two alternator mounting bolts (at upper and lower end) from the front and lift out the alternator in an upward direction.

The installation is a reversal of the removal procedure. The alternator mounting bolts are tightened to 5.7 kgm (41 ft.lb.). Tighten the nut of the smaller terminal (terminal 61/D+) to 0.5 kgm and the nut of the larger terminal (terminal 30/B) to 1.4 kgm (10 ft.lb.). Refit the poly drive belt and the air cleaner housing.

After installation enter the anti-theft code for the radio and reset the time clock.

Diesel Engines

- Disconnect the battery.
- Lift the front of the vehicle to improve the access.
- Slacken the poly drive belt as described in the section dealing with the cooling system and lift off the belt.
- Remove the bottom engine compartment panel at the front. This operation concerns models 639.7/8 from VIN number 055347.
- At the lower end of the alternator located the charge air hose and disconnect it from the turbo charger.
- At the terminal end of the alternator remove the protective cap fitted over the terminal 30 (B+) and disconnect the electrical cable. The nuts is tightened to 1.6 kgm (11.5 ft.lb.) during installation. Also remove the connector plug from the second terminal (terminal 61).
- Remove two bolts to slacken the alternator/turbocharger support and the bolts attaching the alternator to the timing case.
- Remove the alternator from the bottom.

The installation is a reversal of the removal procedure. The bolts securing the turbo charger support to the crankcase are tightened to 2.0 kgm (14.5 ft.lb.). The bolts securing the support to the turbo charger are tightened to 3.0 kgm (22 ft.lb.). Tighten the alternator mounting bolts to 2.0 kgm (14.5 ft.lb.).

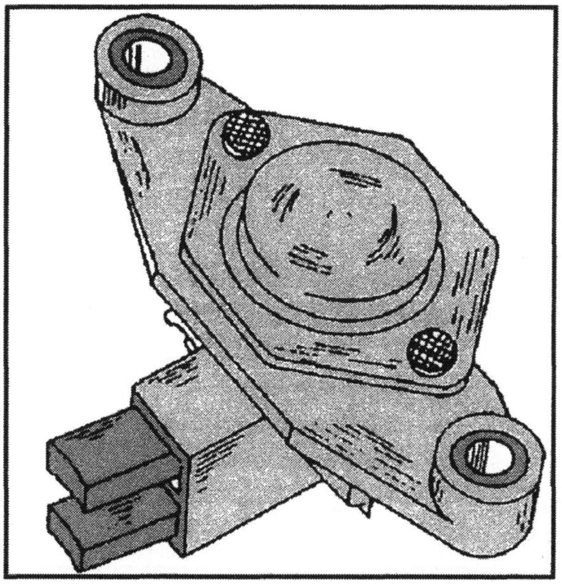

Fig. 9.2 – The ends of the alternator brushes must protrude by at least 5 mm.

9.1.3. SERVICING

A Bosch alternator is used on the engines dealt with in this manual, having an output of 90 amps, but in some cases an alternator of 150 amps can be fitted. We do not recommend that the alternator or control unit should be adjusted or serviced by the owner. Special equipment is required in the way of test instruments and the incorrect application of meters could result in damage to the circuits.

The alternator is fitted with sealed-for-life bearings and no routine attention is required for lubrication. Keep the outside of the alternator clean and do not allow it to be sprayed with water or any solvent.

The alternator brush gear runs in plain slip rings and the brushes have a long life, requiring inspection only after a high mileage has been covered. To inspect the brushes, we recommend the removal of the alternator. Take out the two screws from the brush holder assembly and withdraw for inspection. Measure the length of the brushes, shown in Fig. 9.2. If the protruding length is less than 5.0 mm (0.2 in.) or approaching this length, replace the brushes. New brushes will have to be soldered in position. We would like to point out that it is not an easy operation to guide the brushes over the slip rings when the slip ring cover is being fitted.

9.2. Starter Motor

A Bosch pre-engaged starter motor is fitted to the models covered in this manual.

9.2.0. REMOVAL AND INSTALLATION

The starter motor is shown in Fig. 9.3 in fitted condition. Remove as follows:

- Disconnect the battery earth (ground) cable.

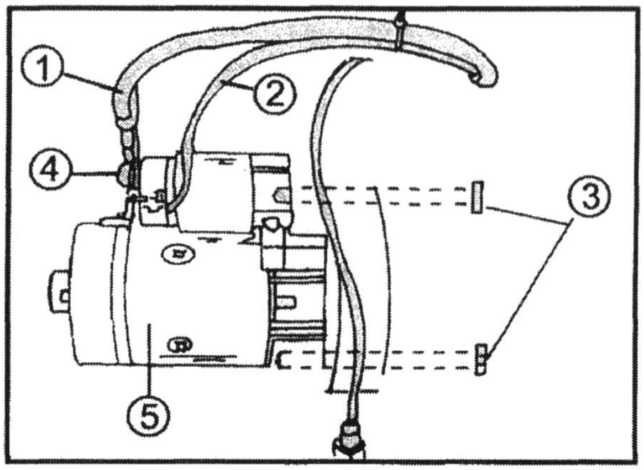

Fig. 9.3 – Fitted starter motor.
1 Cable to terminal "30"
2 Cable to terminal "50"
3 Mounting bolts
4 Contact protector
5 Starter motor

- Disconnect the cables from terminal "30" and "50" and free the cables from the retaining clamps. The contact protector must be removed first from the terminals.
- Place the front end of the vehicle on chassis stands.
- Remove the starter motor mounting bolts and withdraw the unit towards the bottom from the vehicle. Tilt the motor so that the solenoid switch is facing downwards and then remove it sideways.

Install in the reverse sequence to removal. Tighten the starter motor bolts to 4.0 kgm (29 ft.lb.). Make sure that the mating faces are clean before bolting up. Re-connect the wires and the battery terminals. Tighten the nut of the smaller terminal to 0.6 kgm and the nut of the larger terminal (terminal "30") to 1.5 kgm (11 ft.lb.).

9.2.1. SERVICING

It may be of advantage to fit an exchange starter motor if the old one has shown faults. Exchange starter motors carry the same warranty as new units and are therefore a far better proposition.

The following repairs can, however, be attempted without any problems:

Solenoid Switch

- Disconnect the cable between the solenoid switch and the starter motor terminal.
- From the front of the solenoid switch remove the bolts and withdraw the switch from the starter motor, at the same time disengaging the engagement lever from the starter motor drive.
- Refit the new switch in reverse order.

Starter Motor Brush Replacement

- Unscrew and remove the cover from the rear of the starter motor.
- Remove the two screws securing the commutator bearing cover.
- Remove the "C" clip from the rear end of the armature shaft (screwdriver), remove the sundry parts and withdraw the cover from the starter motor yoke. If necessary insert a screwdriver to get the parts apart. Remove the shims.
- Remove the brushes from their holders. The brush leads are either screwed-on or soldered. If they are soldered, you will have to be familiar with a soldering iron. Otherwise take the unit to an auto electrical repair shop to have the bushes replaced.
- Replace the brushes as applicable and re-assemble the starter motor in reverse order.

9.2.2. Starter Motor Faults

Following are a few tips to locate faults in the starting system before the starter motor is replaced:

- *When the key is turned to the starting position, the starter motor turns very slowly, nothing happens and starts to crank the engine and stops:* If the warning lights in the instrument panel are nearly or fully extinguished, check the battery. Loose or corroded cables on the starter motor, bad earth connection are other causes. If the battery has been connected previously, check the terminals for good contact. Use a jumper lead or get a tow-start to start the engine.
- *If the warning lights in the instrument panel remain "on" and a faint click is heard from the area of the starter motor*, it is possible that the engagement plunger of the solenoid switch is sticking. Use a heavy spanner and knock against the outside of the switch. Sometimes this will free the plunger. Other causes are worn starter motor brushes or defective contacts in the solenoid switch. In both cases remove the starter motor to replace the bushes and/or the solenoid switch. If the warning lights remain "on" and no click can be heard, check the cable connections between battery and starter motor.
- *If the starter motor is heard to operate without cranking the engine*, suspect a damaged solenoid switch, or the engagement mechanism, the starter motor pinion of the flywheel teeth are damaged. Try to engage a gear, push the vehicle backwards and forwards a few times and try the starter motor again. Otherwise fit a new starter motor.
- *If the starter motor "spins" after the engine has started and the key has been turned to the normal position*, a sticking solenoid switch is mostly the cause for the fault (contacts fail to open). The ignition switch can also be damaged. *Immediately switch off the engine.*

If the starter motor pinion "spins" after starting the engine (howling noise), switch off the engine immediately. In most cases a new starter motor will be necessary.

9.3. Headlamps

9.3.0. Headlamp/Side Light Bulb - Replacement

All front bulbs are fitted to the rear of a light unit. Fig. 9.4 shows the location of the various bulbs in a headlamp, as seen from the rear. The following instructions must always be observed when you change a bulb, irrespective which of the bulbs is to be replaced:

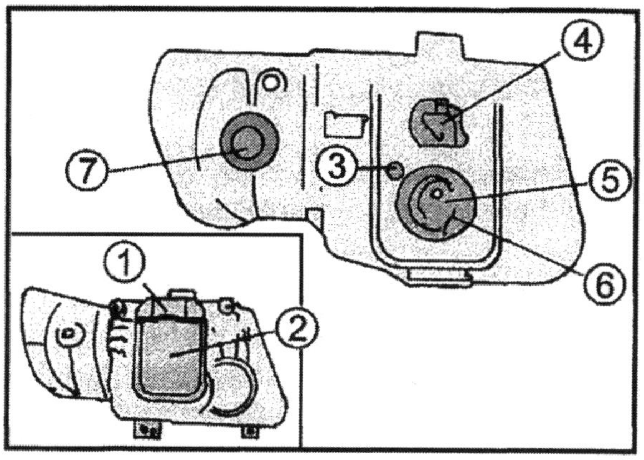

Fig. 9.4 – Details for the replacement of the headlamp bulbs.
1 Retaining clip
2 Headlamp housing cover
3 Parking light (side light) bulb
4 Dipped beam bulb
5 Main beam bulb
6 Fog lamp bulb
7 Flasher lamp bulb

- Switch off all lights to avoid short circuits.
- Only touch new bulbs with tissue paper or a clean cloth. Make sure your fingers are free of oil or grease.
- Only fit bulbs of the same type and the same wattage as the one you replace.
- The headlamp setting must be checked professionally if a headlamp bulb has been replaced.

- The replacement of the bulbs is described in detail in the Operating Instructions manual for your vehicle. The following information is therefore only useful if you have no instruction manual.

Replace the headlamp bulbs as follows:

- Disconnect the battery earth (ground) cable.
- Open the bonnet and from the rear of the headlight unit remove the plastic cover (2) by releasing the spring clip (1) and removing it.
- Withdraw the cable connector plug from the bulb holder in question, snap back the spring retainer securing the bulb and withdraw the bulb.
- Insert the new bulb without touching the glass with the bare fingers. Use a piece of tissue paper. Make sure the bulb is located correctly in its holder.
- All other operations are carried out in reverse order. There should be no need to adjust the headlamp beams if only the bulb has been replaced.

9.3.1. HEADLAMP REPLACEMENT

As the headlamps must be adjusted after replacement of a headlight unit and the adjustment can only be carried out with optical equipment, we expect that you will have the beam alignment checked at your dealer or a service station dealing with headlamp work.

- Disconnect the battery earth cable, open the bonnet and remove the radiator grille and the front bumper as described during the removal of the engine.
- If a headlamp washer system is fitted disconnect the pressure hose, if an outside temperature sender unit is fitted, withdraw the cable connector plug.
- Remove the vacuum hose from the headlamp beam level adjuster.
- Remove the screws from the rear of the headlamp. The flasher lamp, inserted from the outside must be rotated slightly in order to remove the headlamp. Remove the multi-plug connector from the rear of the headlamp unit as soon as it is accessible.

The installation of the headlamp is a reversal of the removal procedure. Provisionally adjust the headlamps as described below, but have the alignment checked at a service station if you are not 100% convinced that the adjustment is correct.

9.3.2. Headlamp Adjustment

The adjustment of the headlamps is carried out from the rear of the headlamp units. You will find two knurled bolts on each headlamp. Use one adjusting screw to move the headlamp beam up or down, and the other adjusting screw to adjust the beams from left to right.

The headlamps should be adjusted with optical equipment and only in emergency should you reset the headlamps.

9.4 Rear Lights

The rear light cluster consists of the lamp carrier, the reflector part and the coloured, outer lamp covers. The cover is fitted with a sealing ring from the outside to the body and secured at the inside together with the reflector by means of nuts. Replacement of bulbs can be carried out with the help of Fig. 9.5.

Replace the bulbs in the conventional manner. The rear light cluster must be removed to change the bulbs. Open the tail gate or the rear door and unscrew three screws securing the rear light cluster. Turn the cluster outwards and pull it out of the locating holes.

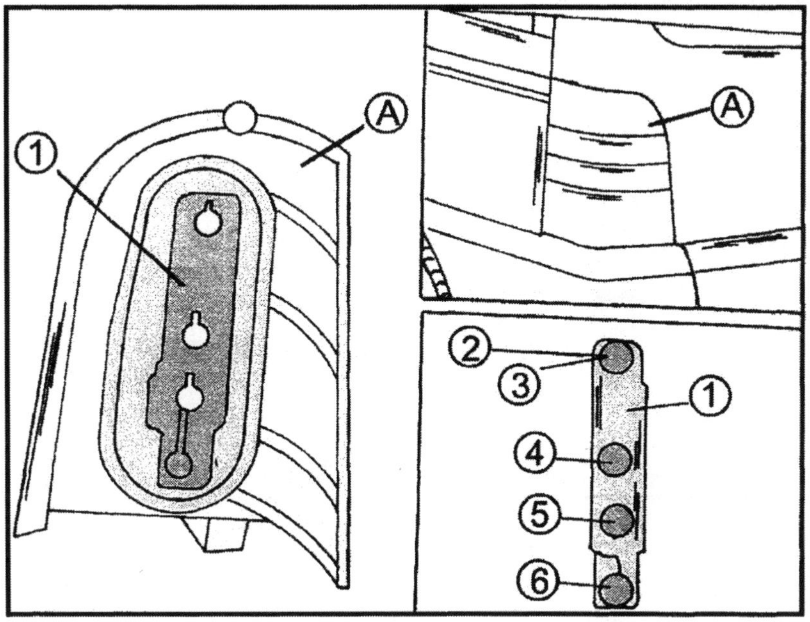

Fig. 9.5 – View of the rear light cluster from the inside and the outside with location of the various bulbs.

1 Bulb carrier
2 Brake light bulb
3 Side light bulb
4 Flasher lamp bulb
5 Rear light bulb
6 Rear fog lamp bulb

Unplug the electrical connector. Depress the four fastening lugs at the side of the bulb holder (two on each side) and remove the bulb holder. The four bulbs, starting at the upper end, are for the brake/tail light, the turn signal light (flasher light), the reversing light and the rear fog light.

9.5. Windscreen Wipers

Wiper linkage and wiper motor can be removed individually. Before the motor is removed you will have to remove the panelling of the centre section of the windscreen as described in Section 5.1 and the wiper linkage.

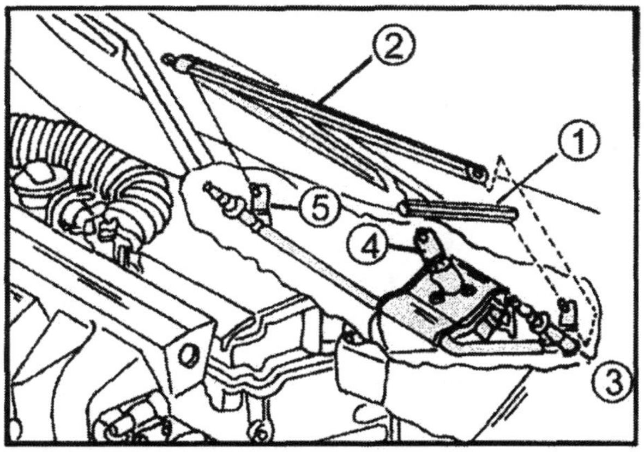

Fig. 9.6 – Details for the removal and installation of the wiper linkage.

1 Short wiper linkage
2 Long wiper linkage
3 Connecting lever
4 Wiper crank lever

Removal and Installation of Wiper Linkage

Fig. 9.6 shows details for the removal and installation.

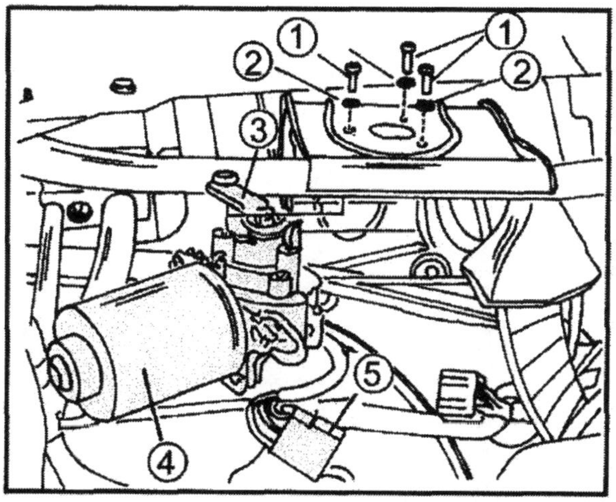

Fig. 9.7 – Fitted windscreen wiper motor.

1 Bolts
2 Washers
3 Crank lever
4 Windscreen wiper motor
5 Connector plug

- Remove the air intake tube.
- Detach the short wiper linkage (1) from the ball joint connection of levers (3) and (5), using a screwdriver without bending the lever or the linkage.

The installation is carried out in reverse order. The linkage is pressed over the ball joint of the levers. After installation switch on the wipers to check for proper operation.

Removal and Installation of Wiper Motor

The wiper motor is shown in fitted position in Fig. 9.7. The motor is located on the bulkhead of the engine compartment. The ignition must be switched off, and the battery disconnected.

- Remove the wiper linkage as described above.
- Remove the bolts (1) and the washers (2).
- Withdraw the cable connector plug (5) from the side of the wiper motor and lift out the motor.

The installation is carried out in reverse order. The linkage is pressed over the ball joint of the levers. After installation switch on the wipers to check for proper operation.

10. PETROL ENGINES

10.0 Main Features

Type:	4-stroke engine, with electronically controlled fuel injection and ignition system.
Fitted Engine:	
- Vito, Viano 119, 122, 123:	112.951 or 112.976
Number of cylinders:	6
Number of valves:	18
Number of camshafts:	2
Capacity:	
- Vito 119, 122 Viano:	3199 c.c.
- Vito 123:	3724 c.c.
Max. Power (typical):	
- Vito 119 (Viano):	190 B.H.P. (140 kW) at 5600 rpm
- Vito 122 (Viano):	218 B.H.P. (160 kW) at 5500 rpm
- Vito 123:	231 B.H.P. (170 kW) at 5600 rpm
Max. Torque:	
- Vito 119 (Viano):	27.0 kgm (270 Nm) at 2750 – 4750 rpm
- Vito 122 (Viano):	30.5 kgm (305 Nm) at 2800 – 4750 rpm
- Vito 123:	34.5 kgm (345 Nm) at 2500 – 4500 rpm
Cylinder Bore:	
- 3.2 litres - Standard:	89.90 mm
- 3.7 litres - Standard:	97.00 mm
Piston stroke:	
- 3.2 litres:	84.70 mm
- 3.7 litres:	84.00 mm
Compression Ratio:	
- 3.2 litres:	10.1 : 1
- 3.7 litres:	10.4 : 1
Max. engine speed under load:	6000 rpm +/- 150 rpm

Fuel injection system:	Electronic fuel injection
Catalytic converter:	Yes
Lambda probe:	Yes
Lead-free petrol:	Yes

General Information

- Two camshafts are fitted. The shaft on the R.H. side, seen in the direction of drive, is responsible for the opening and closing of the exhaust valves, the L.H. shaft is the inlet valve camshaft. The bearings for the camshafts are not machined directly into the cylinder head.
- The valve tappets are inserted between camshafts and valve ends. The cams push against the ends of the tappets to operate the valves. The tappets are known as bucket tappets.

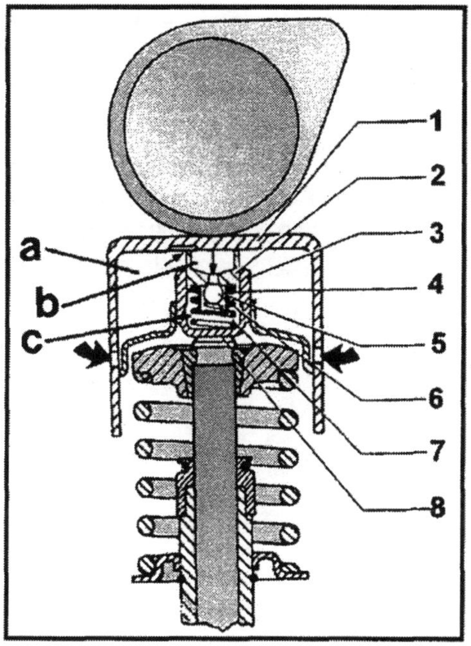

Fig. 10.1 – The operation of the hydraulic tappets. The oil enters via the bore (1) in the tappet into the valve tappet chamber (a), then into the smaller chamber (b) and then via the ball valve into the working chamber (c). The remaining parts are given below.

1 Valve tappet
2 Thrust pin
3 Retaining ring
4 Pressure spring
5 Ball guide
6 Ball
7 Gall guide
8 Guide sleeve

- The adjustment of the valves is no longer necessary on these engines. Hydraulic compensating elements are fitted which will ensure the correct valve clearance at all times. The function of the hydraulic valve clearance compensating elements is to eliminate valve clearance, i.e. the dimensional changes in the valve train (valve lash) due to heat expansion and wear are compensated by the elements. The rocker arm is in constant contact with the cam. The compensating elements cannot be repaired, but can be checked for correct functioning as described below. Fig. 10.1 shows sectional views of a valve with clearance compensation. We will give a short description of the operation. All references refer to Fig. 10.1.

The hydraulic valve compensating element are fitted into the rocker levers and operate the valves directly via a ball socket:

- The thrust pin with oil supply chamber and the return bores and the ball valve (check valve). The ball valve separates the supply chamber from the work chamber.
- The guide sleeve with the work chamber (c), the thrust spring (4) and the closing cap.

When the engine is stopped and the tappet is held under load from the cam, the element can completely retract. The oil displaced from the work chamber (c) flows through an annular gap, i.e. the clearance between the guide sleeve and the thrust pin to the oil supply chamber (b).

When the cam lobe has moved past the valve tappet, the thrust pin will be without load. The thrust spring (4) forces the thrust pin upwards until the valve tappet rests against the cam.

The vacuum resulting from the upward movement of the thrust pin in the work chamber (c) opens the ball valve and the oil can flow from the supply chamber into the work chamber. The ball valve closes when the valve tappet presses against the cam and puts the thrust pin under load. The oil in the work chamber acts as a "hydraulic rigid connection" and opens the valve in question.
When the engine is running and depending on the engine speed and the cam position, the thrust pin is only pushed down slightly.
The oil contained in the oil supply chamber is sufficient to fill the work chamber under all operating conditions of the engine. Oil or leak oil which is not required, as well as air are able to escape via the annular gap between the washer and the rocker lever. The oil ejected from the work chamber flows via the annular gap between the guide sleeve and the thrust pin and the two return bores into the oil supply chamber.

10.1. Engine – Removal and Installation

10.1.0. REMOVAL OF THE POWER UNIT

Note: The following instructions are only valid for models without automatic air conditioning system. The system must be discharged during the removal of the front part (front module) of the vehicle. Discharging and charging of the air conditioning system can only be carried out in a specialist workshop.
The engine and transmission is removed from the vehicle as a complete unit, as shown in Fig. 1.2. Engine and transmission are a heavy unit and a suitable hoist or crane is required to lift out the assembly, remembering that the weight is more than 200 lbs. The following description is a general guide line, as we cannot refer to every possible variation and/or equipment that may be fitted to your vehicle. The engine bonnet must be in an upright position and tied to the vehicle. Note that various operations are only referred to in brief, i.e. remove the propeller shaft, signifies for example that you will have to refer to Section "Propeller Shaft" for further instructions.
The front end (front module) of the vehicle must be removed to gain access to various parts. The operations have already been described during the removal of the diesel engine in section 1.1. As explained, this is a rather complicated operation as many parts must be removed and/or disconnected which you will have to locate. The workshop uses a lifting device, consisting of the parts shown in Fig. 1.1, which is bolted to the inside of the module to lower it to the ground after removal.
Removal of the engine and transmission can now be attempted as follows, after the front module has been removed as described above:

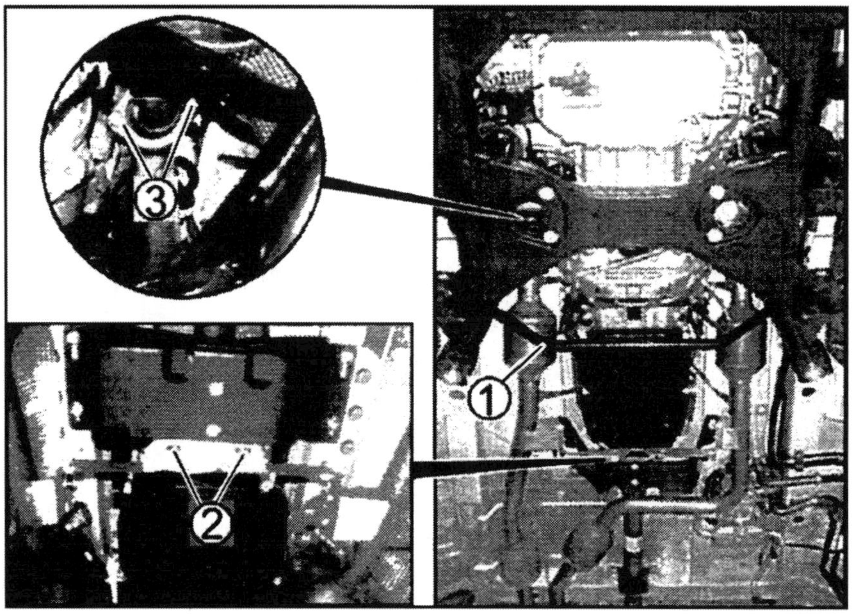

Fig. 10.2 – Details for the removal of the exhaust system. The numbers are referred to in the text.

- Disconnect all hoses and lines from the engine if not already removed.
- Disconnect the exhaust system from the exhaust manifold. The rear engine compartment panelling must be

removed to gain access. Fig., 10.3 shows a view from underneath the items to be removed. Unscrew the bolts (3) on the L.H. and R.H. side of the exhaust manifold. New bolts must be used during installation and are tightened to 2.3 kgm (16.5 ft.lb.). Remove the bolts (2) securing the exhaust system bracket and lower the complete exhaust system onto the stabiliser bar (1).

- Support the transmission from below on a suitable mobile jack and remove the rear engine crossmember. The crossmember is secured by two bolts below the propeller shaft an the L.H. and R.H. side and two further bolts in the centre. Mark the exact position of the crossmember before you take it out. During installation tighten the bolt securing the rear engine mount to the transmission carrier and the bolt securing the transmission carrier to the longitudinal member to 4.0 kgm (29 ft.lb.).
- Remove the propeller shafts as described in the relevant section. After removal push the shaft to one side and tie it up with a piece of wire. The propeller shaft bolts are tightened to 1.5 kgm (11 ft.lb.) and from the final position a further quarter of a turn.
- Remove the two gear selector cables on a model with manual transmission by levering the cables out of the ball socket and remove from the protection. Remove the single selector cable on a model with automatic transmission in a similar manner. The ball joints must be lubricated with long-term grease during installation.
- Disconnect the hydraulic clutch fluid line from the transmission.
- Attach suitable chains to the engine lifting points and the lifting device and lift the engine/transmission assembly until just under tension. Remove the engine mounting bolts on the L.H. and R.H. side of the engine and lift the assembly. Before removing any of the engine mountings ensure that the weight of the assembly is well supported. The mounting bolts are tightened to 5.8 kgm (42 ft.lb.) during installation.
- Lift the engine with the lifting tackle from the subframe and out towards the front. Continuously check that none of the cables, hoses, pipes, etc. have been forgotten and that none of the parts can get caught.

10.1.1. INSTALLATION OF THE POWER UNIT

The installation of the power unit is a reversal of the removal procedure. Here are some of the points, which should be observed:

- Do not connect any of the parts to the engine or transmission until the engine and transmission mountings have been refitted and the engine is fitted free of any load or tension.
- If work has been carried out on the engine or transmission, make sure that the oil drain plug of the respective unit has been tightened and the oil level is correct.
- Keep oil or grease away from all rubber parts.
- All bolts and nuts must be tightened to the correct tightening torque. Some important torque values: Engine carrier to engine mounting 5.8 kgm (42 ft.lb.).
- If the original coolant is refilled to the engine, check the anti-freeze strength. If the coolant has been in the engine for a considerable time, replace it. If possible check the cooling system for leaks before starting the engine.

Note: As the vehicle is fitted with a self-diagnosis and fault memory system it is possible that certain faults have been logged into the memory. The fault memory system can only be checked and cleared in a workshop.

10.2 Engine - Dismantling

Before commencing dismantling of the engine, all exterior surfaces should be cleaned, as far as possible, to remove dirt or grease. Plug the engine openings with clean cloth first to prevent any foreign matter entering the cavities and openings. Detailed information on engine dismantling and assembly is given in the sections dealing with servicing and overhaul (sections commencing at 10.3.) and these should be followed for each of the sub-assemblies or units to be dealt with. The general dismantling instructions given in section 1.2 also apply to the petrol engines.

10.3. Engine – Servicing and Overhaul

10.3.0 CYLINDER HEAD AND VALVES

10.3.0.0 Technical Data

Cylinder Head (all dimensions in mm):

Max. Distortion of cylinder head face:
- In longitudinal direction: .... 0.08 mm
- In transverse direction: .... 0.00 mm

Permissible deviation of faces between upper and
Max. deviation of faces between upper and
lower head sealing faces: .... 0.10 mm
Cylinder head height, face to face: .... 126.85 – 127.15 mm

Valves

Valve Head Diameter (4 in Fig. 10.3):
- Inlet valves (112.951/112.976 engine): .... 36.00 mm/38.00 mm
- Exhaust valves (112.951/112.976 engine): .... 41.00 mm/43.00 mm
- Valve seat angle (inlet and exhaust): .... 45° (+15')

Height of Valve Seat Ring:
- Inlet valves: .... 5.90 mm
- Exhaust valves: .... 7.90 mm

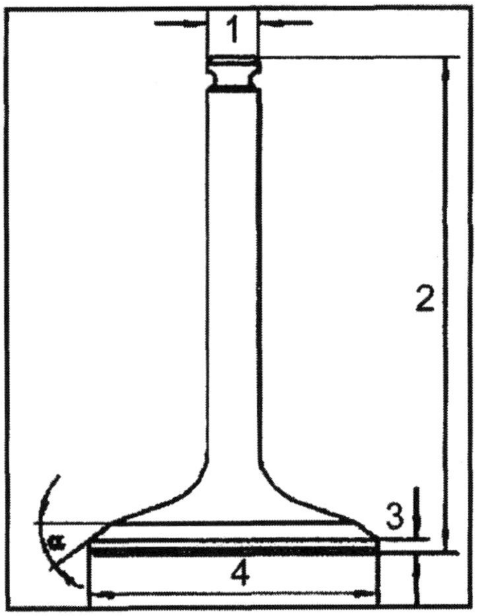

Fig. 10.3 – A valve with the important dimensions, referred to in the technical data.

Valve Stem Diameter (1 in Fig. 10.3):
- Inlet valves: .... 6.975 mm
- Exhaust valves: .... 6.970 mm

Valve identification: .... Quote engine number

Valve seat width:
- Inlet valves and exhaust valves: .... 0.8 – 1.2 mm

Valve Head Height (3 in Fig. 10.3):
- Exhaust valves – New: .... 1.6 mm
- Inlet valves – New : .... 1.2 mm

Valve Length (2 in Fig. 10.3):
- Exhaust valves: .... 122.25 mm
- Inlet valves: .... 119.40 mm

Valve seats

Valve seat width: .... See above

Bore Diameter of Valve Seats in Cylinder Head:
- Exhaust valves (112.951 engine): ...... 42.00 – 42.016 mm (O/S 43.20 – 43.216 mm)
- Exhaust valves (112.976 engine): .... 44.00 – 44.016 mm (O/S not available)
- Inlet valves (112.951 engine): .... 37.00 – 37.016 mm (O/S 38.20 – 38.216 mm)
- Inlet valves (112.976 engine): .... 39.00 – 39.016 mm (O/S not available)

Valve seat angle: .... 45° (+ 15)'
- Upper correction angle: .... 35°
- Lower correction angle: .... 58°

Valve Guides

Inlet Valve Guides:
- Inner diameter : .... 7.00 mm
- Interference fit in guides: .... + 0.12 – 0.031 mm

Bore Diameter in Cylinder Head:
- Standard: .... 12.500 – 12.511 mm
- Oversizes: .... Two available

Camshaft

Camshaft end float: .... 0.05 – 0.15 mm

10.3.0.1. Cylinder Head – Removal and Installation

Fig. 10.4 shows the parts to be removed during the removal of the cylinder head. Some of the operations are described under different headings and must be followed as applicable. Two cylinder heads are fitted and can be removed as applicable.

Certain special tools are used to remove certain parts which will be mentioned during the description. Although the removal and installation may appear difficult at first glance, it can be carried out if you have some experience with engines. Proceed as follows with the removal.

- Disconnect the battery earth cable and drain the cooling system.
- Remove the air cleaner housing and the inlet manifold.
- Detach the exhaust system from the exhaust manifold as described during the removal of the engine. Also see Fig. 10.2.
- Loosen the oil filler pipe for the automatic transmission from the cylinder head cover and the automatic transmission.
- Remove the R.H. cylinder cover (2) and the L.H. cylinder cover (3) as described below under its own heading.
- When removing the R.H. cylinder head remove the Hall sensor for the camshaft (7). Seen from the front you will see the sensor on the L.H. side. Withdraw the connector plug first.

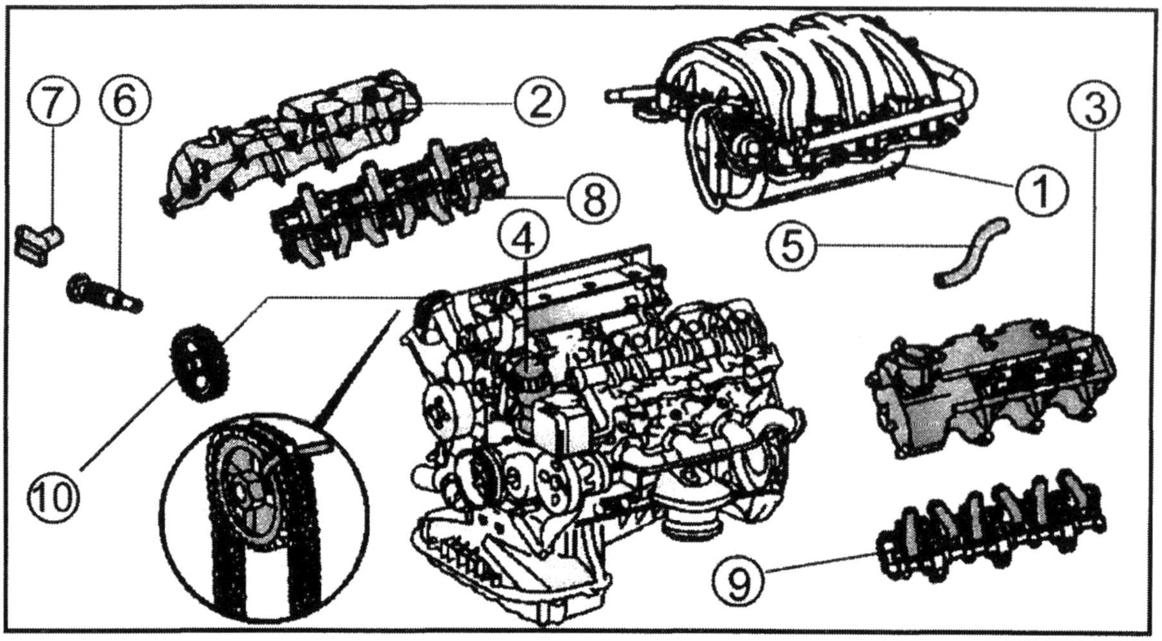

Fig. 10.4 – Details for the removal and installation of the cylinder head (six-cyl.).

1 Inlet manifold
2 R.H. cylinder head cover
3 L.H. cylinder head cover
4 Oil filter housing
5 Coolant hose
6 Chain tensioner
7 Hall sensor for camshaft
8 Camshaft bearing bridge
9 Camshaft bearing bridge
10 Camshaft sprocket

- Remove the oil filter housing (section "Lubrication System") – Refer to page 58.
- Remove the viscous clutch as described in section "Cooling System".
- Remove the Poly V-belt as described in section "Cooling System".
- Remove the coolant hose (5) from the L.H. cylinder head on the bulkhead.
- Remove the two camshafts as described later on in the section in question.

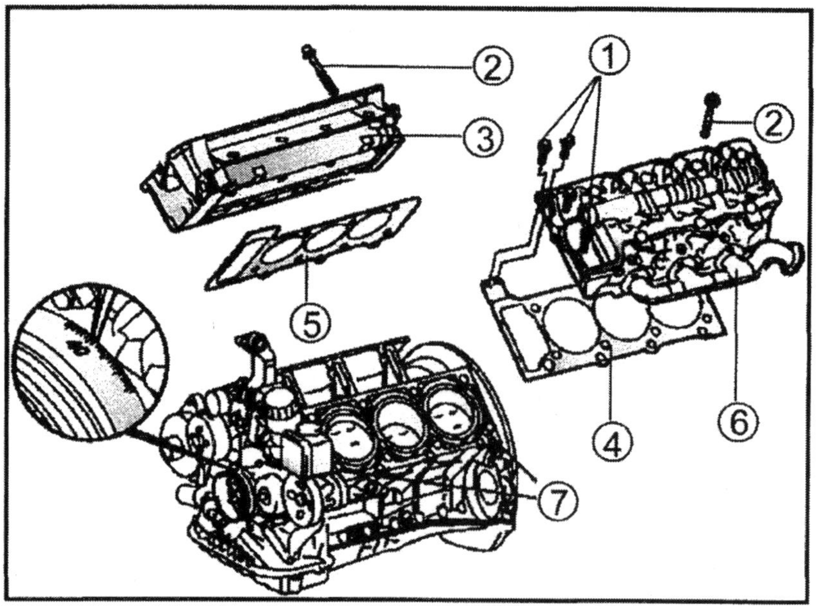

Fig. 10.5 – Details for the removal of the cylinder head. The TDC position of the vibration damper is shown in the circle. The numbers are referred to below.

- Locate the fuel hose connections and unscrew the hose union nuts. The hexagons behind the union nuts must be held with an open-ended spanner.
- Remove the bolts (1) in Fig. 10.5 securing the cylinder head to the timing case cover.
- Slacken the cylinder head bolts (2) in the reverse order shown in Fig. 10.6 in several stages until all bolts have been removed, noting the difference between the L.H. cylinder head (6) and the R.H. cylinder head (3) can now be lifted off the cylinder block. Again a small hand crane or other suitable hoist should be used. Two suitable lifting brackets can be attached to the cylinder head.

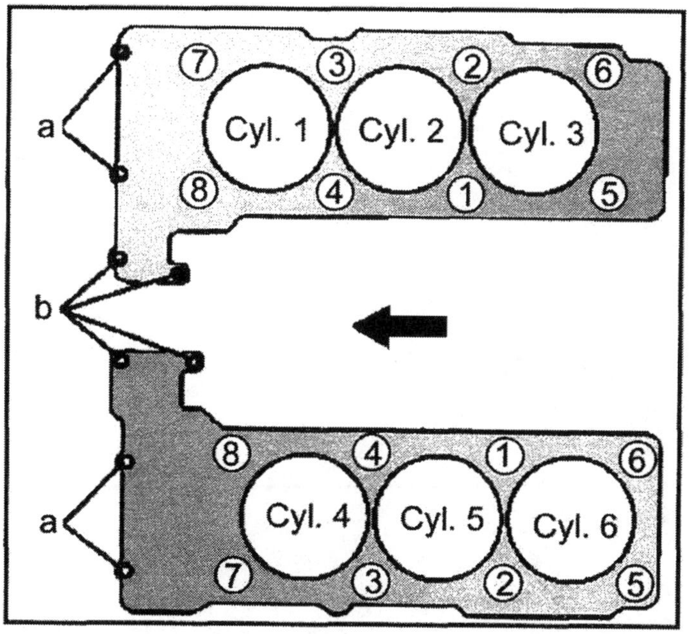

Fig. 10.6 – Tightening order for the cylinder head bolts. Bolts (a) and (b) secure the heads to the timing case.

- Take off the cylinder head gasket (4) or (5) and immediately clean the cylinder head and cylinder block sealing faces. If the guide pins (7) have come out together with the cylinder head transfer them back into the cylinder block. When ordering the new cylinder head gasket specify the vehicle type and model as the engine of type 112 is built in various versions. Gaskets are, for example, different in the diameter of the cylinder bores.

The installation of the cylinder head, although basically a reversal of the removal procedure, is carried out as follows:

- The cylinder head face can be re-ground if necessary, but you should consult an engine shop dealing with this type of V6 engine. **Note:** Re-grinding of the cylinder head face involves re-cutting the valve seats.
- Check that the crankshaft is in the correct position, with the "Zero" in the vibration damper in line with the edge of the adjusting pointer, as shown in the circle in Fig. 10.4. The piston of No. 1 cylinder is now at top dead centre in its compression stroke.
- Place the cylinder head gasket and the cylinder head in position, engaging the dowel pins. Use a plastic or rubber mallet to tap the head well down until seated all round.

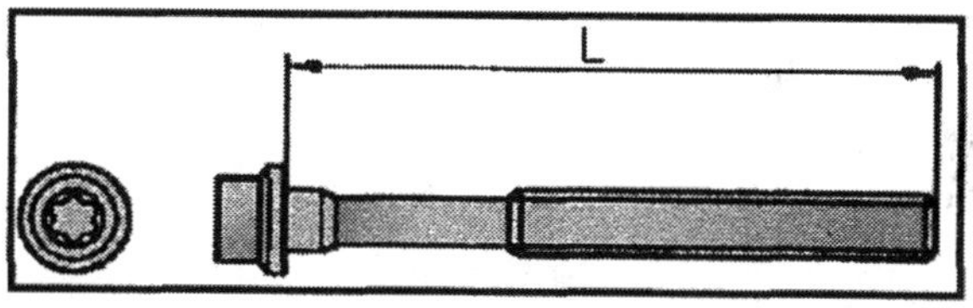

Fig. 10.7 – The cylinder head bolt length (L) is measured as shown.

- Measure the length of each cylinder head bolt (thread diameter M11). Measure from the underside of the bolt head to the end of the bolt thread as shown in Fig. 10.7. All bolts longer than 144.5 mm must be replaced. New bolts have a length of 141.5 mm, i.e., from the dimension(s) obtained you will be able to tell the stretch of each bolt.
- Coat the underside of the cylinder head bolts and the threads with engine oil and insert them one after the other into their bores and tighten them finger-tight.
- Tighten the cylinder head bolts in the order shown in Fig. 10.6 to 1.0 kgm (7 ft.lb.), starting at bolt "1" (on both cylinder heads on the inside). A torx-head insert and a torque wrench must be used. After all bolts have been tightened, again using the torque wrench, tighten all bolts in the order shown to 3.0 kgm (22 ft.lb.). Then take off the torque spanner.
- Tighten each bolt in the order shown in Fig. 10.6 by a quarter of a turn (90°), using the tommy bar only. The tommy bar can, for example be arranged in longitudinal direction of the head and the bolt is tightened until the bar is across the head. After all bolts have been angle-tightened in the manner described, repeat the operation again in the order shown. Tighten the bolts securing the cylinder head to the timing case cover (2.5 kgm/18 ft.lb.).

- The remaining operations are carried out in reverse order. Check the oil level in the sump, irrespective if operations have been carried out on the engine. If the replacement of the cylinder head gasket was necessary because of a "blown" gasket, change the engine oil, as coolant will have entered the oil sump.

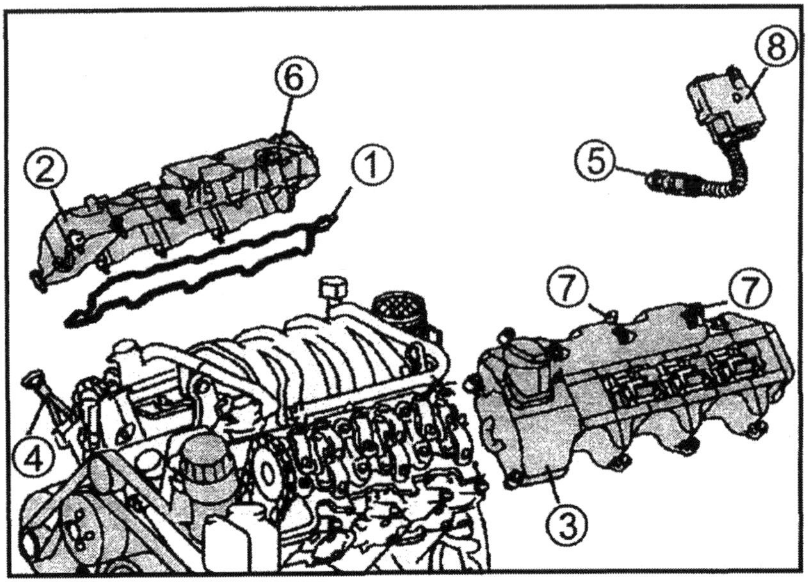

Fig. 10.8 – Details for the removal and installation of the cylinder head covers. The numbers are referred to in the text.

Final Note: An additional re-tightening of the cylinder head bolts is no longer necessary, as the bolts have their final torque setting due to the angle-tightening.

Cylinder Head Covers – Removal and Installation

In the section dealing with the ignition system you will find certain precautions which must always be observed when the cylinder head cover(s) are removed. Removal takes place as follows. Fig. 10.8 will give you some details:

- Remove the air cleaner housing.
- Disconnect the electrical connector plug on the ignition coils of cylinder 1 to 6. Mark the cable connections if not sure.
- Disconnect the spark plug connectors (5) in Fig. 10.8 from the spark plugs. The workshop uses a pair of pliers for this operation.
- Remove the ignition coils (8). Three bolts secure the coils. Tighten them to 0.8 kgm (6 ft.lb.) during installation.
- Remove the refrigerant line on the section side of the engine compartment (vehicle with regulated air conditioning "Tempmatik" or automatic air conditioning).
- If an automatic transmission is fitted remove the oil filler pipe. The pipe is not removed completely. Only unscrew the bolts and rotate the filler pipe to the side.
- Remove the engine venting hose on the fitting (6) for the R.H. cylinder head cover. In the same manner disconnect the venting hose from the fitting (7) on the L.H. cylinder head cover.
- Remove the bolt securing the cylinder head cover on the oil dipstick tube (4). Rotate the dipstick tube to the side to free it of the cover.
- Remove the bolts securing the R.H. cylinder cover (2) and the L.H. cylinder cover (3). The bolts are tightened to 0.8kgm (6 ft.lb.).
- The two covers can now be lifted off.

The installation is a reversal of the removal procedure. Clean the sealing faces. The gasket (1) should always be replaced. Make sure that the gasket is correctly located in the sealing groove of the cylinder head cover, especially at the rear. Follow the tightening torques given above. Make sure that the ignition cables are connected to the correct cylinders.

10.3.0.2. Cylinder Head - Dismantling

The following description assumes that the cylinder head is to be overhauled in an engine shop. If only a top overhaul is asked for, you can remove the valves as described below and can then decide if a complete overhaul of the head is required. The cylinder head must be removed.

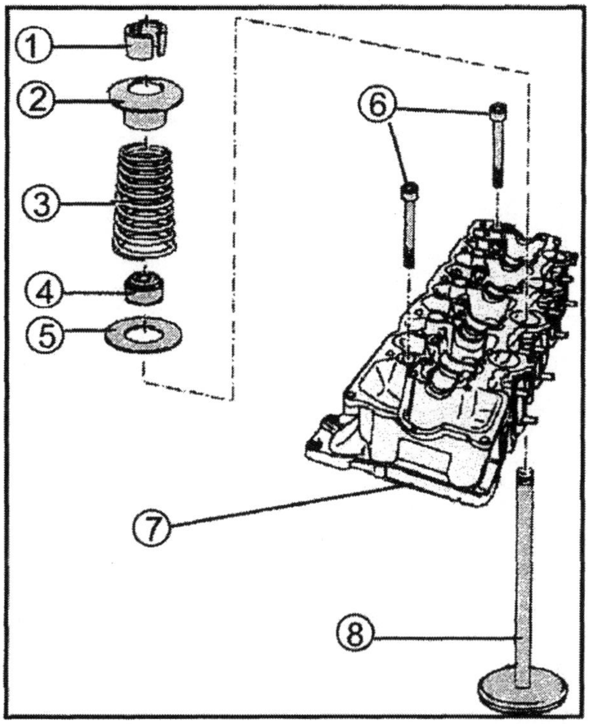

Fig. 10.9 – Valves removed from the cylinder head.

1 Valve cotter halves
2 Upper valve spring cup
3 Valve spring
4 Valve stem seal
5 Lower valve spring retainer
6 Cylinder head bolts
7 Cylinder head
8 Valve

The valve stem oil seals can be replaced with the cylinder head fitted. Signs of worn oil seals are blue exhaust smoke when the vehicle is coasting (gear engaged), when the engine is accelerated after idling for a while or blue smoke when starting the cold engine. If the oil consumption has reached 1 litre per 600 miles, replace the valve stem oil seals as described under a separate heading. The valve stem oil seals can also be replaced with the engine fitted to the vehicle, but various special tools are required. It is therefore a job for the Mercedes dealer.

The valves can be removed as follows. After removal you will have the parts shown in Fig. 10.9.

- Remove all auxiliary parts from the cylinder head, including the exhaust manifold.
- Remove the two camshafts as described under a separate heading.
- A valve spring compressor is required to remove the valves. Fig. 10.10 shows such a compressor. Different types are available, and it must be possible to compress the springs. Valves are held in position by means of valve cotter halves. Compress the springs and remove the valve cotter halves (1) with a pair of pointed pliers or a small magnet (1).

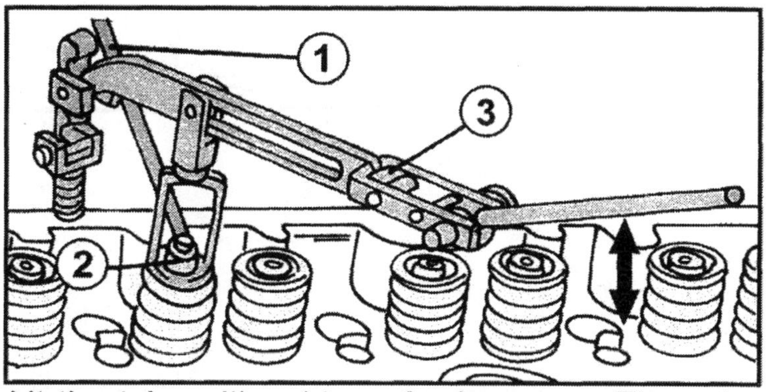

Fig. 10.10 – Removal of valves with a valve lifter (3). The valve cotter halves (2) can be removed with a small magnet (1).

- If a valve spring compressor is not available, it is possible to use a short piece of tube to remove the valve cotter halves. To do this, place the tube over the upper valve spring collar and hit the tube with a blow of a hammer. The valve cotter halves will collect in the inside of the tube and the components can be removed. The valve head must be supported from the other side of the cylinder head. Keep the hammer in close contact with the tube to prevent the cotter halves from flying out.
- Remove the valve spring collar (2) and the valve spring (3) in Fig. 10.9. The valve springs (one spring per valve) are identified with a paint spot and only a spring with a paint spot of the same colour must be fitted.

- Remove valve stem oil seals (1) in the L.H. view of Fig. 10.11 carefully with a screwdriver or a pair of pliers (2).
- Remove the valves one after the other out of the valve guides and pierce them in their fitted order through a piece of cardboard. Write the cylinder number against each valve if they are to be re-used.

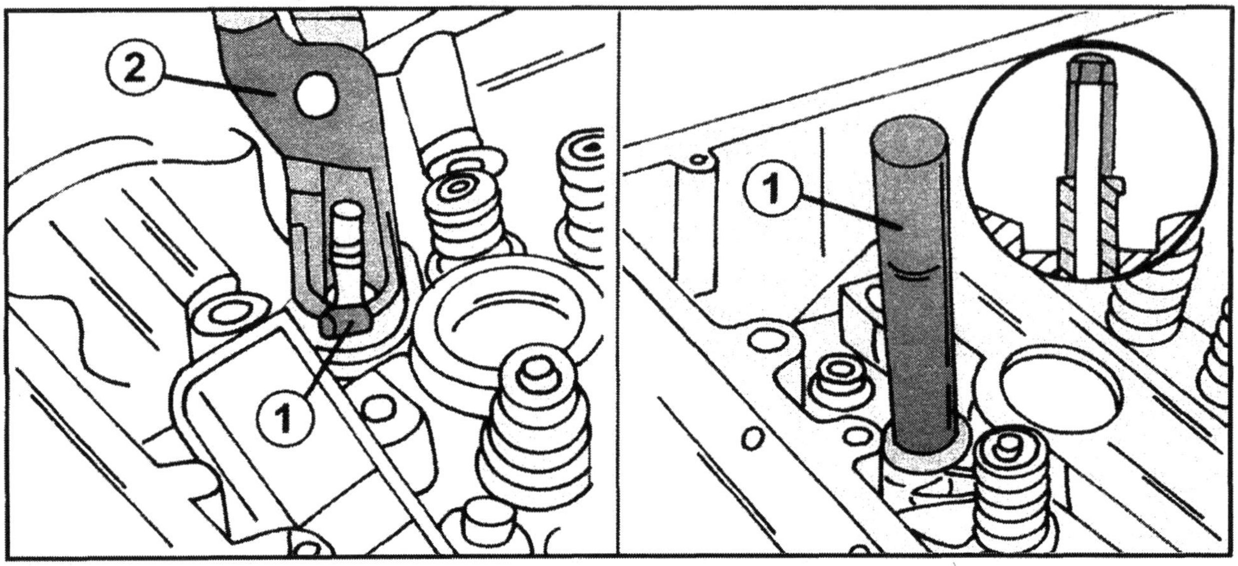

Fig. 10.11 – Removal of valve stem oil seals on the left. Use a pair of pliers (2) to remove the seal (1). The R.H. view shows the installation of a seal with the piece of tube (1).

The camshaft sprockets are marked with timing marks, which indicate the top dead centre position of piston No. 1 when they are aligned. Also check the alignment of the balance shaft, shown in Fig. 10.12. Camshafts are marked with a number. If a shaft is replaced, only fit a shaft with the same number. The parts lists will have the correct camshaft(s) for the engine in question.

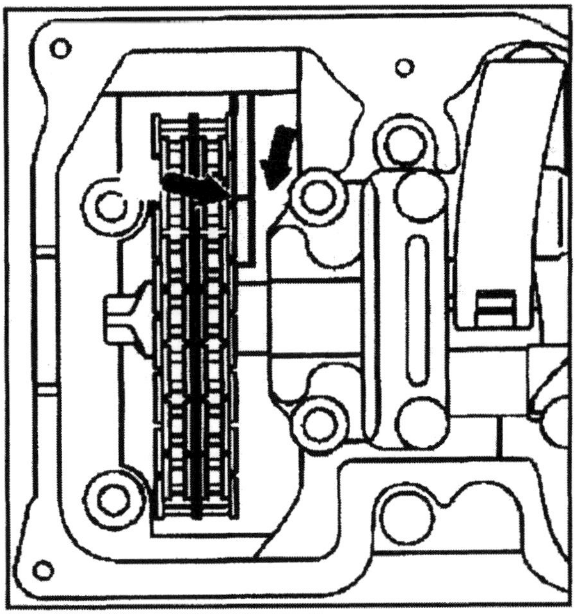

Fig. 10.12 – Alignment of the balance shaft must be as shown.

The following text assumes that the cylinder head is to be replaced. If for example only the valves require attention, ignore the additional operations. The cylinder head must, of course, be removed.

- Remove the thermo switch, sender units, etc. from the cylinder head and unscrew the exhaust manifold. Also unscrew the spark plugs.
- A valve spring compressor is required to remove the valves. Valves are held in position by means of valve cotter halves. Compress the springs and remove the valve cotter halves with a pair of pointed pliers or a small magnet.

If a valve spring compressor is not available, it is possible to use a short piece of tube to remove the valve cotter halves. To do this, place the tube over the upper valve spring collar and hit the tube with a blow of a hammer. The valve cotter halves will collect in the inside of the tube and the components can be removed. The valve head must be supported from the other side of the cylinder head. Keep the hammer in close contact with the tube to prevent the cotter halves from flying out.

Remove the valve spring collar and the valve spring. The valve springs (one spring per valve) are identified with a paint spot and only a spring with a paint spot of the same

colour must be fitted. Remove the valves one after the other and keep them in their numbered order, writing the number in front of each valve.
The following notes refer to the camshafts. The camshafts have an identification number. When a shaft is replaced, only fit a shaft with an identical identification number.
The two camshafts must only be removed when they are set to their basic position. The shafts are sensitive to handling. Never remove the shaft if it is under tension from the valve springs.

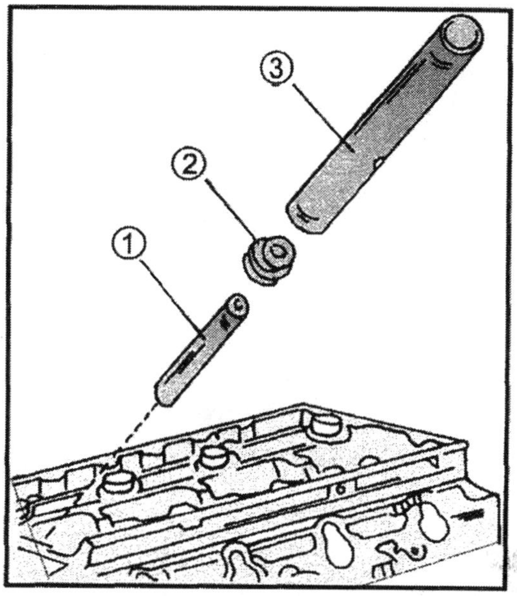

Fig. 10.13 – Fitting the valve guide stem seals. After placing the guide sleeve (1) in position, slide the seal (2) over the sleeve and use a piece of tube (3) to push against the seal. Avoid damage to the seal.

10.3.0.3. Cylinder Head - Overhaul

The cylinder head must be thoroughly cleaned and remains of old gasket material removed. The checks and inspections are to be carried out as required.

Valve Springs: If the engine has a high mileage, always replace the valve springs as a set. To check a valve spring, place the old spring and a new spring end to end over a long bolt (with washer under bolt head) and fit a nut (again with a washer). Tighten the nut until the springs are under tension and measure the length of the two springs. If the old spring is shorter by more than 10%, replace the complete spring set.
The springs must not be distorted. A spring placed with its flat coil on a surface must not deviate at the top by more than 2 mm (0.08 in.).

Valve Guides: Valve guides for inlet and exhaust valves are made of cast iron and the same diameter, irrespective of size. The outer diameter is specified with 12.5 mm. Both guide are fitted with valve stem oil seals, which can be removed with a pair of pliers as shown in Fig. 10.11. The workshop uses a special guide sleeve to fit the valve stem seals. The operation is shown in Fig. 10.13. After placing the sleeve (1) over the valve guide, fit the well-oiled seal (2) and push it in position, using a suitable piece of tube (3). Do not damage the seal.

Clean the inside of the guides by pulling a petrol-soaked cloth through the guides. Valve stems can be cleaned best by means of a rotating wire brush. Measure the inside diameter of the guides. As an inside micrometer is necessary for this operation, which is not always available, you can insert the valve into its guide and withdraw it until the valve head is approx. level with the cylinder head face. Rock the valve to and fro and check for play. Although no exact values are available, it can be assumed that the play should not exceed 1.0 - 1.2 mm (0.04 - 0.047 in.). Mercedes workshops use gauges to check the guides for wear.

Guides are removed with a shouldered mandrel from the combustion chamber side of the cylinder head. If guides with standard dimension can be used, drive them in position, until the retaining ring is resting against the cylinder head. If repair guides are fitted, the locating bores in the cylinder head must be reamed out to take the new guides. As dry ice is required to fit the new valve guides, we recommend to have the work carried out in a workshop.

Before a valve guide is replaced, check the general condition of the cylinder head. The guides must be reamed after installation.
Valves must always be replaced if new valve guides are fitted. The valve seats must be re-cut when a guide has been replaced. If it is obvious that seats cannot be re-ground in the present condition, new valve seat inserts must be fitted.
Again this is an operation for a specialist and the work should be carried out in a workshop.

Valve Seats: If the camshaft bearings are excessively worn, fit a new or exchange cylinder head. In this case there is no need to renovate the valve seats.

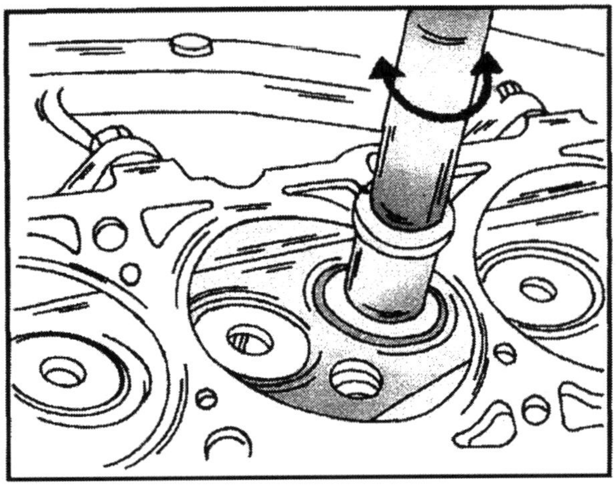

Fig. 10.14 – Grinding-in of valves.

Check all valve seats for signs of pitting or wear. Slight indentations can be removed with a 45° cutter. If this operation is carried out properly, there should be no need to grind-in the valves. Use correction cutters to bring the valve seating area into the centre of the valve seat. Make sure that the valve seat width, given in Section 1.3.0.0. is obtained. This again is achieved by using cutters of different angles (for example 15° and 60°). Valve seat inserts can be fitted to the cylinder head. Replacement of valve seat inserts will require that the old seat insert is removed by machining. The machining must not damage the bottom face of the head recess. As this is a critical operation, we advise you to bring the cylinder head to your Mercedes Dealer who has the necessary equipment and experience to do the job. It may be possible to obtain a reconditioned cylinder head in exchange for the old one to avoid time delay. In this case remove all ancillary parts from the old head and refit them to the new head.
Valves can be ground into their seats in the conventional manner. To do this, coat the valve seat with lapping compound and use a suction tool, as shown in Fig. 10.14. Move the valve backwards and forwards. Every so often, lift the suction tool, move it forward by ¼ of a turn, and continue grinding. Work the seat until an uninterrupted ring is visible around the face of the valve. After grinding-in, clean the cylinder head, and even more important the inside of the valve guide bores thoroughly. Any lapping paste inside the cylinder head will accelerate the wear of the new parts.
Use a pencil and mark across the valve seat closely spaced. Drop the valve into the respective valve guide and turn the valve by 90°, using the suction tool, applying slight pressure to the tool. Remove the valve and if the pencil marks have been removed from the entire circumference. The gap created will indicate the width of the valve seat and can be measured with a ruler or caliper. Otherwise repeat the grinding until this is the case.

Valves: The exhaust valves must not be discarded as ordinary scrap metal. They are filled with sodium, as is the case with other Mercedes engines. Never use such a valve as a drift – it may be tempting. Valves can be cleaned best with a rotating wire brush. Check the valve faces for wear or grooving. If the wear is only slight, valves can be re-ground to their original angle in a valve grinding machine, but make sure that there is enough material left to have an edge on the valve head. The valve head thickness must be at least 0.8 mm. Measure dimension "4" in Fig. 10.15 to check each valve, if in doubt.

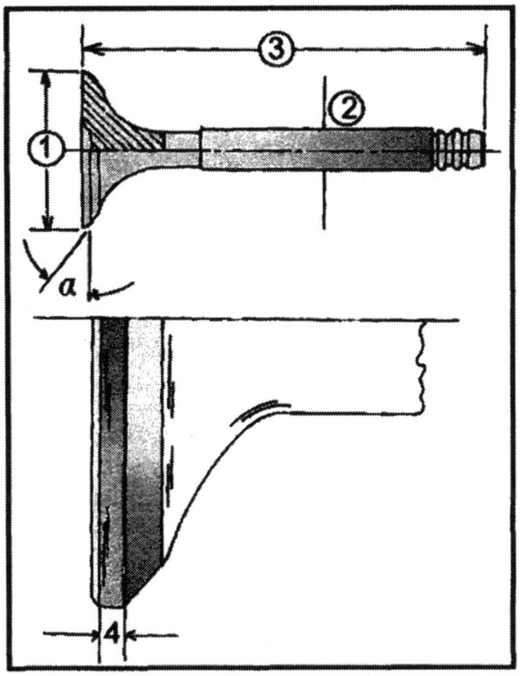

Fig. 10.15 – The principal dimensions of the valves. The values are given in Section 1.4.0.0. The dimension (4) must not be less than 0.8 mm.

1 Valve head diameter
2 Valve stem diameter
3 Valve length
4 Valve head thickness

Measure the valves in accordance with the values given in Section 1.3.0.0. and Fig. 10.15 and check with the data in Section 1.3.0.0. Replace any valve which is outside the limits given.

Check the valve stem diameters and in this connection the inside diameters of the valve guides. If there is a deviation from the nominal values, it may be necessary to replace the valve guides (see above). Also check the end of the valve stems. There should be no visible wear in this area.

Always quote the model year and the engine number when ordering new valves, as different valves are used in the "112" engines. These are marked with "E" (inlet) or "A" (exhaust).

Sometimes it is only required to replace the exhaust valves, if these for example are burnt out at their valve head edges.

Cylinder Head: Thoroughly clean the cylinder head and cylinder block surfaces of old gasket material and check the faces for distortion. To do this, place a steel ruler with a sharp edge over the cylinder head face and measure the gap between ruler and face with feeler gauges. Checks must be carried out in longitudinal and diagonal direction and across the face. If a feeler gauge of more than 0.10 mm (0.004 in.) can be inserted, when the ruler is placed along or across the cylinder head, have the cylinder head face re-ground.

Camshaft: See separate heading.

10.3.0.4. Cylinder Head - Assembly

The assembly of the cylinder head is a reversal of the dismantling procedure. Note the following points:

- Lubricate the valve stems with engine oil and insert the valves into the correct valve guides.
- Valve stem seals must be suitable for the engine in question. Make sure to order the correct seals. The repair kit contains fitting sleeves and these must be used to fit the seals (see last Section).
- The sleeves are fitted over the valve stem before the seal is pushed in position.
- Fit the valve spring and valve spring collar over the valve and use the valve lifter to compress the spring. Insert the valve cotter halves and release the valve spring lifter. Make sure that the cotter halves are in position by tapping the end of the valve stem with a plastic mallet. Place a rag over the valve end - just in case.
- Fit the camshafts as described later on and carry out all other operations in reverse order to the dismantling procedure.

10.3.0.5. Hydraulic Valve Clearance Compensation

The function of the hydraulic valve clearance compensating elements is to eliminate valve clearance, i.e. the dimensional changes in the valve train (valve lash) due to heat expansion and wear are compensated by the elements. The rocker arm is in constant contact with the cam. The compensating elements cannot be repaired, but can be checked for correct functioning as described below. Fig. 10.16 shows sectional views of a valve with clearance compensation. We will give a short description of the operation. All references refer to Fig. 10.16.

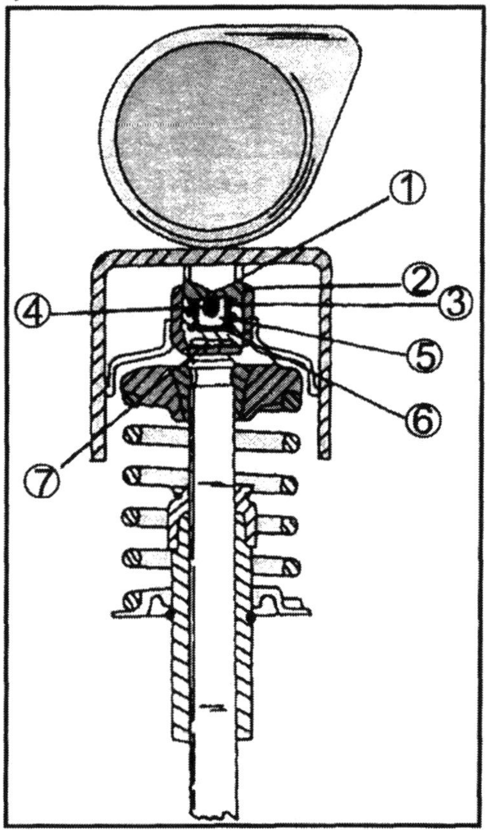

The hydraulic valve compensating element are fitted into the rocker levers and operate the valves directly via a ball socket:

Fig. 10.16 – Sectional view of a valve and a compensating element.

1 Thrust pin
2 Lock Ring
3 Thrust spring
4 Ball guide
5 Ball
6 Thrust spring
7 Guide bush

- The thrust pin with oil supply chamber and the return bores and the ball valve (check valve). The ball valve separates the supply chamber from the work chamber.
- The guide sleeve with the work chamber, the thrust spring (3) and the closing cap.
- When the engine is stopped and the tappet is held under load from the cam, the element can completely retract. The oil displaced from the work chamber flows through an annular gap, i.e. the clearance between the guide sleeve and the thrust pin to the oil supply chamber.
- When the cam lobe has moved past the valve tappet, the thrust pin will be without load. The thrust spring (3) forces the thrust pin upwards until the valve tappet rests against the cam.

The vacuum resulting from the upward movement of the thrust pin in the work chamber (c) opens the ball valve and the oil can flow from the supply chamber into the work chamber. The ball valve closes when the valve tappet presses against the cam and puts the thrust pin under load. The oil in the work chamber acts as a "hydraulic rigid connection" and opens the valve in question.

When the engine is running and depending on the engine speed and the cam position, the thrust pin is only pushed down slightly.

The oil contained in the oil supply chamber is sufficient to fill the work chamber under all operating conditions of the engine. Oil or leak oil which is not required, as well as air are able to escape via the annular gap between the washer and the rocker lever. The oil ejected from the work chamber flows via the annular gap between the guide sleeve and the thrust pin and the two return bores into the oil supply chamber.

If the tappets are removed, note the following points:

- Always keep the tappets in an upright position, i.e. the open side towards the top.
- After removal of a tappet (see below), mark the cylinder number and the compensating element in suitable manner. Always fit original parts in their same locations.

Checking a hydraulic compensating element: As the elements are in continuous contact with the camshaft, you will rarely hear noises from the area of the hydraulic elements. If noises can be heard, check the elements as follows:

- Start the engine and run it approx. 5 minutes at 3000 rpm. The oil level in the sump must be correct.
- Remove the cylinder head cover.

Fig. 10.17 – Checking a hydraulic valve clearance compensating element.

- Rotate the crankshaft until the cam for the tappet to be checked is pointing vertically towards the top.
- Use a drift, as shown in Fig. 10.17 and push the tappet towards the inside, or try to move the tappet with the fingers.
- If the tappet cannot be depressed or excessive clearance can be felt between the tappet and the back of the cam, replace the tappet. The tappet is supplied together with the hydraulic compensating element. Mercedes workshops can reset the tappet to its original position, but this operation is beyond the scope of the home mechanic.

Tappet Removal and Installation:

- Remove the cylinder head covers as previously described.
- Rotate the engine until the engine is at top dead centre, remove the chain tensioner and remove the camshaft bearing bridge from the cylinder head in question. The operations are described later on in connection with the timing mechanism.
- Use a drift of 1 mm in diameter and press the compensating element (2) in Fig. 10.18 on the cylinder in question out of the rocker arm (3). Mark them, if they are to be refitted.

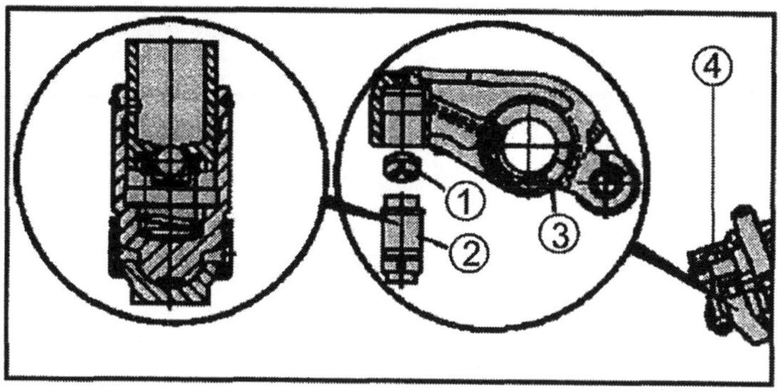

Fig. 10.18 – The hydraulic valve clearance element and associated parts. No. 4 is one of the camshaft bearing bridges.

- Fit the tappets into their original bores, if re-used. The disc (1) in the roller-type rocker arm (3) must point downwards with the slots towards the compensating element (2). Refit the camshaft bearing bridge and the other removed parts in reverse order to the removal procedure.

10.4.1. PISTON AND CONNECTING RODS

10.4.1.0. Technical Data

All dimensions are given in metric units.

Pistons
Available pistons: .... Standard and repair size
Piston Running Clearance :
- Standard (new): .... 0.025 – 0.035 mm
- Wear limit: .... 0.08 mm
Max. weight difference of pistons in
one engine: .... 4 grams (wear limit 10 grams)

Running clearance of piston pins :
- In connecting rod small end: .... 0.007 – 0.017 mm
- In Piston: .... 0.002 – 0.011 mm

Piston Ring Gaps :
- Groove 1: .... 0.20 – 0.35 mm (wear limit 0.80 mm)
- Groove 2: .... 0.20 – 0.40 mm (wear limit 0.80 mm)

Side Clearance of Piston Rings in Piston :
- Groove 1 – 112.951 engine: .... 0.030 – 0.075 mm (wear limit 0.10 mm)
- Groove 1 – 112.976 engine: .... 0.033 – 0.070 mm (wear limit 0.10 mm)
- Groove 2 – 112.951 engine: .... 0.015 – 0.05 mm (wear limit 0.10 mm)
- Groove 2 – 112.976 engine: .... 0.010 – 0.04 mm (wear limit 0.10 mm)

Note: The piston ring in groove 3 must not be removed.

Connecting Rods
Width of con rod at big end bore: .... 19.948 – 20.000 mm
Basic bore diameter of big end bore: .... 55.600 – 55.614 mm

Small End Bush:
- Inner diameter: .... 22.07 – 22.013 mm

Max. twist of connecting rods: .... 0.10 mm per 100 mm
Max. bend of connecting rods: .... 0.015 mm per 100 mm
Max. weight difference in same engine: .... 14 gram (per set)

Connecting rod bearing details: .... See under "Crankshaft"

10.4.1.1. Piston and Connecting Rods – Removal

Pistons and connecting rods are pushed out towards the top of the cylinder bores, using a hammer handle after connecting rod bearing caps and shells have been removed. The engine must be removed to remove the connecting rod piston assemblies. Also remove the cylinder head, the oil sump and the oil pump. If a piston, or all pistons require replacement, we suggest to have the work carried out in an engine shop.

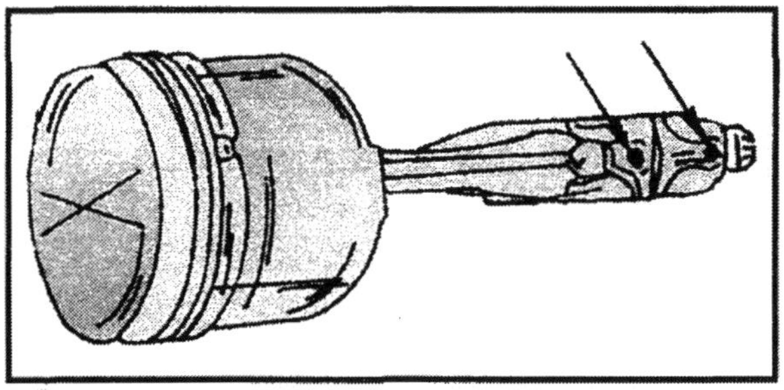

Fig. 10.19 – Big end bearing caps and connecting rods must be marked at opposite points as shown before removal.

Three piston rings are fitted to each piston. But note: The lower ring (oil scraper ring) must not be removed. The two upper rings are the compression rings, i.e. they prevent the pressure above the piston crown to return to the crankcase. The second ring is also acting as oil scraper ring. Its function is to remove excessive oil from the cylinder bore, thereby preventing the entry of oil into the combustion chamber. The three rings are not the same in shape. The upper ring has a taper-faced section, the second ring has a chamfer on the inside. Only the correct fitting of the piston rings will assure a proper piston seal. Before removal of the assemblies note the following points.

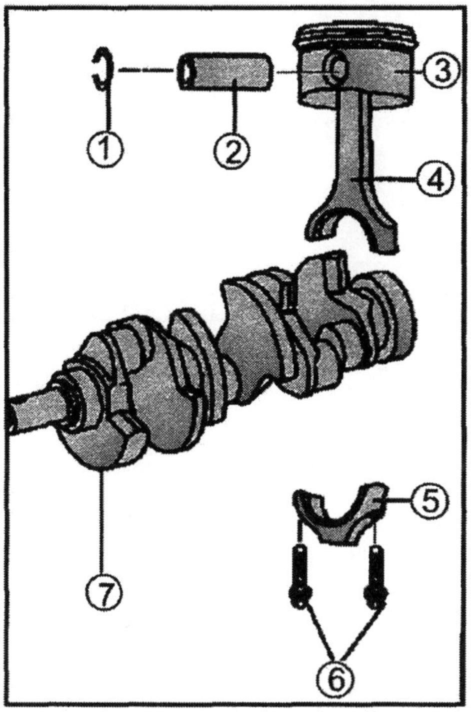

Fig. 10.20 – Component parts of a piston and the connecting rod.
1 Piston pin circlip
2 Piston pin
3 Piston
4 Connecting rod
5 Big end bearing cap
6 Bearing cap bolts

- Piston and cylinder bores have graded diameter classes within specified tolerance groups. If the cylinder block has been re-bored, you will receive the block with the correct pistons.
- Mark each piston and the connecting rod before removal with the cylinder number. This can be carried out by writing the cylinder number with paint onto the piston crown. Also mark an arrow, facing towards the front of the engine (the arrow in the piston crown will be covered by carbon deposits). When removing the connecting rod, note the correct installation of the big end bearing cap. immediately after removal mark the connecting rod and the big end bearing cap on the same side. This is best done with a centre punch (cylinder No. 1 one punch mark, etc., see Fig. 10.19).
- Mark the big end bearing shells with the cylinder number.
- Big end bearing journals can be re-ground to four undersizes (in steps of 0.25 mm between sizes). Corresponding bearing shells are available.
- Remove the bearing caps and the shells and push the assemblies out of the cylinder bore. Any carbon deposits on the upper edge of the bores can be carefully removed with a scraper.
- Remove the piston pin snap rings. A notch in the piston pin bore enables a pointed drift to be inserted, as shown in Fig. 1.28, to remove the rings. Press the piston pins out of the pistons. If necessary heat the piston in boiling water.
- Remove the piston rings one after the other from the pistons, using a piston ring pliers if possible (Fig. 1.29). If the rings are to be re-used, mark them in accordance with their pistons and position.

The installation of a piston and the connecting rod is shown in Fig. 10.20 on the previous page.

10.4.1.2. Measuring the Cylinder Bores

An inside caliper is necessary to measure the diameter of the cylinder bores. The following operations are not possible if none are available or cannot be hired. The instructions given in section 1.3.1.2 for the diesel engine also apply to the petrol engine. Figs. 1.30, 1.31 and 1.32 shows details of the measurement.
To measure the running clearance, determine the piston and cylinder bore diameters as described and calculate the difference between the dimensions.
If the difference is more than 0.12 mm, have the cylinder bores re-bored to fit oversize pistons.

10.4.1.3. Checking Pistons and Connecting Rods

All parts should be thoroughly inspected as described in section 1.3.1.3. for the diesel engine. The data given in section 10.4.0.0. should be referred to. Note the following:

- The piston ring in the bottom groove (oil scraper ring) must not be removed.
- Check the side clearance of each piston ring in its groove as shown in Fig. 1.33.
- Check the piston ring gap by inserting the ring from the bottom into the cylinder bore. Use a piston and carefully push the piston ring approx. 1 in. further into the bore. This will square it up. Insert a feeler gauge between the two piston ring ends to check the ring gap, as shown in Fig. 1.34. Refer to Section 10.4.1.0. for the wear limits. Rings must be replaced, if these are exceeded. New rings should also be checked in the manner described.

The following information concerns the connecting rods:

- Connecting rods which were over-heated due to bearing failure (bluish colour) must not be refitted.
- Connecting and bearing caps are matched to each other and must be fitted accordingly.
- New connecting rods are supplied together with the small end bearing bush and can be fitted as supplied.
- If the piston pin has excessive clearance in the small end bush, a new bush must be fitted. Again this is best left to an engine shop, as the bush must be reamed out to the correct diameter to obtain the correct running clearance for the piston pin.

10.4.1.4. Piston and Connecting Rods - Assembly

If new pistons are fitted, check the piston crown markings to ensure the correct pistons are fitted. If the original pistons are fitted, arrange them in accordance with the cylinder number markings.

- Insert the connecting rod into the piston and align the two bores. Make sure that the arrow in the piston crown is facing the front of the engine.
- Generously lubricate the piston pin with engine oil and insert it into the piston and connecting rod, using thumb pressure only. Never heat the piston to fit the piston pin. Fit the circlips to both sides of the piston, making sure of their engagement around the groove. Move the piston up and down, as indicated in Fig. 10.21 to check for free movement.

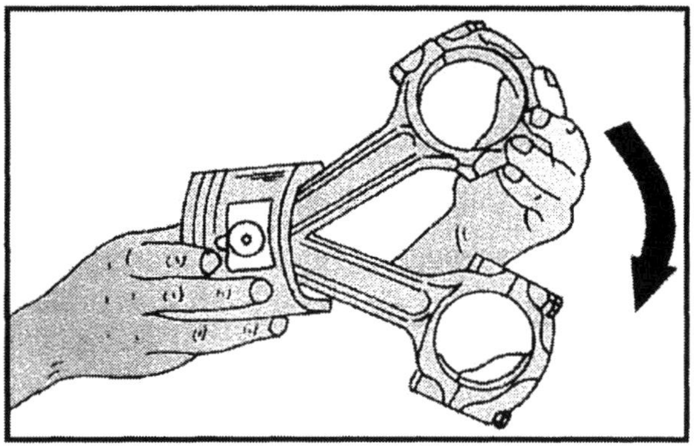

Fig. 10.21 – After assembly move the connecting rod to and fro to check for free movement.

- Using a pair of piston ring pliers (Fig. 1.29), fit the piston rings from the top of the piston, starting with the bottom ring. The two upper rings could be mixed up. Under no circumstances mix-up the upper and lower compression rings.

10.4.1.5. Pistons and Connecting Rods – Installation

- Generously lubricate the cylinder bores with oil. Markings on connecting rods and bearing caps must be opposite each other. The arrows in the piston crowns must face towards the front of the engine.
- Arrange the piston rings at equal spacing of 120° around the circumference of the piston skirt and use a piston ring compressor to push the rings into their grooves.

Check that all rings are fully pushed in. Check that the marking "Top" or the name of the manufacturer can be read from above after installation. Fig. 10.22 shows the section of the piston rings. Make sure that all rings are fitted as shown.

- Rotate the crankshaft until two of the crankpins are at the bottom.

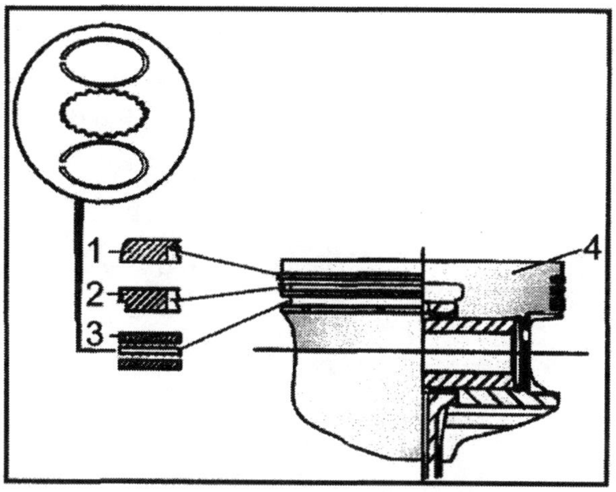

Fig. 10.22 – Sectional view of the pistons and piston rings.
1 Upper compression ring
2 Second compression ring
3 Multi-part oil control ring
4 Piston

- Place a piston ring compressor around the piston rings as shown in Fig. 10.23 and push the piston from above into the cylinder bore as shown in the illustration. The cylinder block can be placed on its side to guide the connecting rod. The big end bearing shell should be in the big end bearing.
- Insert the second bearing shell into the connecting rod bearing cap, with the locating tab and fit the assembly over the connecting rod. Check that connecting rod/cap marks are facing each other.

Fig. 10.23 – Fitting a piston with a piston ring compressor. Push the piston into the bore with a hammer handle.

- Coat the contact areas for the cap bolts with engine oil and fit and tighten the bolts in three stages. First to 0.5 kgm, then to 2.5 kgm and from the final position a further quarter of a turn without using the torque wrench.
- Rotate the crankshaft until the remaining crankpins are at bottom dead centre and fit the two other piston/connecting rod assemblies in the same manner.
- Check the pistons and connecting rods once more for correct installation and that each piston is fitted to its original bore, if the same parts are refitted.
- With a feeler gauge measure the side clearance of each big end bearing cap on the crankpin. The wear limit is 0.50 mm.
- If pistons have been replaced measure the protrusion of the pistons above the cylinder block face. This must be between –0.35 to +0.05 mm. To check the protrusion place the piston to the top dead centre position (must be exact) and measure the protrusion at the points shown by the arrows in the L.H view of Fig. 1.38, i.e. in the direction of the piston pin (same as on diesel engines). You measure the part of the piston above the cylinder block face as shown in the R.H. view of Fig. 1.38. The protrusion will determine the thickness of the cylinder head gasket during assembly, i.e. either a standard thickness gasket or a repair thickness gasket must be used. With the values obtained you will be able to obtain the correct gasket from your parts department.

10.4.2. CYLINDER BLOCK

The cylinder block consists of the crankcase and the actual block with the cylinder bores. Special attention should be given to the cylinder block at each major overhaul of the engine, irrespective of whether the bores have been re-machined or not.
Thoroughly clean all cavities and passages and remove all traces of foreign matter from the joint faces. If any machining of the bores has taken place, it is essential that all swarf is removed before assembly of the engine takes place.
Measurement of the cylinder bores should be left to an engine shop, as they have the proper equipment to measure cylinder bores. It is, however, feasible to check the cylinder block face for distortion, in a similar manner as described for the cylinder head. The max. permissible distortion is 0.05 mm.

10.4.3. CRANKSHAFT AND BEARINGS

10.4.3.0. Technical Data

All dimensions in metric units.
Machining tolerances:

Max. out-of-round of journals:	0.005 mm
Max. taper of main journals:	0.010 mm
Max. taper of crankpins:	0.015 mm
Max. run-out of main journals*:	0.07 mm

* Crankshaft placed with Nos. I and IV journals "V" blocks.

Main Bearing Journal Diameter:
- Nominal: ....63.940 – 63.945 mm
- Repair sizes:..Four sizes available

Basic Bearing Bores:
- For main bearings: ....68.500 – 68.506 mm (standard)
- Con rod big end bores: ....55.600 – 55.614 mm (standard)

Crankpin journal diameter............................Depending on size grading (workshop)

Bearing Running Clearances:
- Main bearings:..0.026 - 0.050 mm
- Big end bearings: ....0.022 - 0.067 mm
- Wear limit:..0.080 mm

Bearing End Float:
- Main bearings: ....0.10 - 0.266 mm
- Big end bearings: .... 0.11 - 0.230 mm
- Wear limit - Main bearings: ....0.30 mm
- Wear limit - Big end bearings:...0.50 mm

Bearing shells: Depending on journal diameter (workshop)

10.4.3.1. Crankshaft - Removal and Installation

Selection of bearing shells is rather complicated, and we advise you to take the cylinder block to an engine shop, if there is a possibility that new bearing shells are necessary.
The engine must be removed to take out the crankshaft. Fig. 10.24 shows the component parts of the crankshaft together with the main bearing shells and caps and the location of the thrust washers.

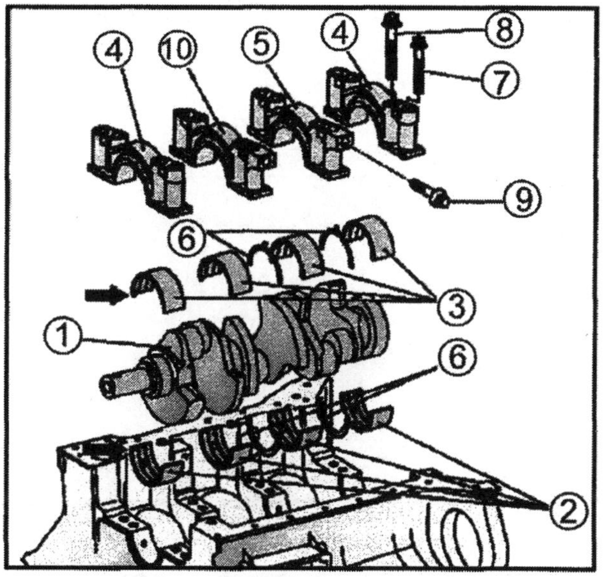

Fig. 10.24 – Crankshaft and bearing parts.

1 Crankshaft
2 Bearing shells in crankcase
3 Bearing shells in main bearing caps
4 Crankshaft bearing cap (end caps)
5 Fitted bearing cap
6 Thrust washers
7 Bearing cap bolt, M8 x 75 mm
8 Bearing cap bolt, M10 x 90 mm
9 Bearing cap side bolt, M8 x 40 mm
10 Crankshaft bearing cap

- Remove the transmission from the engine. Take care not to distort the clutch shaft.
- Counterhold the flywheel in a suitable manner and evenly slacken the clutch securing bolts. Use a centre punch and mark the clutch and flywheel at opposite points. Lift off the clutch plate and the driven plate. Immediately clean the inside of the flywheel and unscrew the flywheel.
- Remove the drive plate for a torque converter of an automatic transmission in the same manner.
- In the following order remove the end cover, the oil sump, the oil pump and the timing case cover. The operations are described under separate headings.
- Remove the cylinder head together with the inlet manifold as described.
- Remove the pistons and connecting rods as described in Section 10.4.1.1.
- The crankshaft end float should be checked before the crankshaft is removed. Refer to page 42 and Fig. 1.42 as described for the diesel engine. If it exceeds 0.30 mm (0.012 in.) replace the thrust washers during assembly, but make sure to fit washers of the correct width. These are located left and right at the centre bearing. Note that only two washers of the same thickness must be fitted.

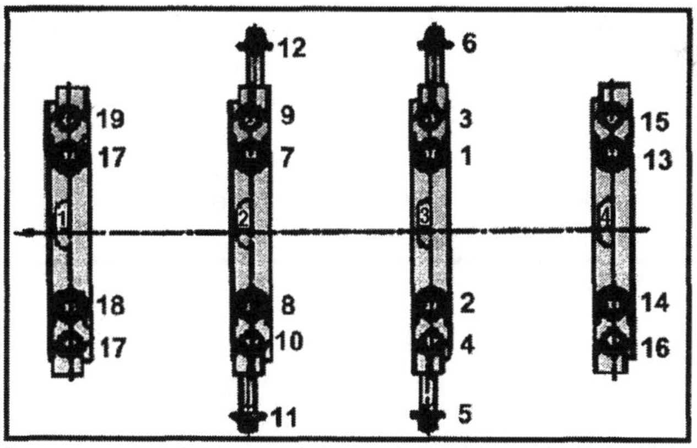

Fig. 10.25 – Tightening order for the crankshaft bearing cap bolts. Note the different diameters and bolt length.

- Unscrew the main bearing bolts in reverse order to the numbering in Fig. 10.25. Note that bolts of different diameter and length are used. Each bearing cap is secured with two bolts on either side from above and the side of each cap is additionally secured by a bolt inserted from the side. Mark each bearing cap and its position on the crankcase.
- Remove the crankshaft bearing caps (4), the bearing cap (10) and the fitted bearing cap (5). Cap (10) and (5) have a tight fit and must be levered out carefully. Remove the bearing shells (2) from the bearing journals (they could also stick to the caps) and immediately mark them on their back faces with the bearing number. The numbers refer to Fig. 10.24.
- Lift the crankshaft (1) out of the cylinder block and remove the remaining thrust washers (6) from the centre bearing location and the remaining bearings shells.
- Keep the shells together with the lower shells and the bearing caps. These shells have an oil bore and a groove and must always be fitted into the crankcase when the crankshaft is installed.

10.4.3.2. Inspection of Crankshaft and Bearings

In general the operations are similar as described for the diesel engine, commencing on page 43. If obvious damage or wear has been detected we recommend to take the crankshaft and the cylinder block to an engine shop, familiar with the new V6 engines.

10.4.3.3. Crankshaft - Installation

Thoroughly clean the bearing bores in the crankcase and lubricate the crankshaft bearing cap (4), bearing cap (10), fitted bearing (5) and the bearing shells in the crankcase (2) and the bearing cap (3) with engine oil. Insert the shells with the drillings into the bearing bores, with the tabs engaging the notches. Fit the thrust washers (6) to the bearing shown, with the oil grooves towards the outside. The numbers refer to Fig. 10.24.

Use the two forefingers as shown in Fig. 1.47 to hold the thrust washers against the bearing cap and fit the cap in position.

Lift the crankshaft in position and fit the bearing caps with the inserted shells (again shells well oiled and locating tabs in notches). Fit the two thrust washers to the centre bearing cap, again with the oil groove towards the outside. Place this cap in position, guiding the two thrust washers in order not to dislodge them. Use the forefingers to hold the washers (Fig. 1.47).

Check the numbering of the bearing caps and fit them in position, with the shells inserted. Use a plastic mallet to tap them down. Caps can only be fitted in one position, but the bearing cap numbers must be observed.

Tighten the bolts in the order shown in Fig. 10.25, but note the diameter and the length of the bolts:

- The M8 bolts (7) are tightened to 2.0 kgm (14.5 ft.lb.) and from the final position a further quarter of a turn (90°).
- The M8 bolts (9) inserted into the side of the bearing caps are tightened to 3.0 kgm (22 ft.lb.).
- The M10 bolts (8) are tightened in three stages. Fist tighten to 0.5 kgm, then re-tighten them to 3.0 kgm (22 ft.lb.) and from the final position a further quarter of a turn (90°). The tightening operation can only be carried out successfully if you countercheck each bolt with Fig. 10.24 to pin-point the exact location of each bolt.
- Rotate the crankshaft a few times to check for binding and re-check the crankshaft end float as described during the removal.
- The remaining operations are carried out in reverse order to the removal procedure. The various sections give detailed description of the relevant operations, i.e. piston and connecting rods, rear oil seal flange, timing mechanism, flywheel and clutch or drive plate, oil pump, oil sump and cylinder head.

10.4.3.4. Flywheel or Drive Plate (Automatic)

The flywheel is treated in the same manner as described for the diesel engine on pages 45 and 46. Not shown in Fig. 1.50 is a washer which is fitted against the flywheel. The tightening torque is identical as specified.

10.4.3.5. Crankshaft Pulley and Vibration Damper

The engine is fitted with a crankshaft pulley and a vibration damper.

Remove the parts as follows, noting that a puller may be necessary to withdraw the hub:

- Remove the air cleaner housing, the viscous fan and the Poly V-belt as described in the section in question (fan and V-belt in section "Cooling System). The belt must only be lifted off the crankshaft pulley.
- Rotate the engine until the piston of No. 1 cylinder is at TDC, i.e. the long line in the crankshaft pulley must be opposite the pointer, as shown in Fig. 10.5, bottom left.
- Engage a gear and apply the handbrake to lock the engine against rotation. In the case of a vehicle with automatic transmission remove the starter motor and lock the starter motor ring gear in suitable manner. The same can be carried out when a manual transmission is fitted and the bolt cannot be removed by engaging a gear.
- Unscrew the bolt securing the crankshaft pulley with the vibration damper. The bolt is tightened to a very high torque (20 kgm/144 ft.lb.), i.e. you will need the necessary tools to remove the bolt.
- Withdraw the vibration damper, but as already mentioned you may need a puller.

The installation of the crankshaft pulley and the vibration damper proceed as follows:

- Rotate the crankshaft until the Woodruff key is visible and slide the combination crankshaft pulley/vibration damper with the key way over the key and the shaft end. Make sure that the key has engaged.
- Coat the centre bolt threads with engine oil. Tighten the bolt to 20.0 kgm (144 ft.lb.) and from the final position a further 95° (just a little more than a quarter of a turn).
- The remaining operations are carried out in reverse order.

10.4.3.6. Rear Crankshaft Oil Seal and End Cover

The rear crankshaft oil seal is located inside an end cover flange (oil seal carrier), which is bolted to the rear of the crankcase. Two dowels, one on each side of the cover, locate the flange correctly in relation to the crankshaft centre. The flange is secured by means of bolts, as shown in Fig. 10.26 and is fitted with sealing compound ("Loctite").

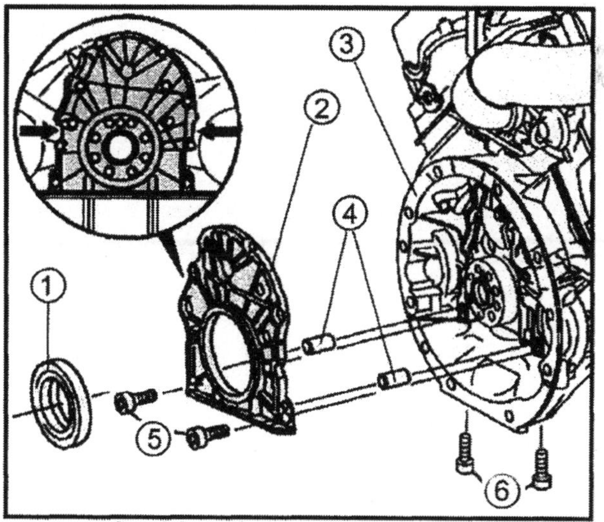

Fig. 10.26 – The arrows show the location of the oil seal carrier securing bolts. Tighten the bolts as specified in the text.

1 Rear crankshaft oil seal
2 Rear cover (oil seal carrier)
3 Crankcase
4 Dowel sleeves
5 Bolts, 1.4 kgm
6 Bolts from below, note torque value

Transmission and flywheel and/or drive plate must be removed to replace the oil seal. If the oil seal carrier is to be removed, remove the bolts around the outside edge and two bolts from below. The two lower bolts secure the cover to the oil sump. Apply two screwdrivers underneath the lugs on each side of the cover and carefully lever the carrier off the crankcase. The oil seal can now be removed from the inside.

Clean the carrier and crankcase faces. The sealing lip of the new oil seal is offset to prevent it from running on the same crankshaft area. Fill the space between the sealing lip and the dust protection lip with grease.

Coat the carrier face with sealing compound and fit it to the block, with the dowel pins engaged. Slightly tap the carrier in position. Great care must be taken during installation in order not to damage the oil seal.

Fit the bolts. First tighten the two lower bolts and then the other bolts. The torque for the lower bolts depend on the thread diameter. M6 bolts to 0.9 kgm (7 ft.lb), M8 bolts to 2.0 kgm (14.5 ft.lb.).
Carry out all other operations in reverse order.

10.4.3.7. Front Crankshaft Oil Seal

The front crankshaft oil seal is located on the inside of the crankshaft vibration damper. Oil leaks at this position can also be caused by a leaking timing cover gasket. Check before replacing the oil seal. The front of the vehicle must be resting on secure chassis stands.
The vibration damper and the crankshaft pulley must be removed as already described before the oil seal can be replaced. The seal can be carefully removed with a screwdriver. Screw a self-tapping screw into the outside of the seal and apply the screwdriver plate under the screw head.
Thoroughly clean the surrounding parts. Burrs on the timing cover bore can be removed with a scraper. Fill the space between sealing lip and dust protection lip with grease and carefully drive a new oil seal into the timing cover and over the crankshaft until the ring outer face is flush. Refit the vibration damper as described in Section 10.4.3.5.

10.4.4. TIMING COVER – REMOVAL AND INSTALLATION

The removal of the timing cover is obviously a time consuming and comprehensive operation as all parts in front of the cover must be removed. Some experience is required to remove the cover. Depending on the reason for the removal it may be better to remove the engine. The following description refers to an engine fitted to the vehicle, if for example the cover gasket shows oil leaks around the cover. Most operations are described under different headings and must be referred to. One additional operation is the removal of the air pump which is described below. On vehicles with A/C system do not disconnect any refrigerant pipes/hoses. First remove the front module (front part) of the vehicle as described during the removal of the engine. Then proceed as follows, commencing with the removal of the air pump.

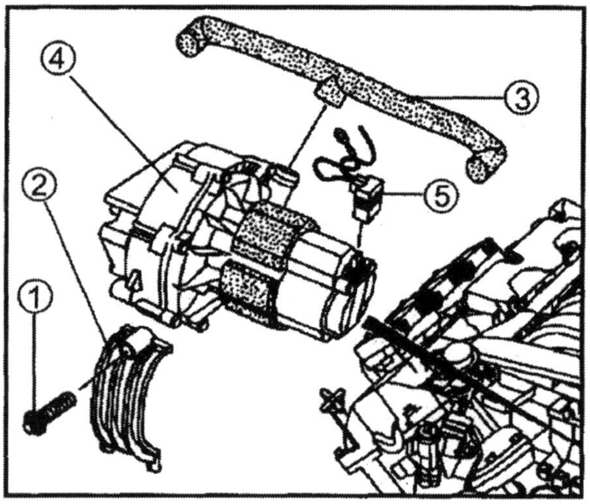

Fig. 10.27 – Details for the removal and installation of the air pump. The numbers are referred to below.

- Remove the air cleaner housing and then refer to Fig. 10.27. Remove the bolt (1) and take off the mounting clamp (2) and place the earth cable to one side. Tighten the bolt to 0.8 kgm (6 ft.lb.). Disconnect the air hose (3) from the air pump (4). On the R.H. side withdraw the connector plug (5) and take off the pump. Below the pump you will find the coolant temperature sensor which is to be removed.
- In the following order remove the viscous fan, the Poly V-belt, the alternator, the V-belt tensioning pulley (in the centre between the crankshaft and the steering pump pulley), the crankshaft pulley, the coolant pump, the thermostat housing, the cylinder head covers, the oil filter housing, the power steering pump, the A/C compressor (if fitted), the chain tensioner and the oil sump. Locate the timing cover securing bolts and take off the timing cover.

The installation is a reversal of the removal procedure. A new gasket must always be fitted. Clean the cover sealing faces and apply "Loctite 5970" to the cover face. Tighten all bolts (crankcase and cylinder head) to 2.0 kgm (14.5 ft.lb.).

10.4.5. CAMSHAFTS

10.4.5.1. Checking the Camshaft Timing Position

As already mentioned: Camshafts and associated parts can only be removed when the camshafts are in their basic timing position. The workshop uses a locating plate which is inserted into a groove in the camshaft, but by carefully turning the crankshaft this can be avoided. The cylinder head cover must be removed. Proceed as follows:

- Rotate the crankshaft until the 40° mark in the vibration damper (pulley) is opposite the pointer, as shown in the circle in Fig. 10.5 (after TDC of cyl. 1). Only turn the engine in direction of rotation.
- Check with Fig. 10.12 that the pin on the crankcase and a notch in the balancing weight are aligned as shown.

10.4.5.2. Adjusting the Camshaft Timing Position

In order to set the two camshafts to the basic timing position they can be rotated without the danger of the pistons hitting the valves, provided that the piston of No. 1 cylinder is in the top dead centre position as already described (see also Figs. 10.5 and 10.12). If a new adjustment is required:

- Remove the chain tensioner and the camshaft Hall sensor.
- Remove the R.H. and L.H. camshaft timing gears (sprockets). The camshafts are prevented from rotating with an open-ended spanner. Lift the timing chain from two timing sprockets. Do not allow the chain to drop into the timing cover (tie it up).
- Rotate the camshafts into their correct position as described in the previous section.
- Place the timing chain over the two camshaft sprockets and fit the sprockets to the camshafts. The bolts must always be replaced. Tighten them to 5.0 kgm (37 ft.lb.) and then a further quarter of a turn.
- Refit the Hall sensor and the chain tensioner and then proceed as described in Section 10.4.5.1 to set the camshafts to their basic timing position.

10.4.5.3. Camshafts – Removal and Installation

The removal of a camshaft involves the removal of the bearing bridge before the camshaft(s) can be removed. After removal of the bearing bridge in question, the camshaft is lifted out.

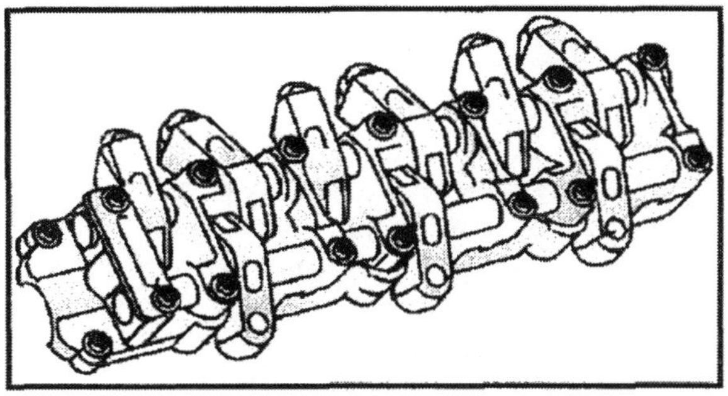

Fig. 10.28 – One of the camshaft bearing bridges.

Removal and Installation of Bearing Bridges

Fig. 10.28 shows one of the bearing bridges. The following should be noted:

- Lubricate the camshaft bearings before installation. The camshafts must not be replaced individually.
- If the bearing bridge is damaged, replace the cylinder head with the camshaft bearing bridge. The M7 bolts must always be replaced. Tighten the bolts in accordance with the diagram shown in Fig. 10.29.

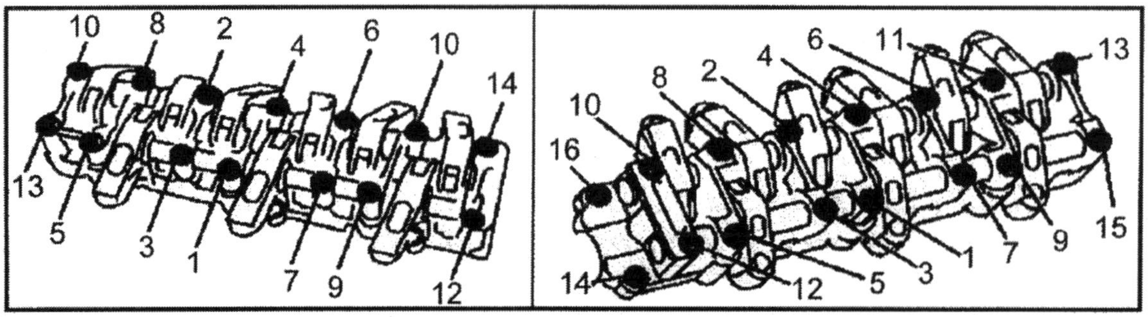

Fig. 10.29 – Tightening sequence for the camshaft bearing bridges. Note the different number of bolts used on the L.H. and R.H. bridge.

The removal of a bearing bridge is carried out after removal of the cylinder head cover. Set the No. 1 cylinder to top dead centre position (40° after ignition timing point, as shown in the circle of Fig. 10.5). After removal of the chain tensioner slacken the bolts in reverse order to the sequence shown in Fig. 10.29. Note that some bolts are 45 mm long, others 84 mm. Mark where they have been removed from.
The installation is a reversal of the removal procedure. Insert the bolts in accordance with the length and tighten them, using the diagrams shown in Fig. 10.29, depending which of the bridges has been removed. Note that one bridge has 14 bolts and the other one 16 bolts. Count the number of bolts and use the diagram accordingly.

Removal and Installation of Camshafts

- Remove the camshaft sprockets of the L.H. and R.H. camshaft. An open-ended spanner is used to prevent the camshaft from rotating. Use a piece of wire to attach the timing chain to the sprockets and remove the sprockets from the camshafts.
- Remove the bearing bridges in the manner described above and lift out the camshaft(s). If necessary replace both camshafts. It is not possible to replace camshafts individually.

The installation is a reversal of the removal procedure. Tighten the camshaft sprocket bolts to 5.0 kgm (36 ft.lb.) and from the final position a further quarter of a turn without a torque wrench.

10.4.6. TIMING MECHANISM (includes balance shaft)

10.4.6.1. Chain Tensioner – Removal and Installation

The chain tensioner is fitted to the timing cover, seen from the front end of the engine on the L.H. side of the pulley group. The air filter housing, the V-belt and the alternator must be removed to unscrew the tensioner. A sealing ring underneath the tensioner must be replaced during installation. Tighten the tensioner to 8.0 kgm (58 ft.lb.) during installation. The alternator bolts are tightened to 5.7 kgm (41 ft.lb.).

10.4.6.2. Timing Chain – Removal and Installation

We cannot recommend the replacement of the timing chain as a DIY operation when the engine is fitted to the vehicle. Apart from special tools, to separate the chain links, fitting a new chain, etc. which cannot be substituted by make-shift tools and a hand grinding machine, you will need the experience to deal with timing chains of a two camshaft Mercedes-Benz engine. The timing chain has a long service life and may not need replacing.
If the engine is removed you can replace the chain in accordance with the instructions given, as the new chain is placed in position over the various sprockets. The remaining operations can be taken from the various operations described in this section.

10.4.6.3. Balance Shaft – Removal and Installation

Fig. 10.30 shows details for the removal and installation of the balance shaft, fitted to these engines. The engine must be removed to remove and install the balance shaft. Proceed as follows:

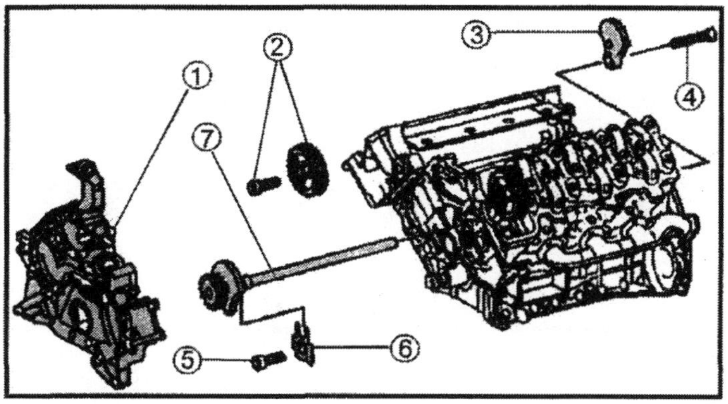

Fig. 10.30 – Details for the removal and installation of the balance shaft. The numbers are referred to in the text.

- Remove in the following order the flywheel, the end cover, the oil sump and the timing cover (1) as described earlier on.
- Set the engine to the top dead centre position. The alignment of the balance shaft must be as shown in Fig. 10.31.

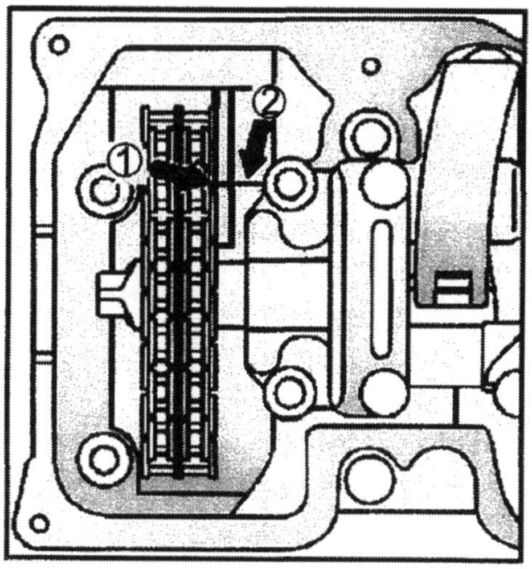

Fig. 10.31 – Alignment of the balance shaft.

- Remove the R.H. camshaft sprocket (2) as described during the removal of the camshafts. Use a piece of wire to attach the timing chain to the sprocket. The engine must not be rotated with the sprocket removed. The sprocket bolt is tightened to 5.0 kgm + 90°.
- Remove the bolt (4) at the rear end of the engine, securing the counterweight to the shaft. A hole in the counterweight enables you to insert a drift to prevent the shaft from rotating. Withdraw the counterweight (3). The bolt is tightened to 2.0 kgm (14.5 ft.lb.) and then a further quarter of a turn.
- At the front of the engine remove the bolt (5), take off the locking plate (6) and remove the shaft towards the front. The bolt is tightened to 1.5 kgm (10 ft.lb.) during installation.

The installation is a reversal of the removal procedure. After installation check the timing alignment with Fig. 10.31. The markings of the pin on the crankcase (2) and the notch (1) in the balance weight must be aligned.

10.5. Tightening Torque Values

All tightening torques are specified in the various removal/installation instructions.

11. PETROL FUEL INJECTION

The engine is fitted with a fuel injection system known as "ME-SF5". The system operates in conjunction with the ignition system. The system has six injection valves, which inject the fuel during starting speed up to 600 rpm and during acceleration at the same time into all six cylinders.

The electronic control unit calculates the duration of the injection in accordance with the operating conditions of the engine. The injected amount of fuel depends therefore on the opening time of the injection valves, which can be between 1.5 seconds and 130 milli-seconds. The vacuum in the inlet manifold, the engine speed, the coolant temperature and the intake air temperature are factors which contribute to the injection amount.

The engine speed is controlled by a speed limiter. If the engine speed exceeds a certain speed, the fuel is cut-off to the cylinders. The control unit will detect when sixth gear is engaged by comparing the driving speed with the engine speed.

The fuel supply is also interrupted when the vehicle is coasting, i.e. the engine speed is higher than specified by the fitted gearbox, the idle speed switch is closed and a cruise control system (if fitted) is switched off. When the engine speed is again above 1100 rpm, fuel supply will resume and the fuel is injected once more into the two cylinders mentioned above.

General Note: Any work on the component parts of the injection system will require special instruments and equipment to adjust, measure and check important values. It is therefore recommended to leave any work to a dealer. The following description should be considered for information only.

11.1.0. PRECAUTIONS WHEN WORKING ON THE SYSTEM

Always note the points listed below when carrying out any operations on the injection system and in this connection on the ignition system:

Attention: We advise that persons with a heart pacemaker should not work on the injection system.

- Always switch off the ignition and disconnect the battery. When we refer to disconnecting the battery, always disconnect the negative cable and push it well away.
- Never touch any connections, cable, leads, etc. when the engine is running, even when the engine is switched off, but the ignition is on.
- Never disconnect or connect any electrical cables when the engine is running.

11.1.1. Possible Work to be carried out

The following operations can be carried out with a little knowledge and are therefore included, but note that some special tools are listed.

Lambda probe – Removal and Installation

The workshop uses a special wrench (No. 000 589 71 03 00), which is a long socket with a slot on one side, so it can be pushed passed the cable of the Lambda probe. Each catalytic converter is fitted with two Lambda probes, one before the converter and one after the converter. Remove as follows:

- Disconnect the battery.
- Follow the routing of the cable to the connection on the Lambda probe and separate the cable plugs. Four connector plugs must be separated.
- Unscrew the Lambda probe with wrench mentioned above or use a spanner which can be applied to the hexagon. Remove the probe.
- Coat the thread of the new Lambda probe with heat-resistant grease. You may be able to obtain a finger-full from a workshop. Tighten the probe to 4.5 kgm.
- No problem with the special wrench, as a torque spanner can be applied. If you use an ordinary wrench, you will have to estimate the torque.
- The remaining operations are carried out in reverse order.

Fuel Distribution Rail

Fig. 10.32 shows the parts in question. The injection system must be free of pressure. The workshop uses the service valve (1) to de-pressurise the fuel system. The actual removal presents no major problems. More difficult is the location of the various items to be disconnected and/or remove. We will try to explain where the various parts are connected. The engine is seen from the side when the oil filler cap is on the R.H. side during the following instructions. Remove as follows:

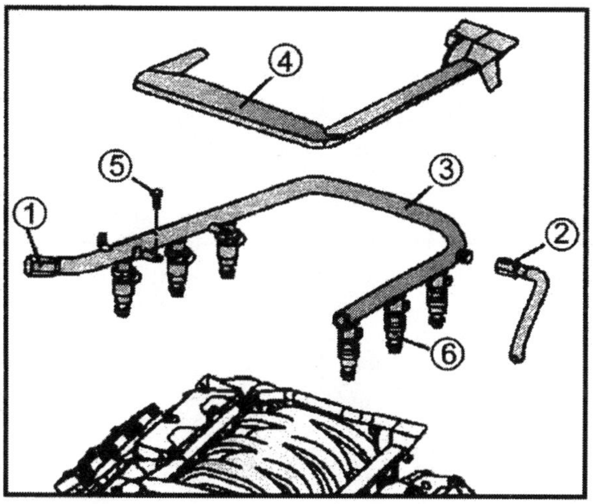

Fig. 10.32 – Details for the removal and installation of the fuel distribution rail. The numbers are referred to below.

- Disconnect the battery.
- Remove the air cleaner housing.
- Loosen the clamp screw and disconnect the intake hose from the mass air flow meter.
- Disconnect the fuel feed line. The union nut is tightened to 2.4 kgm (18 ft.lb.) during installation.
- Disconnect the electrical connector plug from the ignition coil (L.H. side of engine).
- Remove the engine vent hose on the R.H. cylinder cover and on the L.H. cylinder cover.
- Disconnect the connector plug from the camshaft Hall sensor. Looking at the engine you will find it in the L.H. bottom corner. In the same area you will find the connector on the air injection valve and further to the right the connector for the coolant temperature sensor and the intake manifold pressure sensor. Remove the plugs. A further plug must be disconnected on the R.H. side next to the vacuum unit with the connected hose (for the intake switchover valve) and at the L.H. rear of the engine (for the vacuum transducer). The latter is located near the vent hose (top left on the engine).
- Disconnect the cable tie from the fuel distributor rail (3) – new tie must be fitted
- Unlock the cable duct (4) from the fuel distributor rail (3).
- Remove the bolts (5) securing the fuel distributor rail (3) to the intake manifold (tighten to 1.0 kgm) and remove the rail together with the injection valves (6) in an upwards direction. Disconnect the connector plugs from the injection valves and fully remove the rail. If a new distributor rail is fitted, you can transfer the injection valves from the old rail to the new rail.

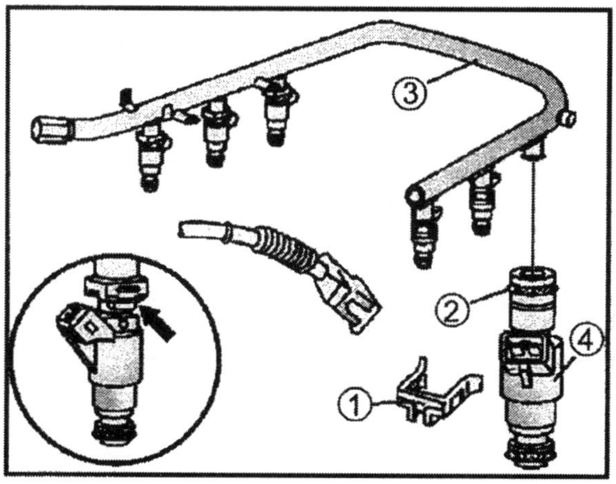

Fig. 10.33 – The injection valves are fitted to the fuel distribution rail (1) with the retaining plates (2), a type of clip, as shown. The arrows show where the sealing rings must be fitted.

The installation is a reversal of the removal procedure, noting any torques given above.

Injection Valves – Replacement

The valve are secured by means of retaining clips. The fuel distributor rail must be removed as described to replace a valve. Fig. 10.33 shows details. After removal of the clip (1) withdraw the valve (4) from the distribution rail (3). A sealing washer (2) is fitted to the end of the valves, which must always be replaced. During installation lubricate the sealing rings with engine oil and push the valve(s) into the bore(s). The twist lock at the position shown by the arrow must engage into the cut-out of the fuel distributor rail.

Air Cleaner and Air Cleaner Element

The removal of the air cleaner housing is carried out in a similar manner as described for the diesel engine. The intake hose must be removed after slackening the clamp screw on the air filter housing. Remove the hose carefully to prevent the intake pipe unlocking from the sealing ring in the inlet manifold.

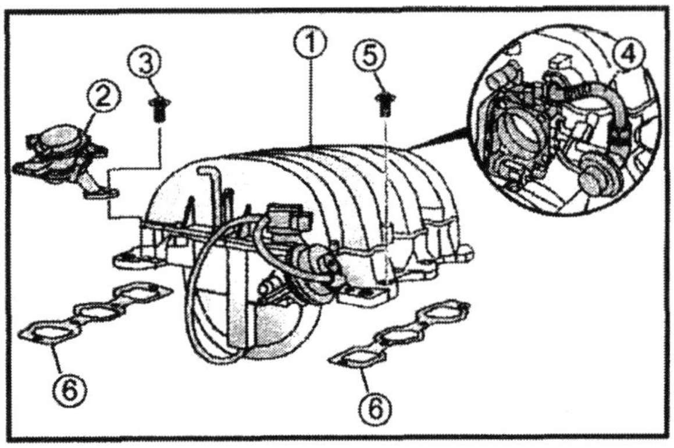

Fig. 10.34 – The attachment of the inlet manifold.

1 Inlet manifold
2 Aspirator shut-Off valve
3 Bolt
4 Fuel feed pipe, 2.4 kgm
5 Bolt, 2.0 kgm
6 Inlet manifold gaskets

Inlet Manifold – Removal and Installation

This is obviously a time-consuming and complicated operation. As, however, the inlet manifold must be removed sometimes for other operations, we give you in brief the necessary work to be carried out. Fig. 10.34 shows the attachment of the manifold and will help you. Again the fuel system must be free of pressure. The battery must be disconnected. As the fuel distributor rail and the injection valves must be removed, most preliminary operations have already been carried out. Further remove the hot film mass air flow sensor, the intake pipe, the R.H. air filter housing bracket, any hoses connected to the inlet manifold and the exhaust gas re-circulating valve line.

The manifold can now be removed by referring to Fig. 10.34. Tighten the inlet manifold bolts to 2.0 kgm (14.5 ft.lb.), the exhaust gas re-circulating line to 4.0 kgm (22 ft.lb). The manifold gaskets (6) must be replaced.

11.1.2. IGNITION SYSTEM

11.1.2.0. Precaution when working on the Ignition System

If work is carried out on the ignition and fuel injection system, note the following precautions:

- *Persons with a heart pacemaker should not work on engines with fuel injection and electronic ignition – so we have been told.*
- Never disconnect the battery with the engine running.
- Always disconnect both battery cables before the battery is re-charged.
- Never use a battery of more than 12 volts to start the engine.
- Never disconnect the control unit with the ignition switched on.
- Before connecting a connector plug check the condition of the terminals. Also check the retaining springs of connectors for good fit.
- Always make sure that the battery is disconnected when electric welding work is carried out on the vehicle.
- No test lamps must be used to check electrical circuits.
- Never use wires to bridge any of the connections to earth.
- Make sure that the battery terminals are properly tightened before the engine is started (damage to electronic component parts).

11.1.2.1. Ignition Coils – Removing and Installation

The coils are located on the side of the engine and are secured by means of three bolts. Each coil is connected with the three spark plugs on the respective side of the engine. The bolts are tightened to 0.8 kgm (6 ft.lb.). The coil can be removed after removal of the air cleaner housing and disconnecting the electrical connectors and the ignition cables. Separate the tie strap if necessary. During installation make sure that each ignition cable is connected to the correct connector. Mark the cables before removal if not sure.

11.1.2.2. Spark Plug – Removing and Installation

Fig. 10.35 shows one side of the engine with two of the plugs (2) and their connectors (1). If spark plugs are removed, make sure that the surrounding area is clean to prevent foreign matter from falling into the plug holes as soon as the plugs are unscrewed. We must point out that a special plug spanner (No. 112 589 01 09 00) is used to remove the plugs in a workshop. Clean the plug face with a wire brush and check the electrodes for wear or burns.

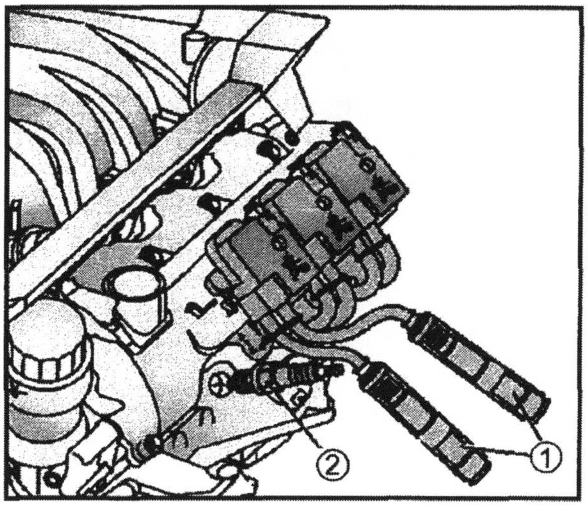

Fig. 10.35 – Remove the spark plug connectors (1) before removal of the plugs (2).

Use a feeler gauge to check the electrode gap and close the gap if necessary, by tapping the outside electrode with the handle of a screwdriver. To open up a plug gap, insert the blade of a small screwdriver and bend up the side electrode. Never bend the centre electrode in order to correct plug gaps as this will damage the insulator. Either Beru, Bosch, NGK or Champion plugs are fitted during production. The electrode gap on Champion plugs is 1.1 mm, on the other plugs 1.0 mm.

Note: Spark plugs are sometimes changed. Always ask for the latest plugs if buying new ones. Tighten the spark plugs, if possible, with a torque wrench to a torque setting of 2.8 kgm (20 ft.lb.).

Remove the plugs as follows, referring to Fig. 10.35 for details of the installation.

- Disconnect the battery.
- Remove the air cleaner housing.
- Remove the spark plug connectors (1). Mark their position if not sure.
- Unscrew the plugs with the special wrench specified above or another suitable plug spanner and an extension.

During installation fit the spark plug by hand as far as they will go and then tighten them to 2.8 kgm (20 ft.lb.). Re-connect the connector plugs. All other operations are carried out in reverse order.

Inspect the condition of the insulator tip and the electrodes. Following there are a few examples to interpret the condition of plugs removed from the engine.

Normal Plug Face: The colour of the insulator should appear greyish - brown or tan coloured. The electrodes should be black or sooted. These are signs of a plug which has been used under normal conditions with alternative short and long driving periods. White or yellow deposits mean that the car has been used for long periods at high speeds and can be ignored.

Worn Plug Appearance: Insulator tip and electrodes are burnt off. All plugs which show this condition must be replaced. Always replace the whole set and make sure to fit the correct plug.

Oiled-up Appearance: Normally this condition is recognised by wet oil deposits which have been left by excessive ingress of oil into the combustion chamber (worn piston rings or pistons, inlet valves or valve guides, worn bearings, etc.). Hotter plugs are normally able to overcome the fault, but in serious cases an overhaul of the engine is necessary.

Burnt or Overheated Appearance: Burnt or overheated plugs can normally be recognised by their electrodes being coloured white or being burnt, or by the presence of blisters on the insulator or the electrodes. Electrodes may also be burnt off. Faults can be traced to the cooling system or improper ignition timing.

12. SERVICING AND MAINTENANCE

Most of the maintenance operations can be carried out without much difficulty. In many cases it is, however, better to have certain maintenance operations carried out in a workshop as experience and special equipment, for example test instruments, are required to carry out a certain job. Most important are the regular inspections and checks which are described below. Operations to be carried out after a certain mileage are described later in the Maintenance Booklet you received with your vehicle. We will advise when specific jobs should be left to a Mercedes Dealer.

12.1. Regular Maintenance you can do

Oil Level Check: Check the engine oil level every 500 miles. With the vehicle standing on level ground, remove the oil dipstick and wipe it clean with a clean rag or a piece of tissue paper. Re-insert the oil dipstick and remove once more. The oil level must be visible between the upper and the lower mark on the dipstick. If the oil level is below the lower mark, top-up with engine oil of the correct viscosity. Never overfill the engine - the level must never be above the upper dipstick mark.

Checking the Brake Fluid Level: - The brake fluid reservoir is in the engine compartment on the drivers side. The reservoir is transparent and it is easy to check whether the fluid level is between the "Min" and "Max" mark. If necessary, top-up to the "Max" mark with the correct brake fluid.

Checking the Brake Lights: The operation of the brake lights can either be checked with the help of another person or you can check it by yourself by driving the vehicle backwards near the garage door. Operate the brake pedal and check if the reflection of the brake lights can be seen on the garage door by looking through the rear view mirror.

Checking the Vehicle Lights: In turn check every vehicle light, including the horn and the hazard warning light system. Rear lights and reversing lights can be checked in the dark in front of a garage door, without leaving the vehicle.

Checking the Tyre Pressures: Check the tyre pressures at a petrol station. Pressures are different for the various models. Either your Operators Manual or tyre charts will give you the correct pressures.

If continuous speeds of more than 100 mph are anticipated, increase the tyre pressure by 0.2 kg/sq.cm. (3 psi.).

Checking the Coolant Level: See Section "Cooling System". Never open the radiator filler cap when the engine is hot.

Checking the Fluid Level in the Automatic Transmission: As specified in your Maintenance Booklet (workshop).

Changing the Engine Oil and Oil Filter: Some petrol stations will carry out an oil change free of charge – You only pay for the oil. The same applies to the oil filter (there may be a small extra charge), but not every petrol station will be able to obtain a Mercedes filter. To change the filter yourself, refer to the relevant page.

Lubrication Jobs: Apart from the engine lubrication there are further lubrication points which should be attended to. These include the throttle linkage and shafts (only grease the swivel points), the engine bonnet catch and the hinges (use a drop of engine oil) and perhaps the door mechanism.

Air Filter Service: Remove the air filter element for cleaning.

Checking the Brake System: If no trouble has been experienced with the brake system, there is little need to carry out extensive checks. To safeguard for the next 6000 miles, however, follow the brake pipes underneath the vehicle. No rust or corrosion must be visible. Dark deposits near the pipe ends point to leaking joints. Brake hoses must show no signs of chafing or breaks. All rubber dust caps must be in position on the bleed valves of the calipers. Insert a finger underneath the master cylinder, where it is fitted to the brake servo unit. Moisture indicates a slightly leaking cylinder.
The brake pads must be checked for the remaining material thickness as has been described in Section "Brakes" for the front and rear brakes.

Adjusting the Parking Brake: Adjust the parking brake as described in Section "Brakes" under the relevant heading.

Brake Test: A brake test is recommended at this interval. Your will decide yourself if the brakes perform as you expect them to. Otherwise have the brakes tested on a dynamometer. The read-out of the meter will show you the efficiency of the brake system on all four wheels.

Checking the Wheel Suspension and Steering: In the case of the front suspension remove both wheels and check the shock absorbers for signs of moisture, indicating fluid leaks.
Check the free play of the steering wheel. If the steering wheel can be moved by more than 25 mm (1 in.) before the front wheels respond, have the steering checked professionally.
Check the rubber dust boots of the track rod and suspension ball joints. Although rubber boots can be replaced individually, dirt may have entered the joints already. In this case replace the ball joint end piece or the suspension ball joint.
Check the fluid level in the reservoir for the power-assisted steering. Refer to the "Steering" section for details. If steering fluid is always missing after the 12,000 miles check, suspect a leak somewhere in the system - See your dealer.

Tyre Check: Jack up the vehicle and check all tyres for uneven wear. Tyres should be evenly worn on the entire surface. Uneven wear at the inner or outer edge of front tyres points to misalignment of the front wheel geometry. Have the geometry measured at your dealer. Make sure that a tread depth of 1.6 mm is still visible to remain within the legal requirements. Make sure to fit tyres suitable for your model, mainly if you buy them from an independent tyre company.

Re-tighten Wheel Bolts: Re-tighten the wheel bolts to the value given in the tightening torque sections (front or rear suspension). Tighten every second bolt in turn until all bolts have been re-tightened.

Checking the Cooling System: Check all coolant hoses for cuts, chafing and other damage. Check the radiator for leaks, normally indicated by a deposit, left by the leaking anti-freeze. Slight radiator leaks can be stopped with one of the proprietary sealants available for this purpose.

Checking the Clutch: Check the clutch operation. The fluid reservoir should be full. If it is suspected that the clutch linings are worn near their limit, take the vehicle to a dealer. A gauge is used to check the protrusion of the slave cylinder push rod. It is a quick check and may prevent the clutch driven plate to wear down to the rivets.

Checking the Anti-freeze: The strength of the anti-freeze should be checked every 12,000 miles. Petrol stations normally have a hydrometer to carry out this check. Make sure that only anti-freeze suitable for Mercedes engines is used.

Checking the Manual Transmission Fluid Level: Refer to Section "Gearbox".

FAULT FINDING SECTION

The following section lists some of the more common faults that can develop in a motor car, both for petrol and diesel engines. For the purpose of this manual, references to diesel engines are of course, first and foremost, as the detection faults in a petrol engine is a job for a special workshop, dealing with fuel injection system. The section is divided into various categories and it should be possible to locale faults or damage by referring to the assembly group of the vehicle in question.
The faults are listed In no particular order and their causes are given a number. By referring to this number it is possible to read off the possible cause and to carry out the necessary remedies, if this is within the scope of your facilities.

ENGINE FAULTS

Engine will not crank:	1, 2, 3, 4
Engine cranks, but will not start:	5, 6, 7, 8
Engine cranks very slowly:	1, 2, 3
Engine starts, but cuts out:	5, 6, 9, 10
Engine misfires in the lower speed ranges:	5, 6, 9, 11
Engine misfires in the higher speed ranges:	5, 6, 11, 12
Continuous misfiring:	5, 6, 7, 10 to 15, 21, 22
Max. revs not obtained:	5, 6, 12, 22
Faulty idling:	5, 6, 8 to 11, 13, 15, 16, 21 and 22
Lack of power:	3, 5 to 11,13 to 15, 22
Lack of acceleration:	5 to 8, 12, 14 to 16
Lack of max. speed:	5 to 8, 10, 12, 13 to 15 ,22
Excessive fuel consumption:	3, 5, 6, 15 ,16
Excessive oil consumption:	16 to 19
Low compression:	7, 11 to 13, 16, 20 to 22

Causes and Remedies

1. Fault in the starter motor or its connection. Refer to "Electrical Faults".
2. Engine oil too thick. This can be caused by using the wrong oil, low temperatures or using oil not suitable for the prevailing climates. Depress the clutch whilst starting (models with manual transmission). Otherwise refill the engine with the correct oil grade, suitable for diesel engines.
3. Moveable parts of the engine not run-in. This fault may be noticed when the engine has been overhauled. It may be possible to free the engine by adding oil to the fuel for a while.
4. Mechanical fault. This may be due to seizure of the piston(s), broken crankshaft,

connecting rods, clutch or other moveable parts of the engine. The engine must be stripped for inspection.

5. Faults in the glow plug system. Refer to "Glow Plug Faults".
6. Faults in the fuel system. Refer to "Fuel Faults".
7. Incorrect valve timing. This will only be noticed after the engine has been re-assembled after overhaul and the timing belt has been replaced incorrectly. Re-dismantle the engine and check the timing marks on the timing gear wheels.
8. Compression leak due to faulty closing of valves. See also under (7) or leakage past worn piston rings or pistons. Cylinder head gasket blown.
9. Entry of air at inlet manifold, due to split manifold or damaged gasket.
10. Restriction in exhaust system, due to damaged exhaust pipes, dirt in end of exhaust pipe(s), kinked pipe(s), or collapsed silencer. Repair as necessary.
11. Worn valves or valve seats, no longer closing the valves properly. Top overhaul of engine is asked for.
12. Sticking valves due to excessive carbon deposits or weak valve springs. Top overhaul is asked for.
13. Cylinder head gasket blown. Replace gasket and check block and head surfaces for distortion.
14. Camshaft worn, not opening or closing one of the valves properly, preventing proper combustion. Check and if necessary fit new camshaft.
15. Incorrect valve (tappet) clearance. There could be a fault in the hydraulic tappets.
16. Cylinder bores, pistons or piston rings worn. Overhaul is the only cure. Fault may be corrected for a while by adding "Piston Seal Liquid" into the cylinders, but will re-develop.
17. Worn valve guides and/or valve stems. Top overhaul is asked for.
18. Damaged valve stem seals. Top overhaul is asked for.
19. Leaking crankshaft oil seal, worn piston rings or pistons, worn cylinders. Correct as necessary.
20. Loose glow plugs, gas escaping past thread or plug sealing washer damaged. Correct.
21. Cracked cylinder or cylinder block. Dismantle, investigate and replace block, if necessary.
22. Broken, weak or collapsed valve spring(s). Top overhaul is asked for.

GLOW PLUG FAULTS

Check a suspect glow plug as follows:

- Remove the glow plug lead from the rear glow plug and from the remaining plugs the bus bars.
- Connect a 12 volts test lamp to the plus terminal of the battery and with the other lead of the lamp touch in turn the connecting threads of each glow plug. The faulty plug is detected when the test lamp does not light up.

Further faults in the glow plug system can develop in the glow plug relay. Check as follows:

- Disconnect the electrical lead from the glow plug on the flywheel end and connect a test lamp between the lead and a good earthing point.
- Disconnect the electrical lead from the coolant temperature sender unit and move away from earth.
- Turn the ignition key to the "glowing" position and observe the test lamp. The lamp should light up for approx. 25 - 30 sec. and then switch off.
- Turn the ignition switch "off" and then again to the "glowing" position. The test lamp must light up once more.
- Hold the disconnected lead from the temperature sender unit against earth. The test lamp must switch off.

- Turn the ignition switch "off" and then again to the "glowing" position. The test lamp should light up.
- Operate the starter motor and check that the test lamp remains "on".
- If the above tests cannot be carried out satisfactory, see your dealer.

LUBRICATION SYSTEM FAULTS

The only problem the lubrication system should give is excessive oil consumption or low oil pressure, or the oil warning light not going off.
Excessive oil consumption can be caused by worn cylinder bores, pistons and/or piston rings, worn valve guides, worn valves stem seals or a damaged crankshaft oil seal or leaking gasket on any of the engine parts. In most cases the engine must be dismantled to locate the fault.
Low oil pressure can be caused by a faulty oil pressure gauge, sender unit or wiring, a defective relief valve, low oil level, blocked oil pick-up pipe for the oil pump, worn oil pump or damaged main or big end bearings, In most cases it is logical to check the oil level first. All other causes require the dismantling and repair of the engine. If the oil warning light stays on, switch off the engine IMMEDIATELY, as delay could cause complete seizure within minutes.

COOLING SYSTEM FAULTS

Common faults are: Overheating, loss of coolant and slow warming-up of the engine:

Overheating:

1. ***Lack of coolant:*** Open the radiator cap with care to avoid injuries. Never pour cold water in to an overheated engine. Wait until engine cools down and pour in coolant whilst engine is running.
2. ***Radiator core obstructed by leaves, insects, etc.***: Blow with air line from the back of the radiator or with a water hose to clean.
3. ***Cooling fan not operating:*** Check fan for proper cut-in and cut-out temperature. If necessary change the temperature switch or see your Dealer.
4. ***Thermostat sticking:*** If sticking in the closed position, coolant can only circulate within the cylinder head or block. Remove thermostat and check as described in section "Cooling".
5. ***Water hose split:*** Identified by rising steam from the engine compartment or the front of the vehicle. Slight splits can be repaired with insulation tape. Drive without expansion tank cap to keep the pressure in the system down, to the nearest service station.
6. ***Water pump belt torn:*** Replace and tension belt.
7. ***Water pump inoperative:*** Replace water pump.
8. ***Cylinder head gasket blown:*** Replace the cylinder head gasket.

Loss of Coolant:

1. ***Radiator leaks:*** Slight leaks may be stopped by using radiator sealing compound (follow the instructions of the manufacturer). In emergency a egg can be cracked open and poured into the radiator filler neck.
2. Hose leaks: See under 5, "Overheating".
3. Water pump leaks: Check the gasket for proper sealing or replace the pump.

Long Warming-up periods:

1. Thermostat sticking in the open position: Remove thermostat, check and if necessary replace.

DIESEL FUEL SYSTEM FAULTS

The following faults are given in general, not necessarily for the CDI engine only.

Engine is difficult to start or does not start:	1 to 13
Engine starts, but stop soon afterwards:	14 to 20

Engine misfires continuously:	1 to 13
Bad idling:	14 to 20
Black, white or blue exhaust smoke:	21 to 29
Lack of power:	30 to 39
Excessive fuel consumption:	40 to 47

Causes and Remedies

1. Fuel tank empty. Refuel.
2. Pre-glowing time too short. Operate until warning light goes "off".
3. Cold starting device not operated. Pull cable and push in after approx. 1 min.
4. Glow plug system inoperative. Refer to "Glow Plug Faults".
5. Electro-magnetic cut-off device, loose or no current. Check cable to cut-off at top of injection pump. Ask a second person to operate ignition key and check if a "click" is heard. Either interrupted current supply or defective cut-off device.
6. Air in fuel system. Operate starter motor until fuel is delivered.
7. Fuel supply faulty. Slacken the injection pipes at injectors, and check if fuel is running out. Other faults: kinked, blocked or leaking injection pipes blocked fuel filter, tank breathing system blocked. Wrong fuel for cold temperatures.

& Injection pipes refitted in wrong order after repair.

9. Injection timing of pump out of phase: Have the adjustment checked and corrected.
10. One or more injectors faulty, dirty or incorrect injection pressure. Have injectors repaired or replace them.
11. Injection pump not operating properly. Fit an exchange pump or have it repaired.
12. Valves not opening properly.
13. Compression pressures too low. See item "8" under "Engine Faults".
14. Idle speed not properly adjusted. Adjust.
15. Throttle cable not properly adjusted or sticking. Re- adjust or free-off.
16. Fuel hose between filter and pump not tightened properly. Tighten connections.
17. Rear mounting of injection pump loose or cracked. Tighten or replace.
18. See items 6, 7, 9, 11, 12 and 13.
19. Engine mounting not tightened properly or worn. Tighten or replace.
20. Sticking accelerator pedal. Free-off pedal.
21. Engine not at operating temperature. Check exhaust smoke colour again when engine is warm.
22. Too much acceleration at low revs. Use individual gears in accordance with acceleration.
23. Air cleaner contaminated. Clean or replace.
24. Fuel filter contaminated. Replace.
25. Max. speed adjustment incorrect. Re-adjust.
26. Injectors are dripping. Have them checked or replace faulty ones.
27. Injector nozzles sticking or broken. Replace injector.
28. Injection pressure too low. Have injectors checked and adjusted.
29. See items 9, 11, 12 and 13.
30. Throttle cable travel restricted. Re-adjust. Check that floor mats cannot obstruct pedal movement.
31. Throttle cable not correctly adjusted. Re-adjust.
32. Operating lever loose on pump. Re-tighten.
33. Max. speed not obtained. Re-adjust max. speed or have it adjusted.
34. Injector pipes restricted in diameter (near connections). Disconnect pipes and check that diameter is at least 2.0 mm (0.08 in.).
35. Heat protection sealing gaskets under injectors not sealing or damaged. Remove injectors and check. Replace if necessary. Fit the washers correctly.
36. Injection pressure of injectors wrong. Have them re-adjusted.
37. See items 6, 7, 9, 11 and 13.

38. See item 20.
39. See items 23, 24, 26 and 27.
40. Road wheels dragging. Brakes seized or wheel bearings not running freely.
41. Engine not running "free". Refers to new or overhauled engine.
42. Fuel system leaking. Check hoses, pipes, filter, injection pump, etc. for leaks.
43. Fuel return line blocked. Clean with compressed air if possible.
44. Idle speed too high. Re-adjust.
45. Max. speed too high. Re-adjust.
46. See items 10, 11, 12 and 13.
47. See items 24, 26, 27 and 28.

CLUTCH FAULTS

Clutch slipping:	1, 2, 3, 4, 5
Clutch will not disengage fully:	4, 6 to 12, 14
Whining from clutch when pedal is depressed:	13
Clutch judder:	1, 2, 7, 10 to 13
Clutch noise when idling:	2, 3
Clutch noise during engagement:	2

Causes and Remedies

1. Insufficient clutch free play at pedal.
2. Clutch disc linings worn, hardened, oiled-up, loose or broken. Disc distorted or hub loose. Clutch disc must be replaced.
3. Pressure plate faulty. Replace clutch.
4. Air in hydraulic system. Low fluid level in clutch cylinder reservoir.
5. Insufficient play at clutch pedal and clutch release linkage. Rectify as described.
6. Excessive free play in release linkage (only for cable operated clutch, not applicable). Adjust or replace worn parts.
7. Misalignment of clutch housing. Very rare fault, but possible on transmissions with separate clutch housings. Re-align to correct.
8. Clutch disc hub binding on splines of main drive shaft (clutch shaft) due to dirt or burrs on splines. Remove clutch and clean and check splines.
9. Clutch disc linings loose or broken. Replace disc.
10. Pressure plate distorted. Replace clutch.
11. Clutch cover distorted. Replace clutch.
12. Fault in transmission or loose engine mountings.
13. Release bearing defective. Remove clutch and replace bearing.
14. A bent clutch release lever. Check lever and replace or straighten, if possible.

- The above faults and remedies are for hydraulic and mechanical clutch operation and should be read as applicable to the model in question, as the clutch fault finding section is written for all types of clutch operation.

STEERING FAULTS

Steering very heavy:	1 to 6
Steering very loose:	5, 7 to 9, 11 to 13
Steering wheel wobbles:	4, 5, 7 to 9, 11 to 16
Vehicle pulls to one side:	1, 4, 8, 10, 14 to 18
Steering wheel does not return to centre position:	1 to 6, 18
Abnormal tyre wear:	1, 4, 7 to 9, 14 to 19
Knocking noise in column:	6, 7, 11, 12

Causes and Remedies

1. Tyre pressures not correct or uneven. Correct.
2. Lack of lubricant in steering.

3. Stiff steering linkage ball joints. Replace ball joints in question.
4. Incorrect steering wheel alignment. Correct as necessary.
5 Steering needs adjustment. See your dealer for advice.
6. Steering column bearings too tight or seized or steering column bent. Correct as necessary.
7. Steering linkage joints loose or worn. Check and replace joints as necessary.
8. Front wheel bearings worn, damaged or loose. Replace bearing.
9. Front suspension parts loose. Check and correct.
10. Wheel nuts loose. Re-tighten.
11. Steering wheel loose. Re-tighten nut.
12. Steering gear mounting loose. Check and tighten.
13. Steering gear worn. Replace the steering gear.
14. Steering track rods defective or loose.
15. Wheels not properly balanced or tyre pressures uneven. Correct pressures or balance wheels.
16. Suspension springs weak or broken. Replace spring in question or both.
17. Brakes are pulling to one side. See under "Brake Faults".
18. Suspension out of alignment. Have the complete suspension checked by a dealer.
19. Improper driving. We don't intend to tell you how to drive and are quite sure that this is not the cause of the fault.

BRAKE FAULTS

Brake Failure: Brake shoe linings or pads excessively worn, incorrect brake fluid (after overhaul), insufficient brake fluid, fluid leak, master cylinder defective, wheel cylinder or caliper failure. Remedies are obvious in each instance.

Brakes Ineffective: Shoe linings or pads worn, incorrect lining material or brake fluid, linings contaminated, fluid level low, air in brake system (bleed brakes), leak in pipes or cylinders, master cylinder defective. Remedies are obvious in each instance.

Brakes pull to one side: Shoes or linings worn, incorrect linings or pads, contaminated linings, drums or discs scored, fluid pipe blocked, unequal tyre pressures, brake back plate or caliper mounting loose, wheel bearings not properly adjusted, wheel cylinder seized. Rectify as necessary.

Brake pedal spongy: Air in hydraulic system. System must be bled of air.

Pedal travel too far: Linings or pads worn, drums or discs scored, master cylinder or wheel cylinders defective, system needs bleeding. Rectify as necessary.

Loss of brake pressure: Fluid leak, air in system, leak in master or wheel cylinders, brake servo not operating (vacuum hose disconnected or exhauster pump not operating). Place vehicle on dry ground and depress brake pedal. Check where fluid runs out and rectify as necessary.

Brakes binding: Incorrect brake fluid (boiling), weak shoe return springs, basic brake adjustment incorrect (after fitting new rear shoes), piston in caliper of wheel cylinder seized, push rod play on master cylinder insufficient (compensation port obstructed), handbrake adjusted too tightly. Rectify as necessary. Swelling of cylinder cups through use of incorrect brake fluid could be another reason.

Handbrake ineffective: Brake shoe linings worn, linings contaminated, operating lever on brake shoe seized, brake shoes or handbrake need adjustment. Rectify as necessary.

Excessive pedal pressure required: Brake shoe linings or pads worn, linings or pads contaminated, brake servo vacuum hose (for brake servo) disconnected or wheel cylinders seized. Exhauster pump not operating (diesel). Rectify as necessary.

Brakes squealing: Brake shoe linings or pads worn so far that metal is grinding against drum or disc. Inside of drum is full of lining dust. Remove and replace, or clean out the drum(s). Do not inhale brake dust.

Note: Any operation on the steering and brake systems must be carried out with the necessary care and attention. Always think of your safety and the safety of other road users. Make sure to use the correct fluid for the power-assisted steering and the correct brake fluid.

Faults in an ABS system should be investigated by a dealer.

ELECTRICAL FAULTS

Starter motor failure:	2 to 5, 8, 9
No starter motor drive:	1 to 3, 5 to 7
Slow cranking speed:	1 to 3
Charge warning light remains on:	3, 10, 12
Charge warning light does not come on:	2, 3, 9, 11, 13
Headlamp failure:	2, 3, 11, 13, 14
Battery needs frequent topping-up:	11
Direction indicators not working properly:	2, 3, 9, 13, 14
Battery frequently discharged:	3, 10, 11, 12

Causes and Remedies

1. Tight engine. Check and rectify.
2. Battery discharged or defective. Re-charge battery or replace if older than approx. 2 years.
3. Interrupted connection in circuit. Trace and rectify.
4. Starter motor pinion jammed in flywheel. Release.
5. Also 6, 7 and 8. Starter motor defective, no engagement in flywheel, pinion or flywheel worn or solenoid switch defective. Correct as necessary.
9. Ignition/starter switch inoperative. Replace.
10. Drive belt loose or broken. Adjust or replace.
11. Regulator defective. Adjust or replace.
12. Generator inoperative. Overhaul or replace.
13. Bulb burnt out. Replace bulb.
14. Flasher unit defective. Replace unit.

WIRING DIAGRAMS – Wiring diagram index

Cable Colour Code

bl	=	blue	nf	=	natural colour
br	=	brown	rs	=	pink
ge	=	yellow	rt	=	red
gn	=	green	sw	=	black
gr	=	grey	vi	=	violet
			ws	=	white

Cable identification: a = size, square mm, b = basic colour, c = second colour

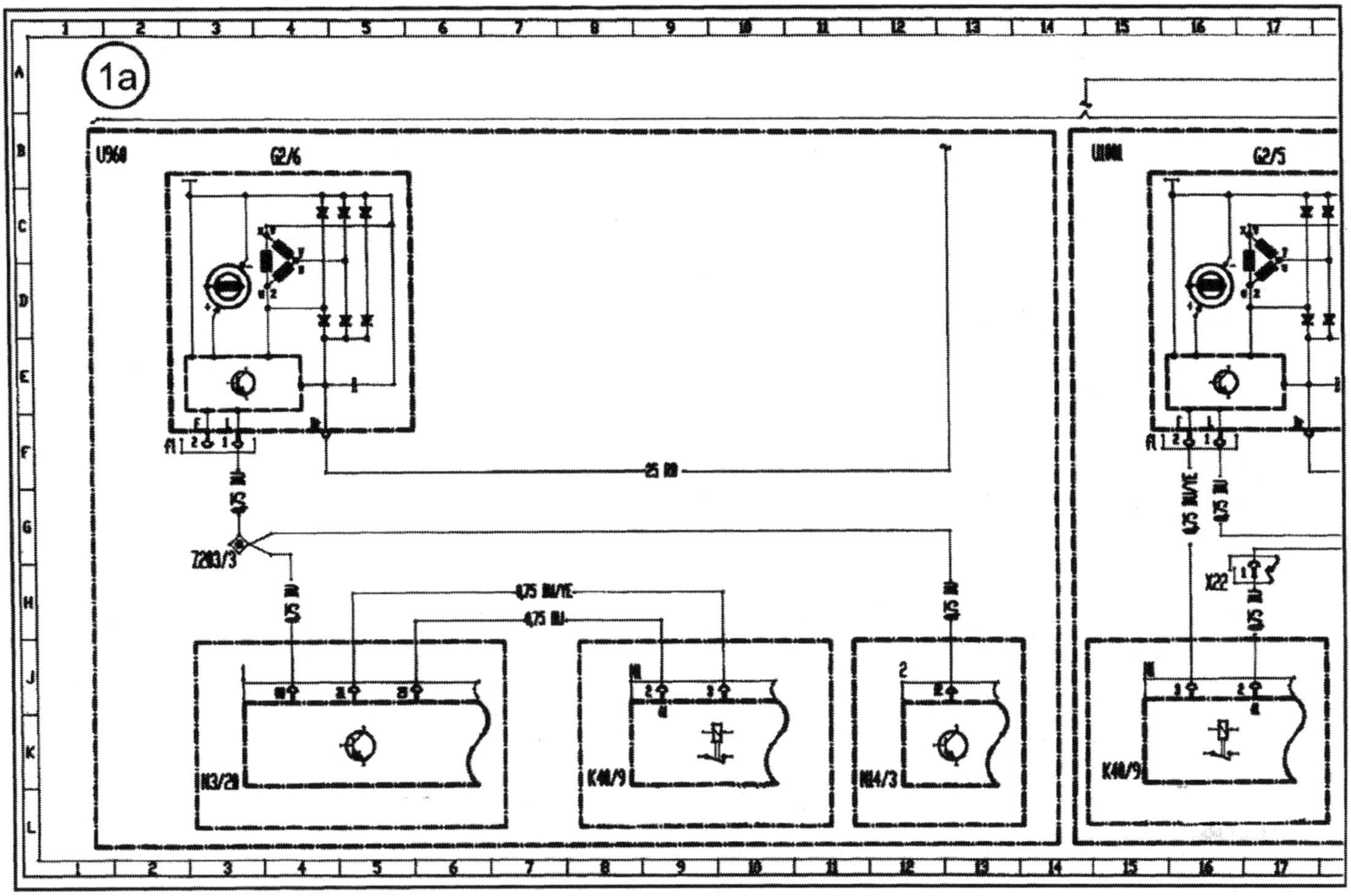
1a
U960
G2/6
Z203/3
N3/20
K40/9
N14/3
G2/5
X22
K40/9

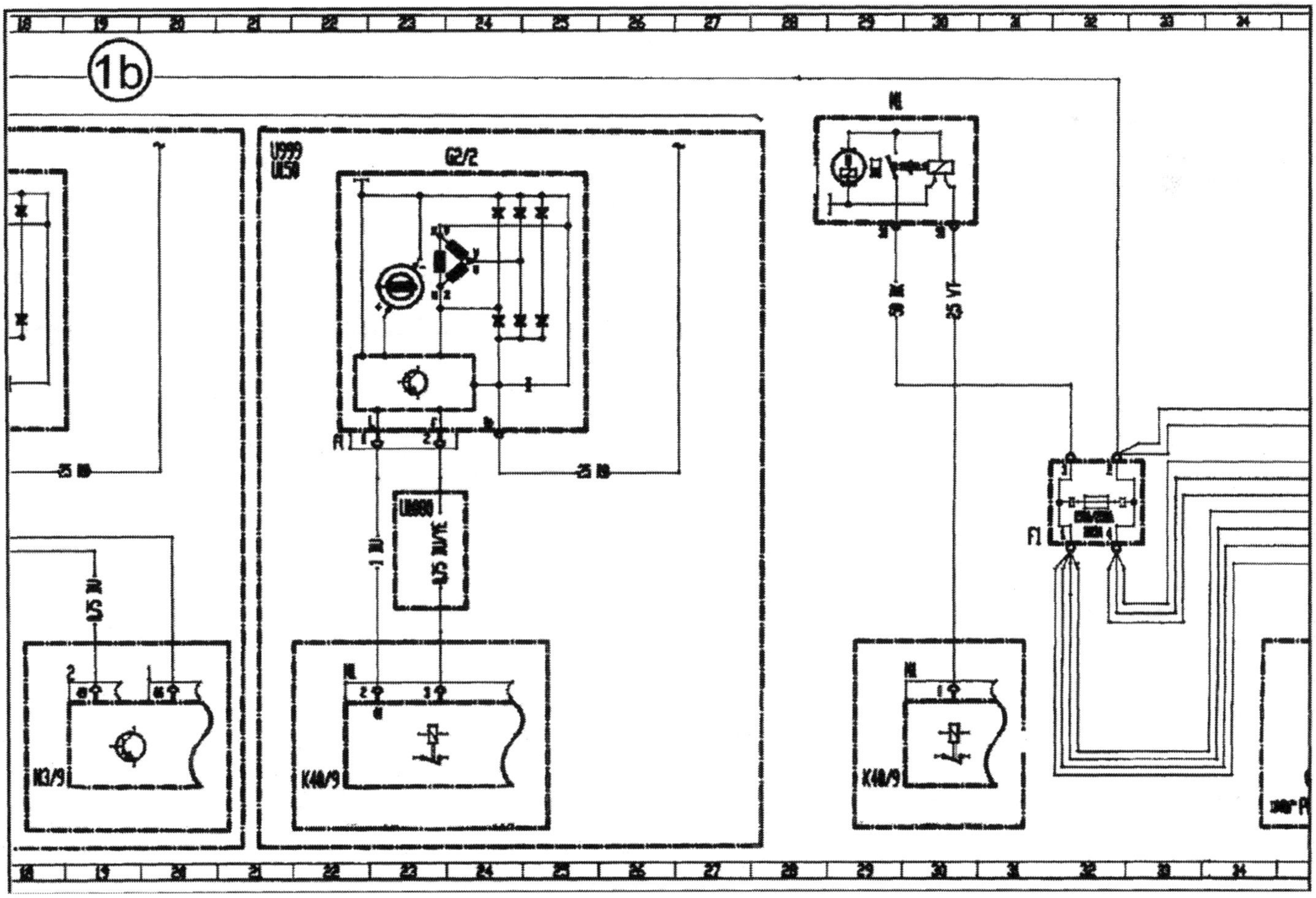
1b
G2/2
N3/9
K40/9
K40/9
F1

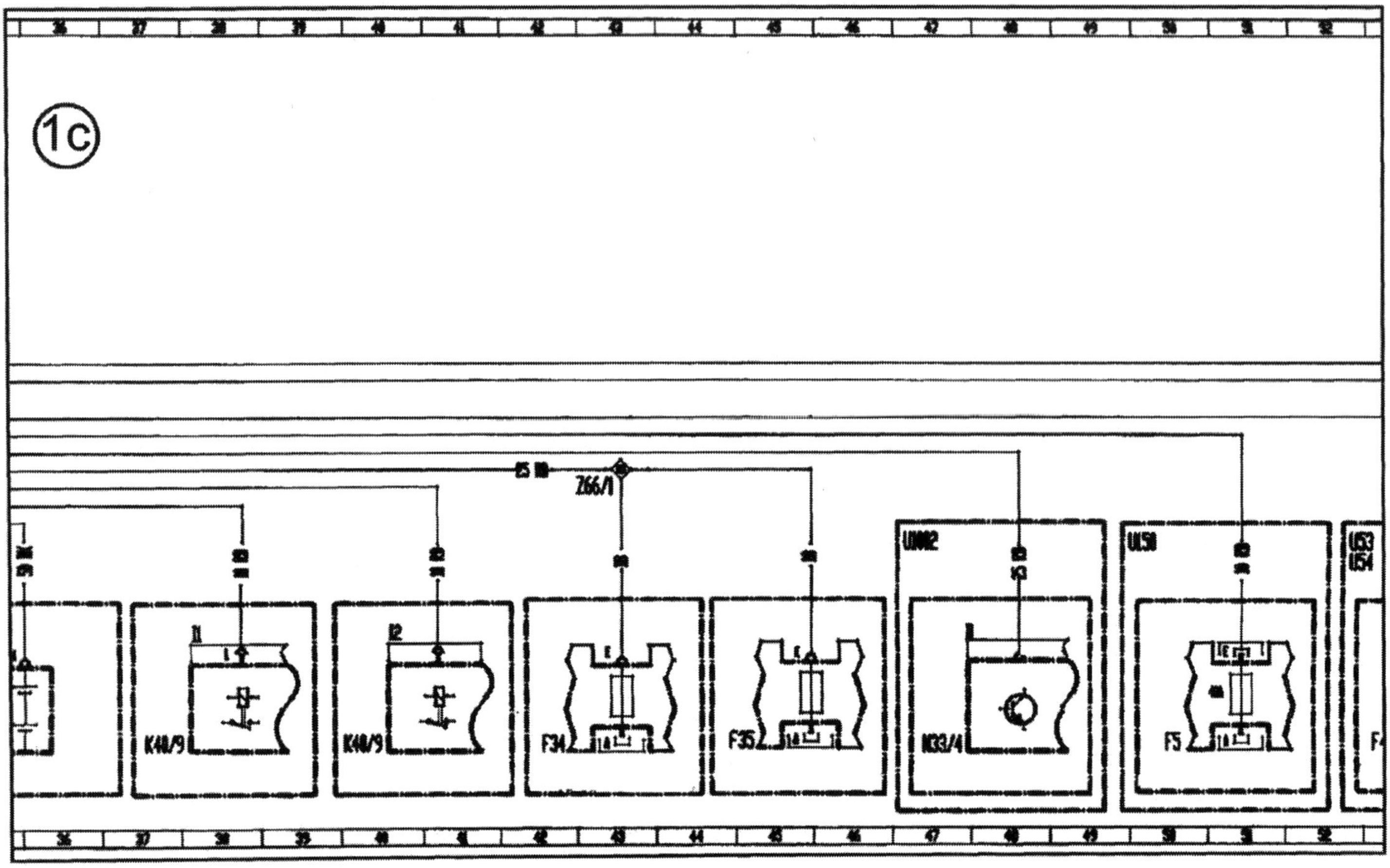
1c
Z66/1
K40/9
K40/9
F34
F35
N33/4
F5

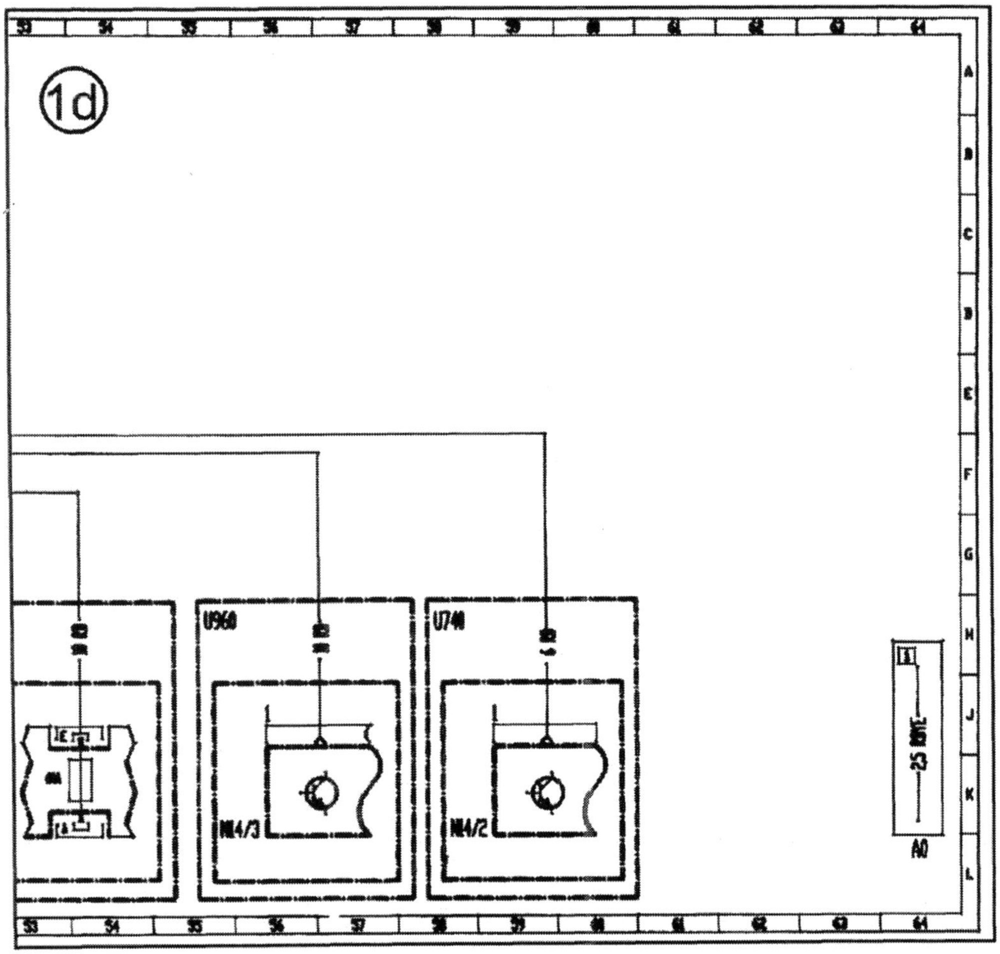
1d
U960
U740
N14/3
N14/2
A0

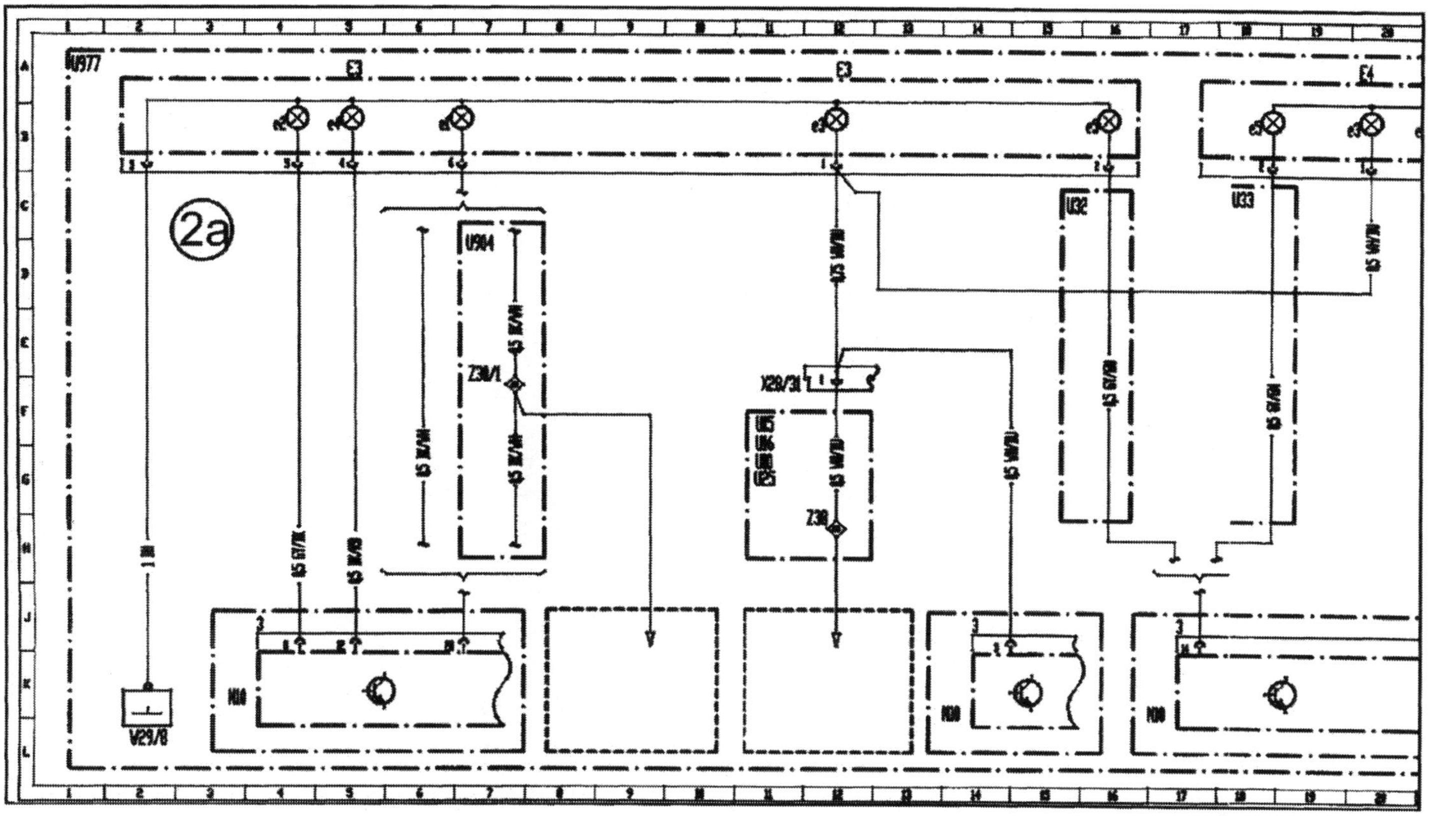
2a
U977
E3
E4
U904
Z30/1
X29/31
Z30
U32
U33
V29/8

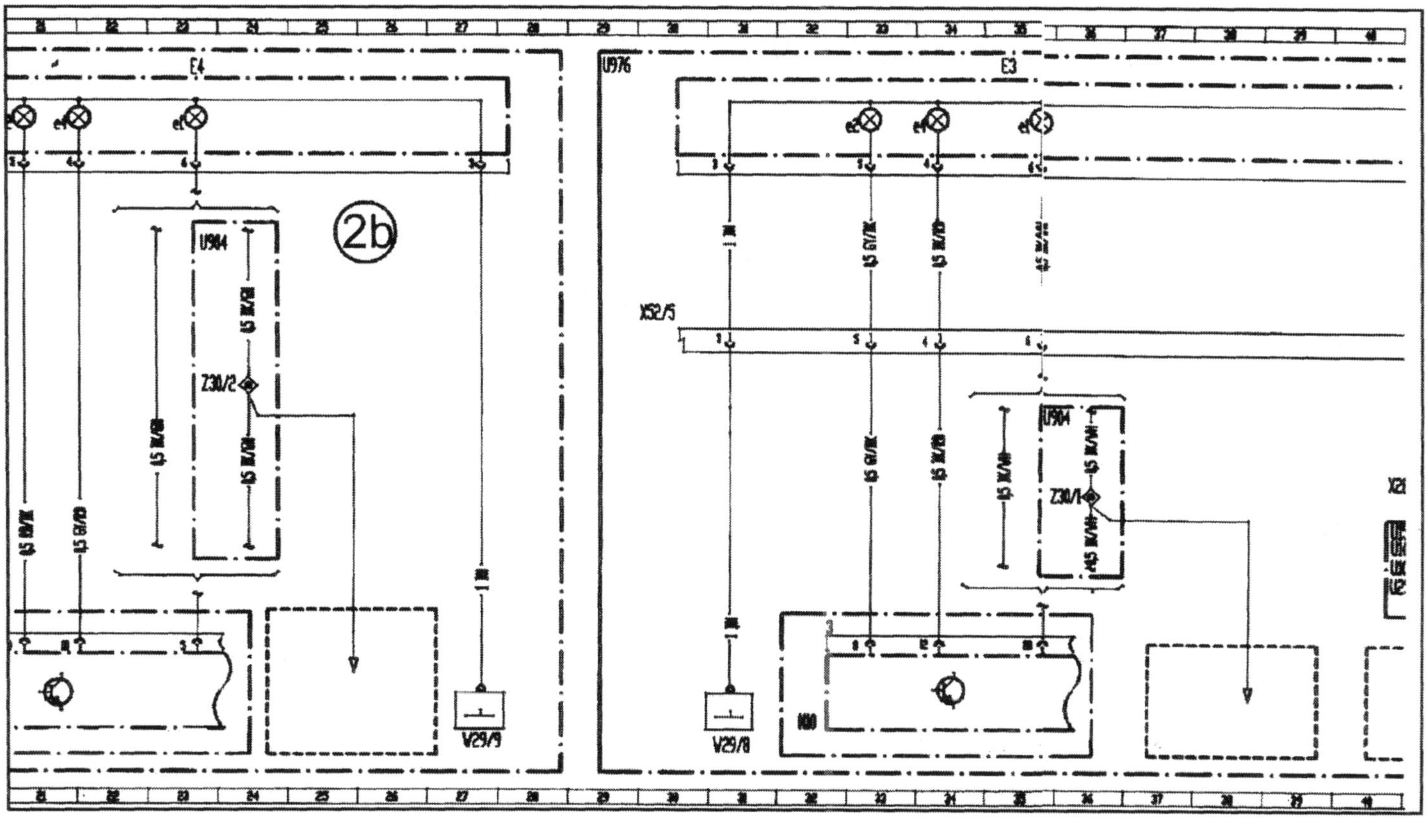
2b
E4
U904
Z30/2
V29/9
U976
E3
X52/5
Z30/1
V29/8
X21

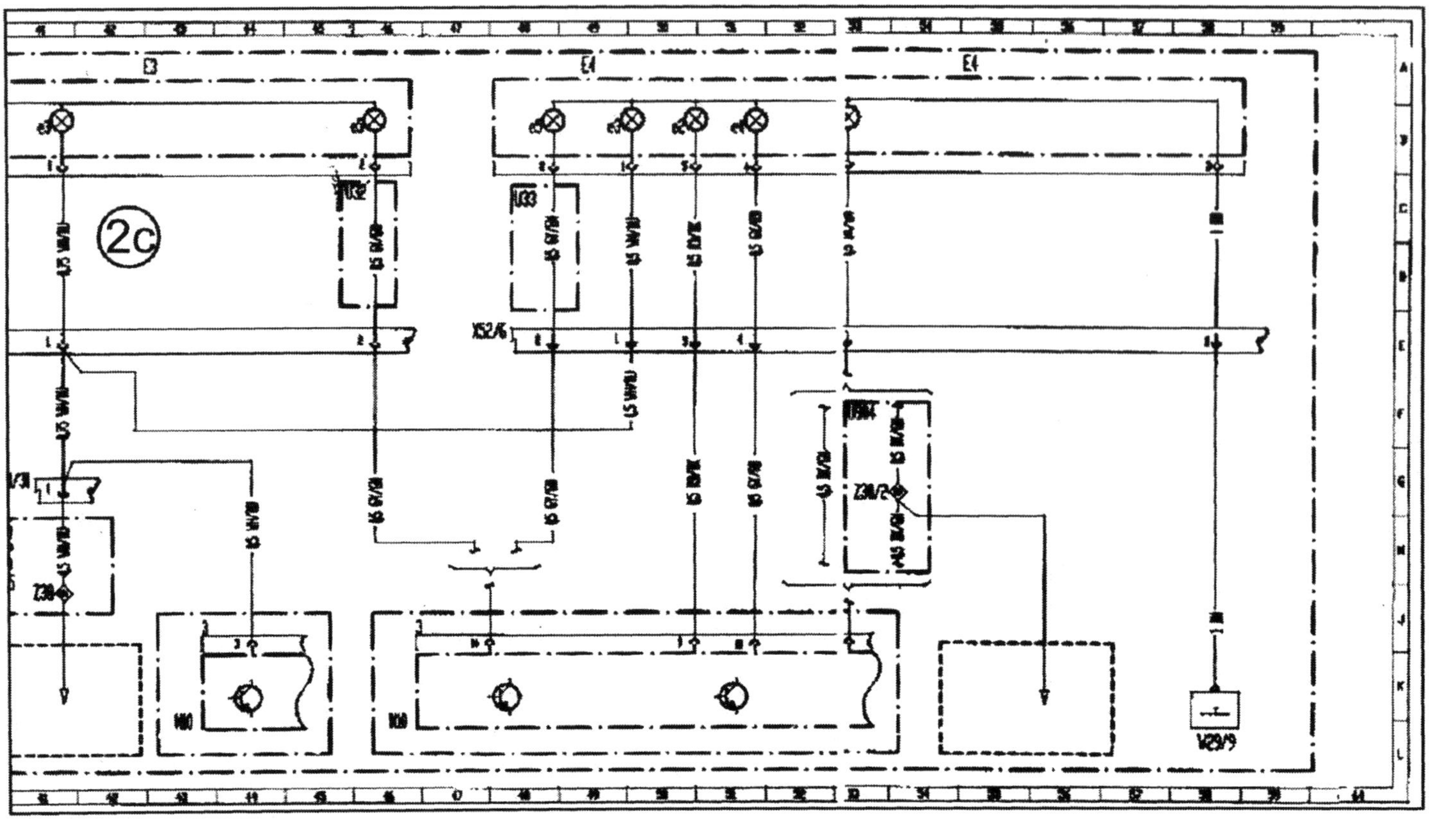
2c
E3
E4
X52/6
Z30/2
Z30
W29/9

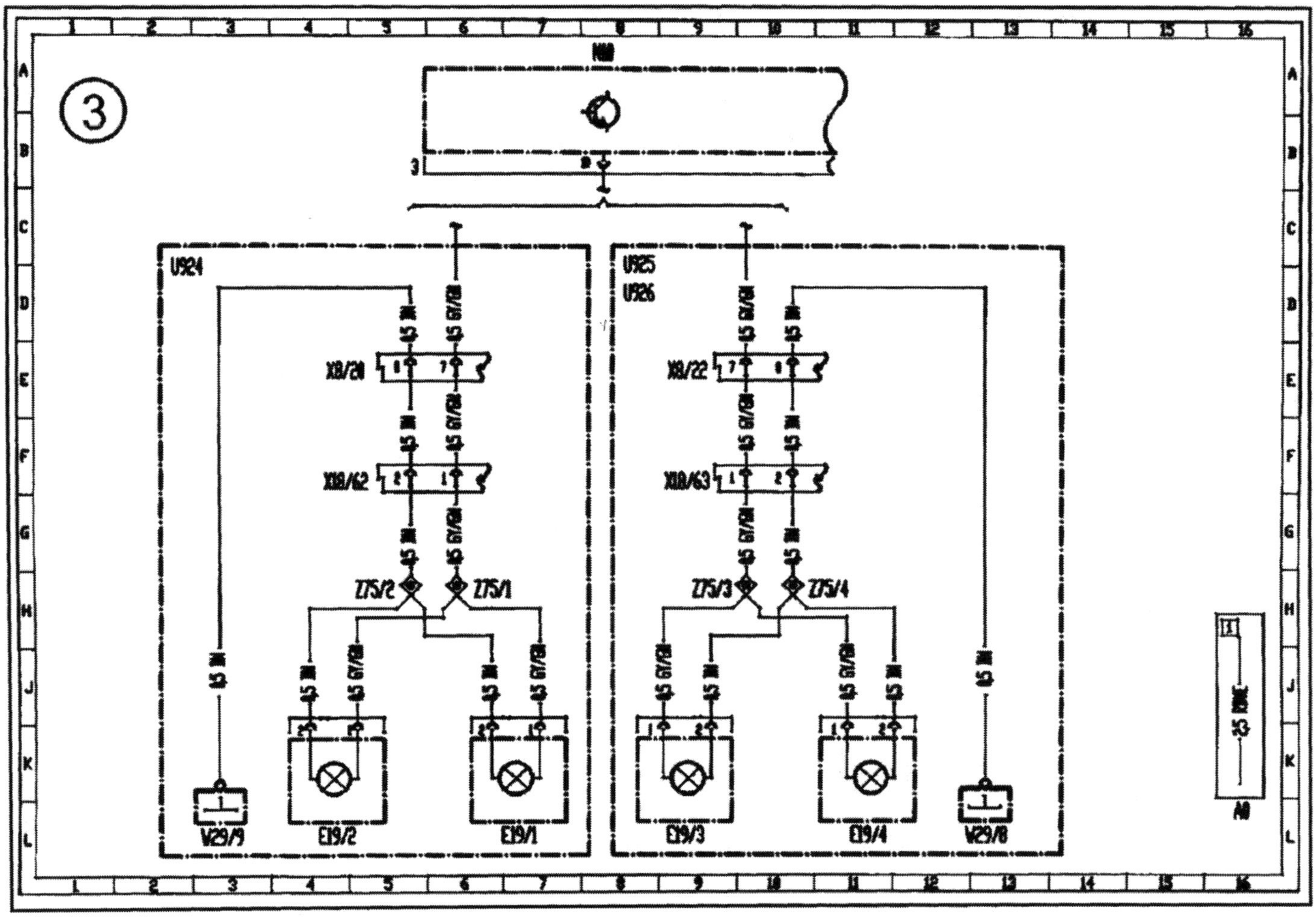
3
U924
U925
U926
X8/20
X8/22
X18/62
X18/63
Z75/2
Z75/1
Z75/3
Z75/4
W29/9
E19/2
E19/1
E19/3
E19/4
W29/8

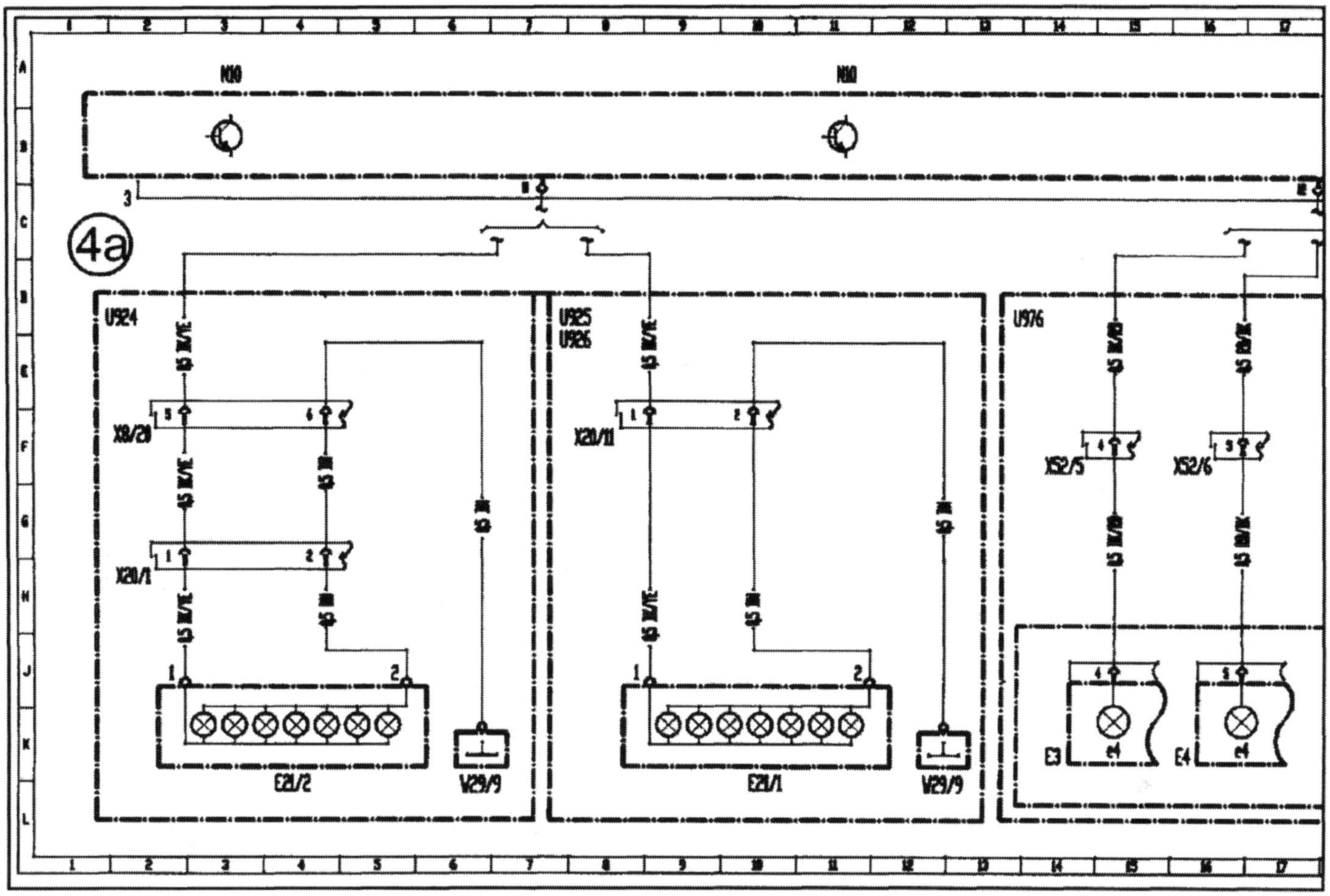
4a
N10
U924
U925
U926
U976
X8/20
X20/1
X20/11
X52/5
X52/6
E21/2
E21/1
V29/9
E3
E4

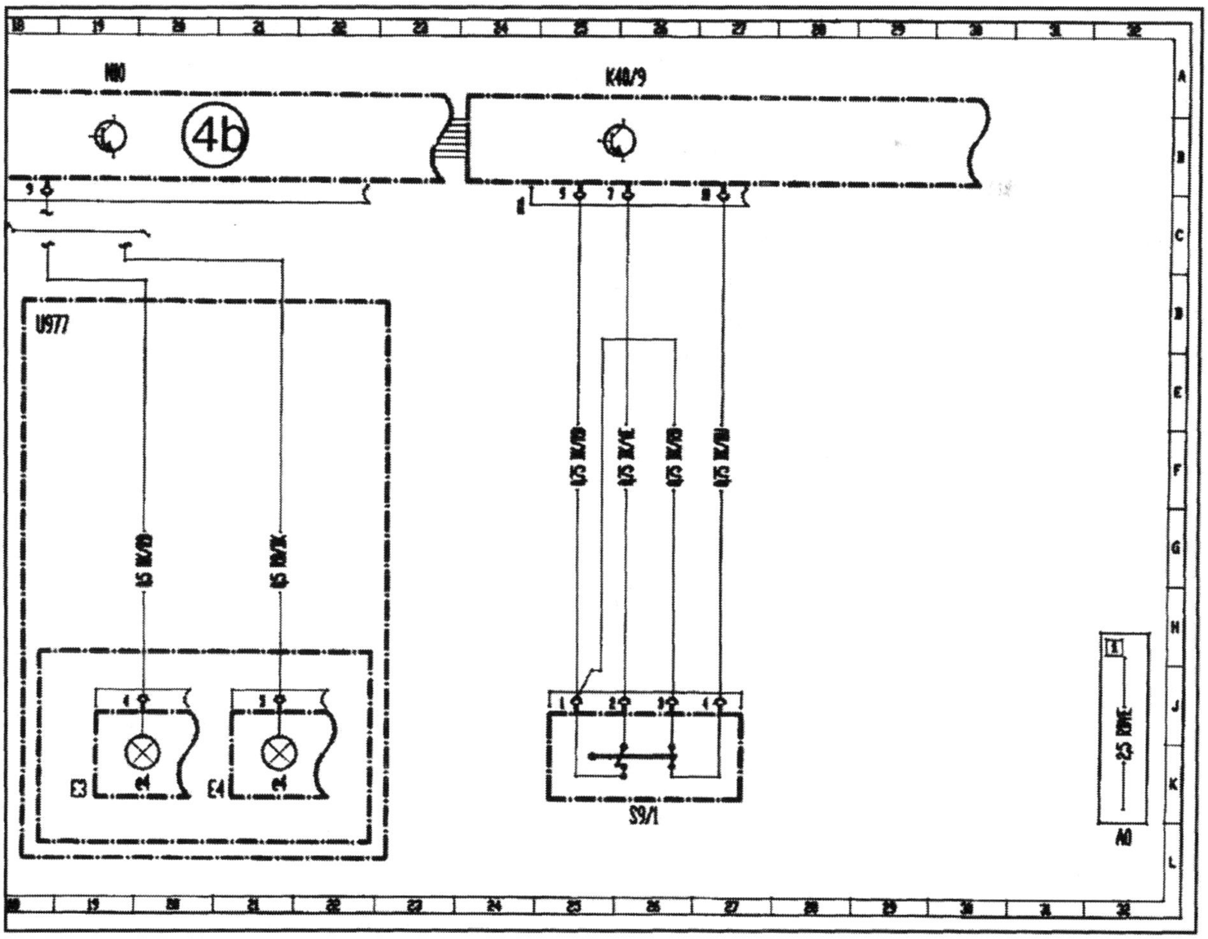
4b
N10
K40/9
U977
E3
E4
S9/1
A0

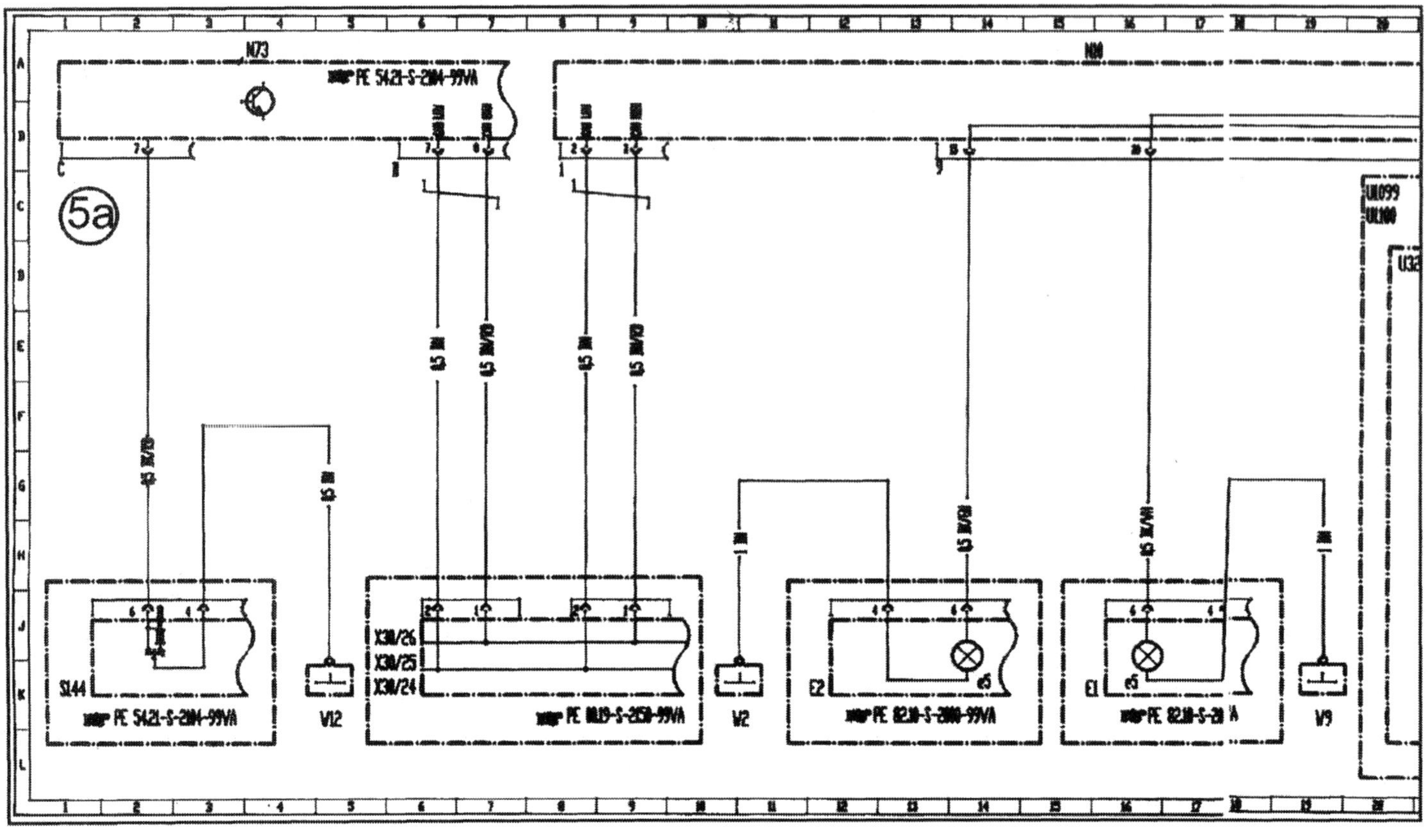
5a
N73
S144
V12
X30/26
X30/25
X30/24
V2
E2
E1
V9
U32

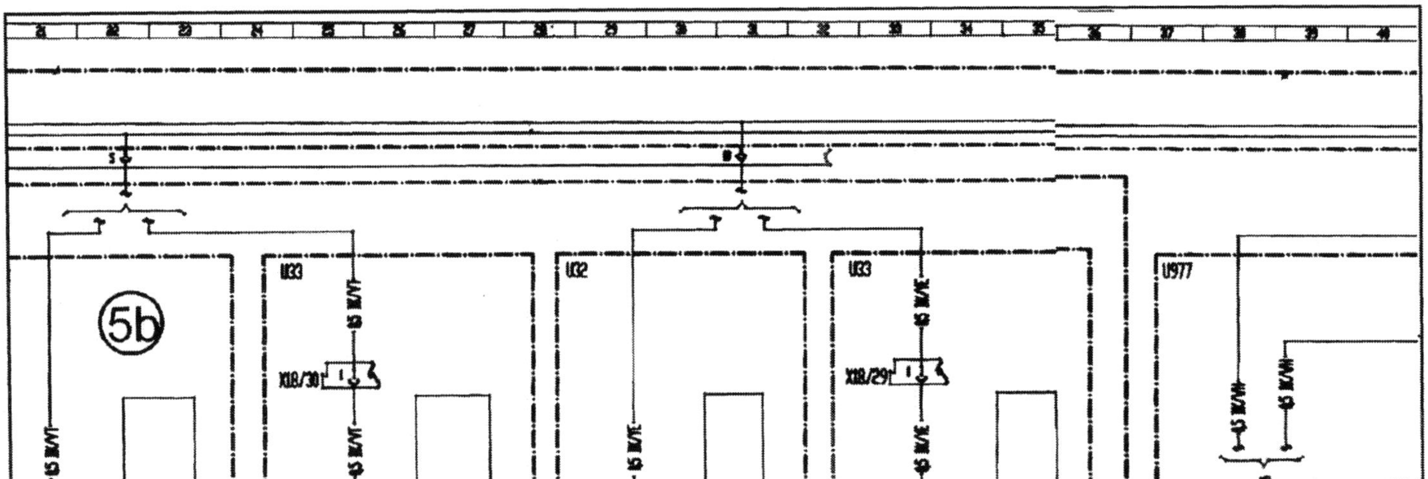
5b
U33
U32
U977
X18/30
X18/29
E22/2
V2
V2/6
E22/1
V9
E3

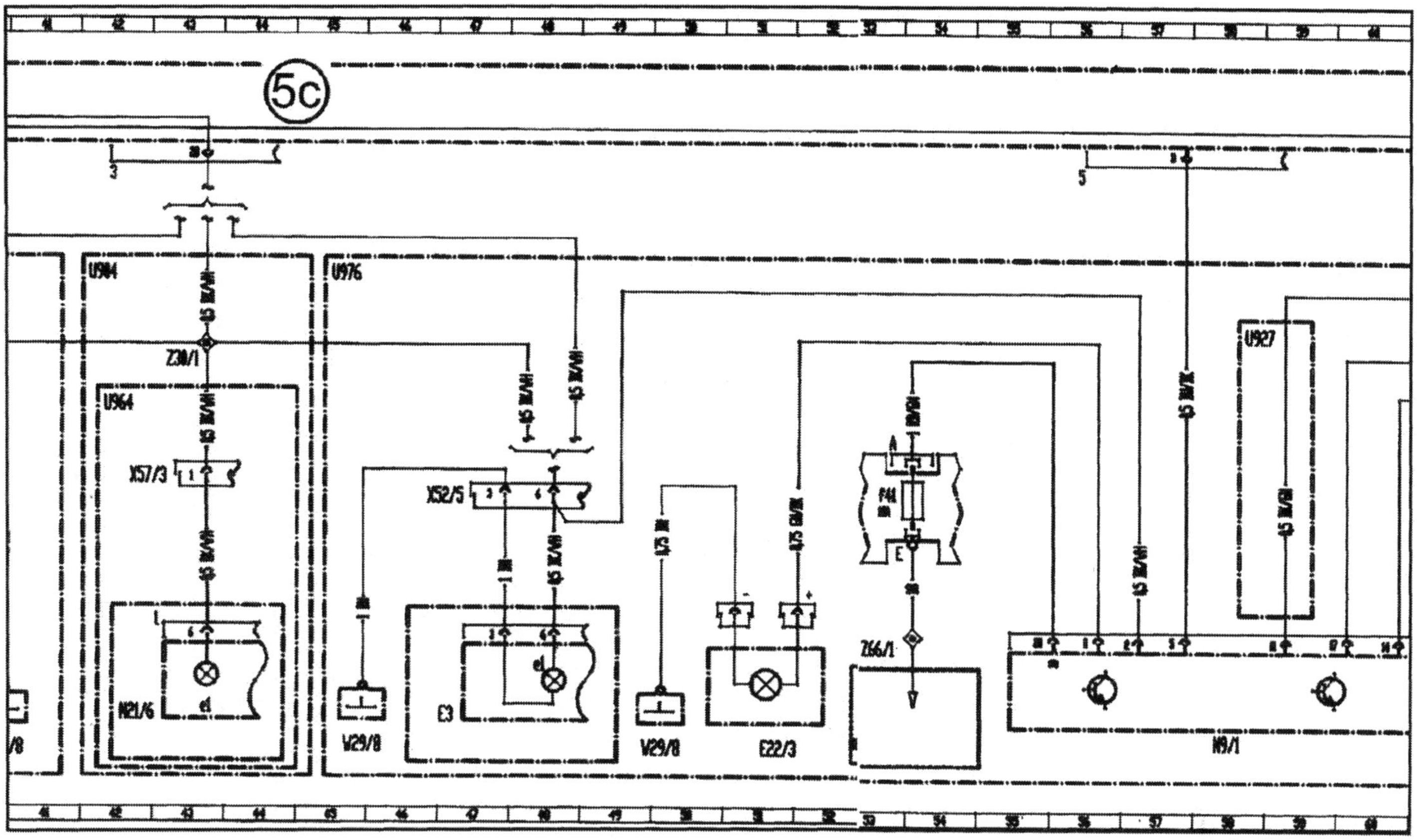

5c
U904
U976
Z30/1
U964
X57/3
X52/5
N21/6
E3
V29/8
E22/3
F44
Z66/1
U927
N9/1

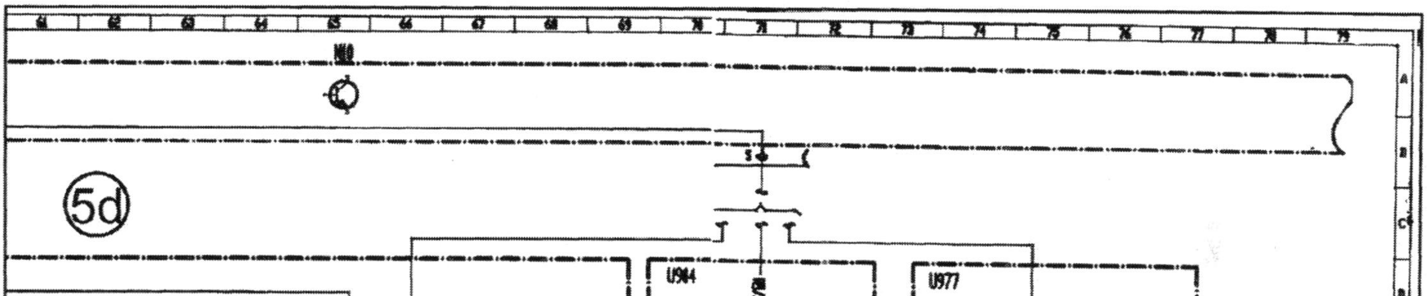

N10
5d
X52/6
U904
30/2
U964
X5/4
U977
E22/4
V29/9
E4

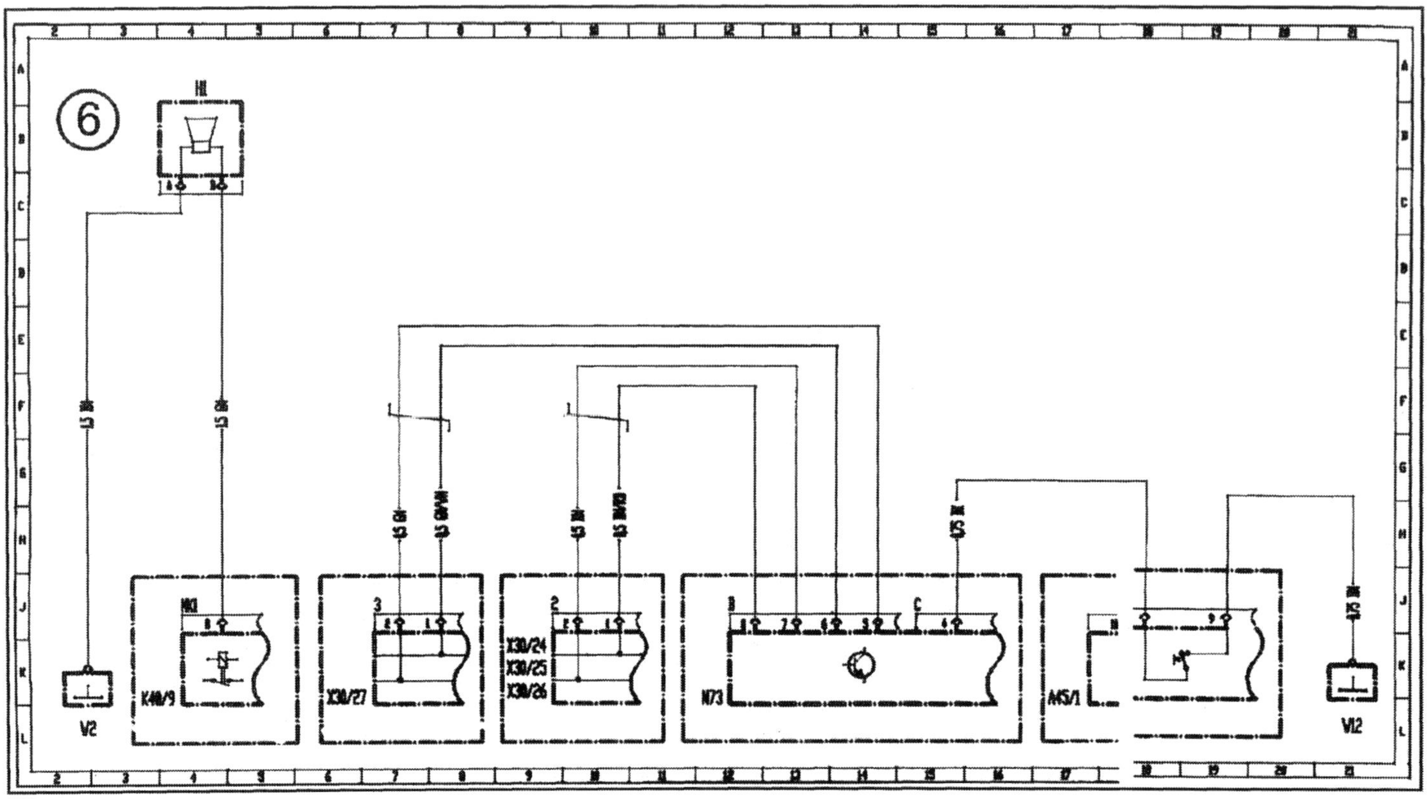
6
H1
K40/9
X30/27
X30/24
X30/25
X30/26
N73
A45/1
V2
V12

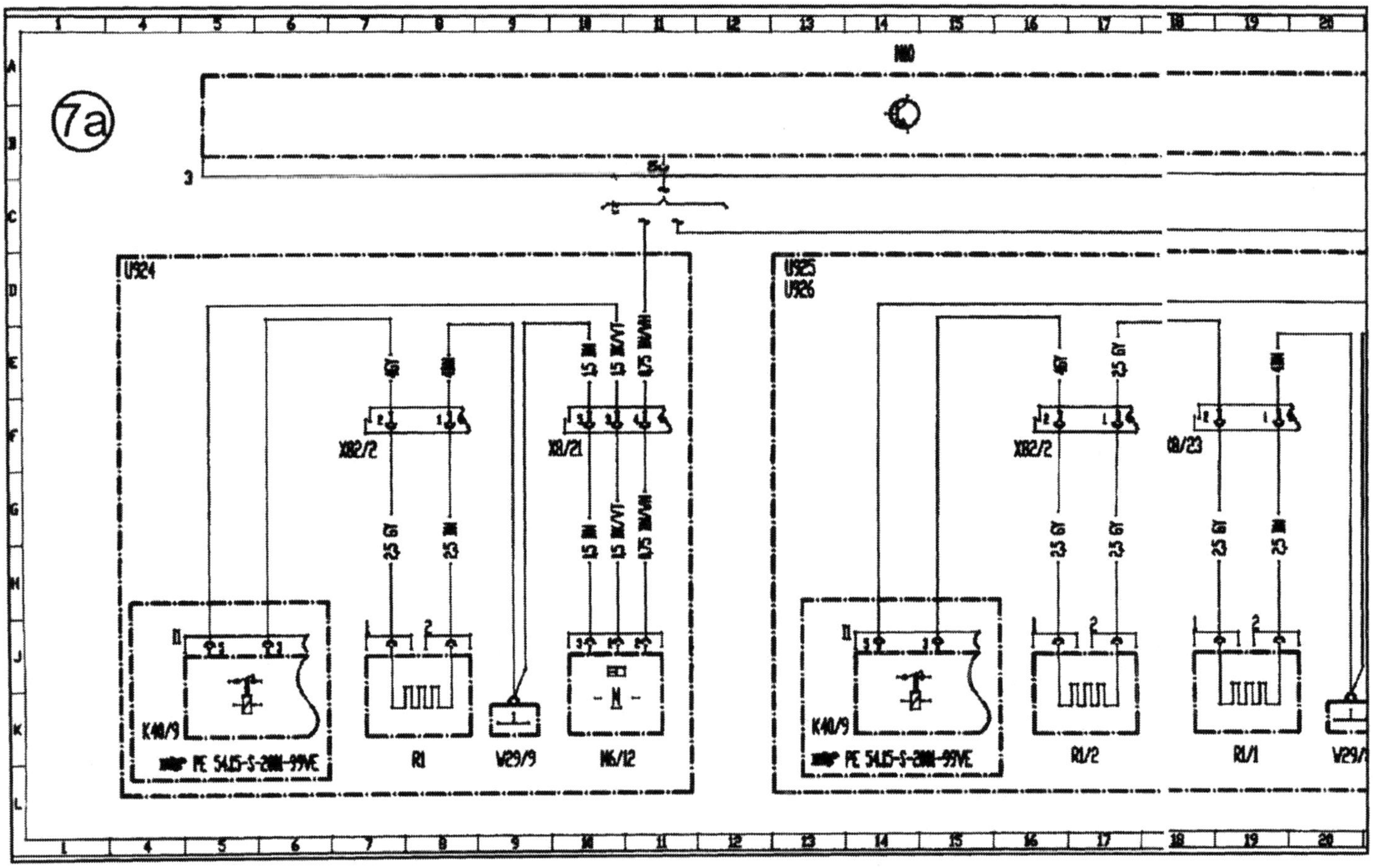
7a
N10
U924
U925
U926
X82/2
X8/21
K40/9
R1
W29/9
N6/12
R1/2
R1/1

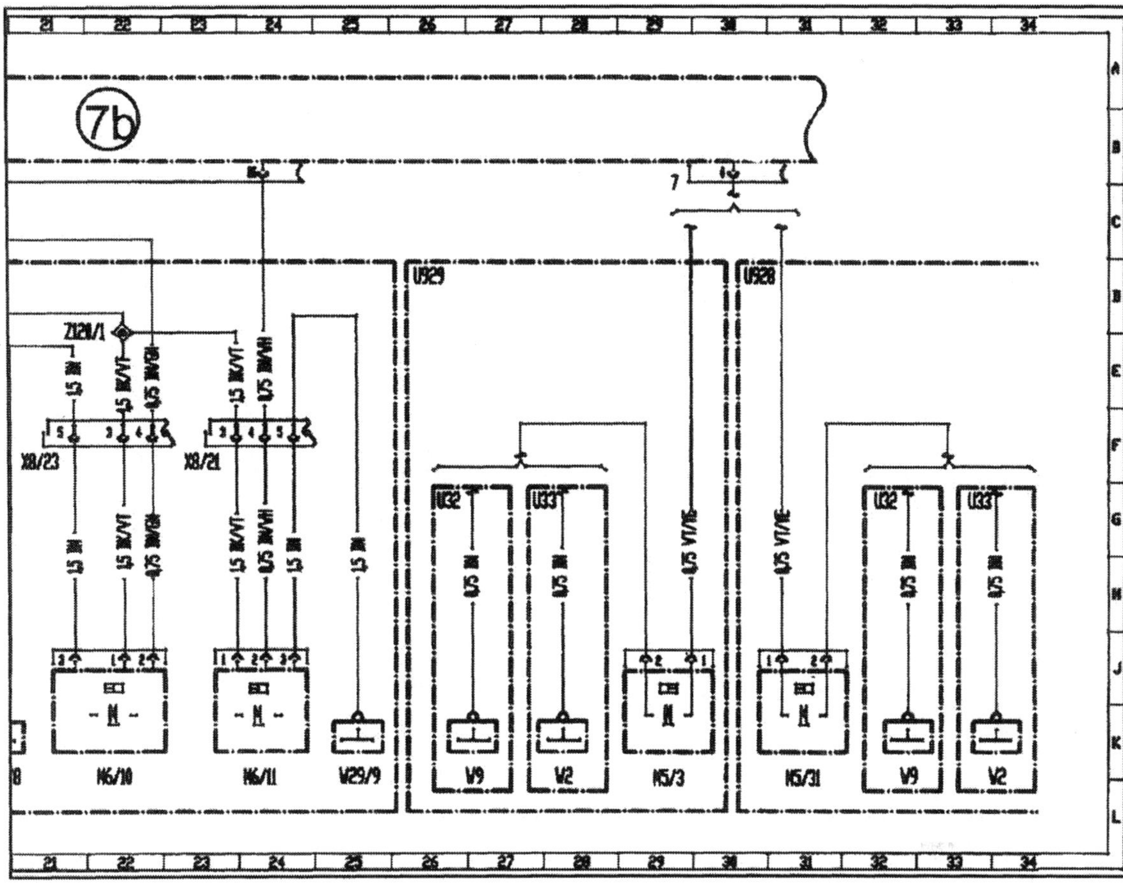
7b
Z120/1
X8/23
X8/21
U929
U928
U32
U33
N6/10
N6/11
W29/9
V9
V2
N5/3
N5/31

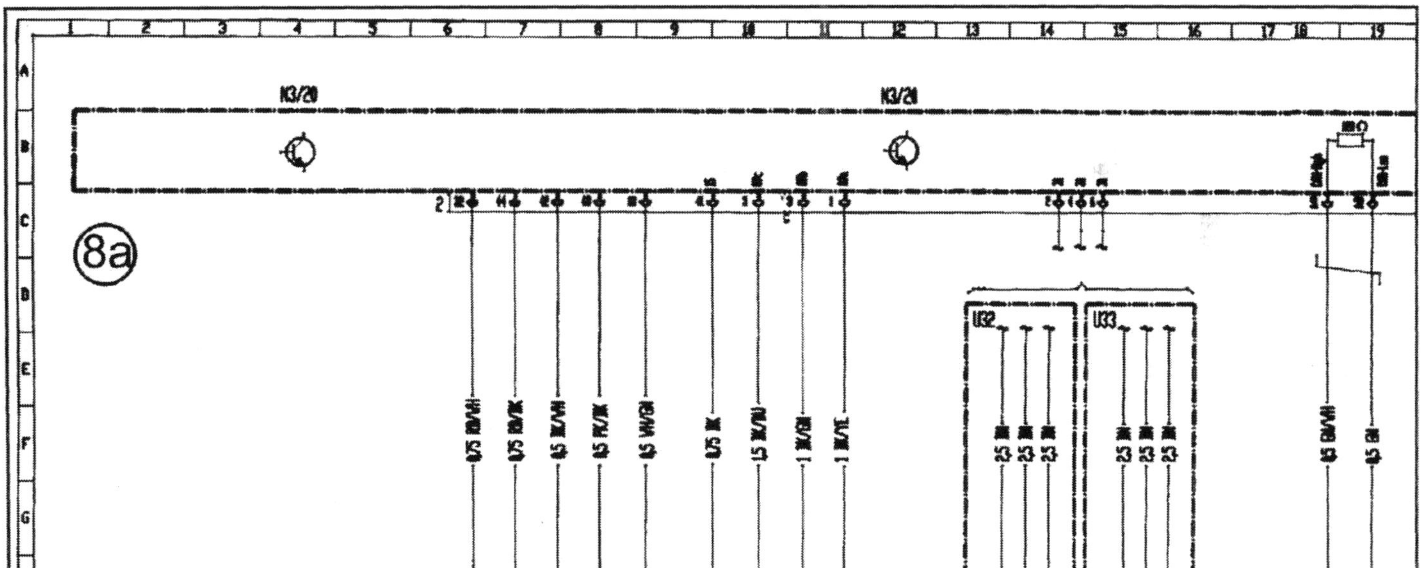
8a
N3/20
0,5 PK/BK
X11
K40/9
U32
U33
V9
V2

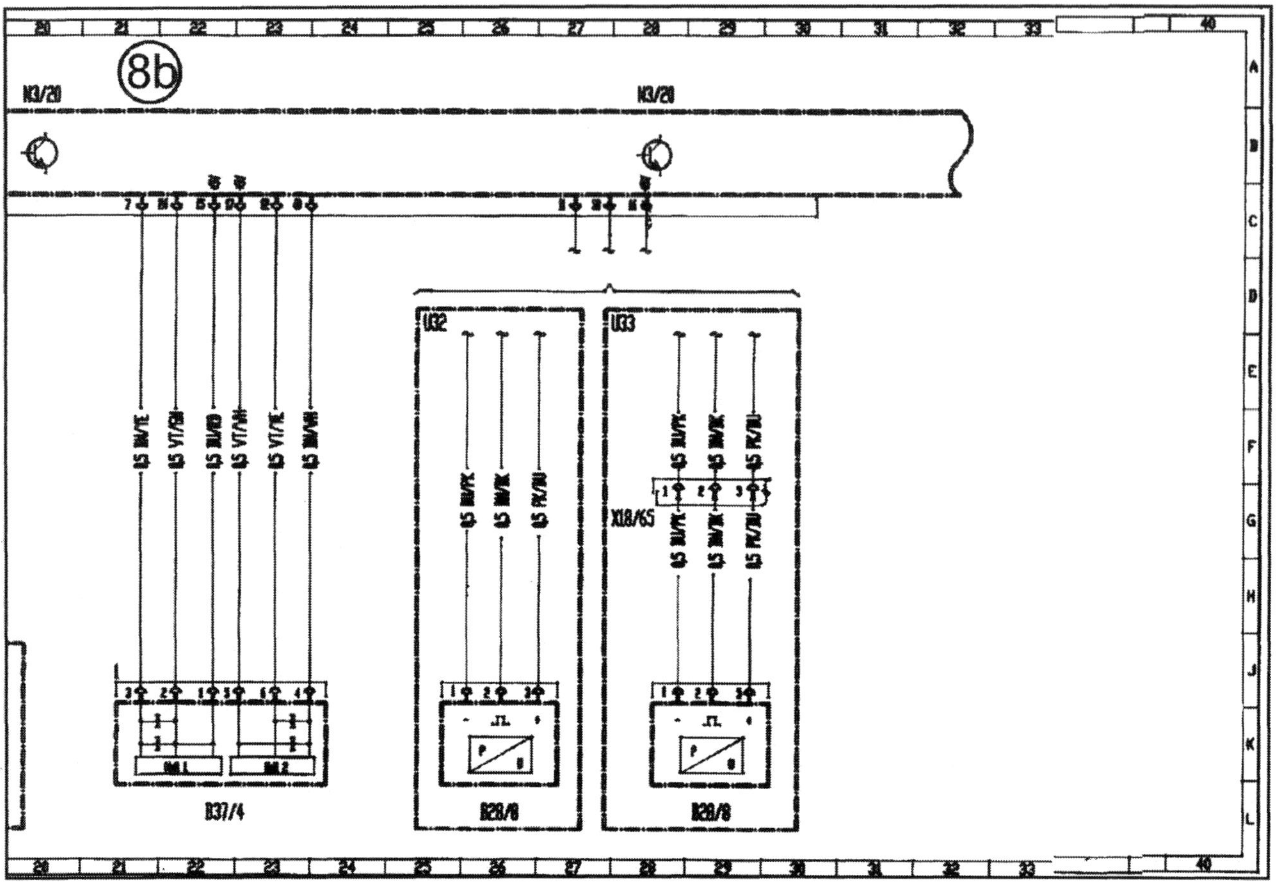
8b
N3/20
N3/20
U32
U33
X18/65
B37/4
B28/8
B28/8

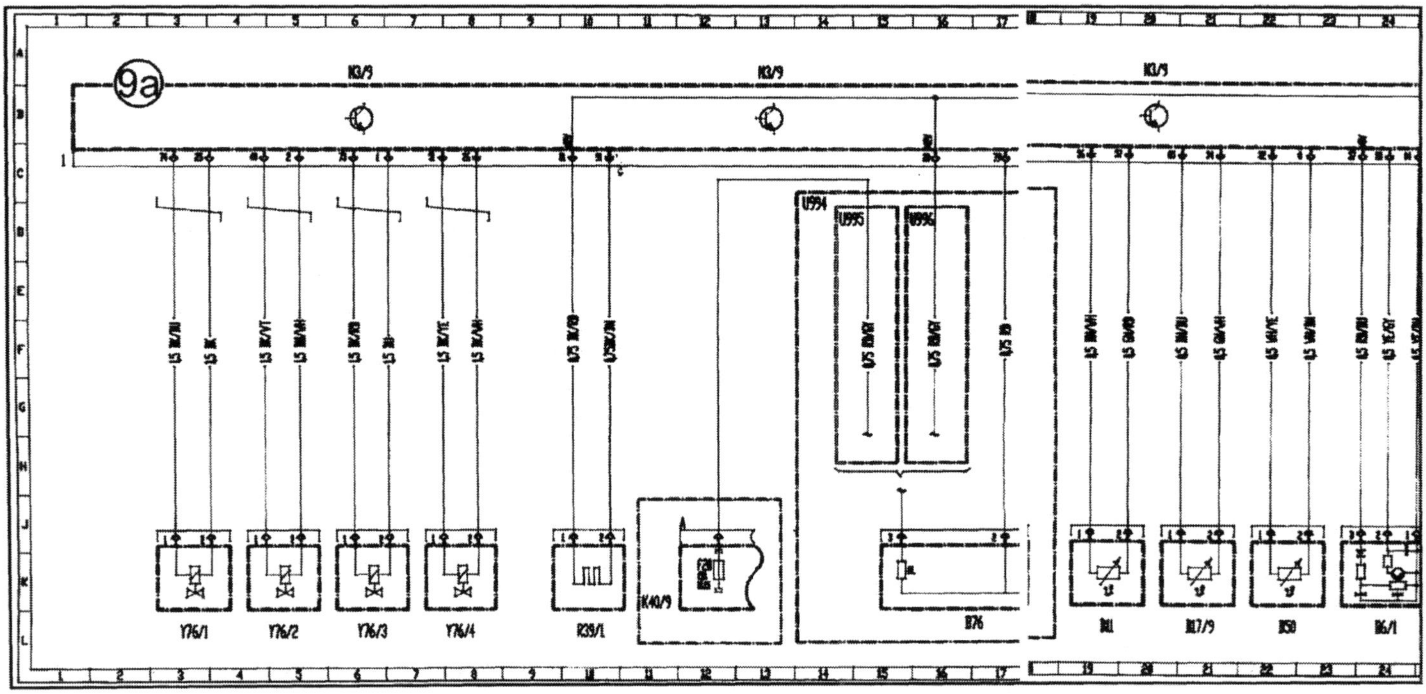
9a
N3/9
N3/9
N3/9
U994
U995
U996
Y76/1
Y76/2
Y76/3
Y76/4
B39/1
K40/9
B76
B11
B17/9
B50
B6/1

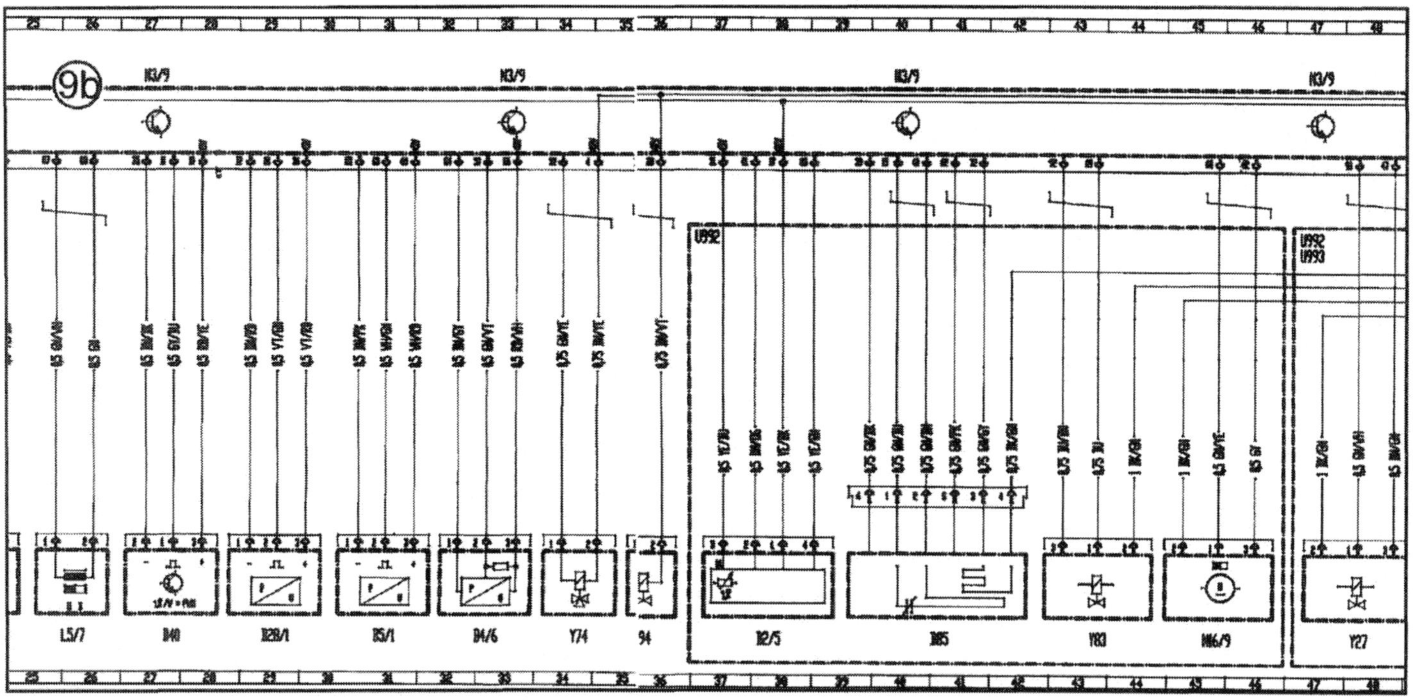
9b
N3/9
U992
U993
L5/7
B40
B28/1
B5/1
B4/6
Y74
B2/5
B65
Y83
M16/9
Y27

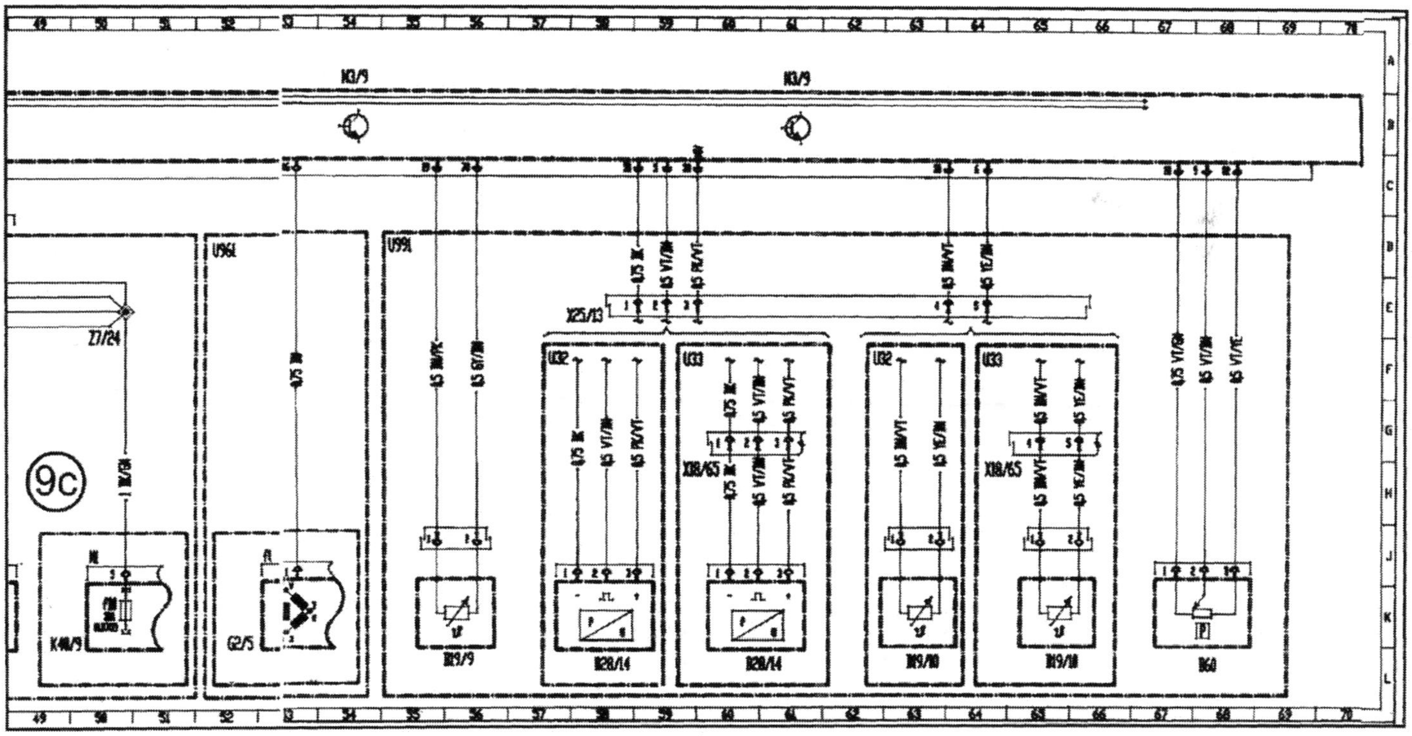
9c
N3/9
U961
U991
Z7/24
K40/9
G2/5
X25/13
U32
U33
X18/65
B19/9
B28/14
B19/10
B60

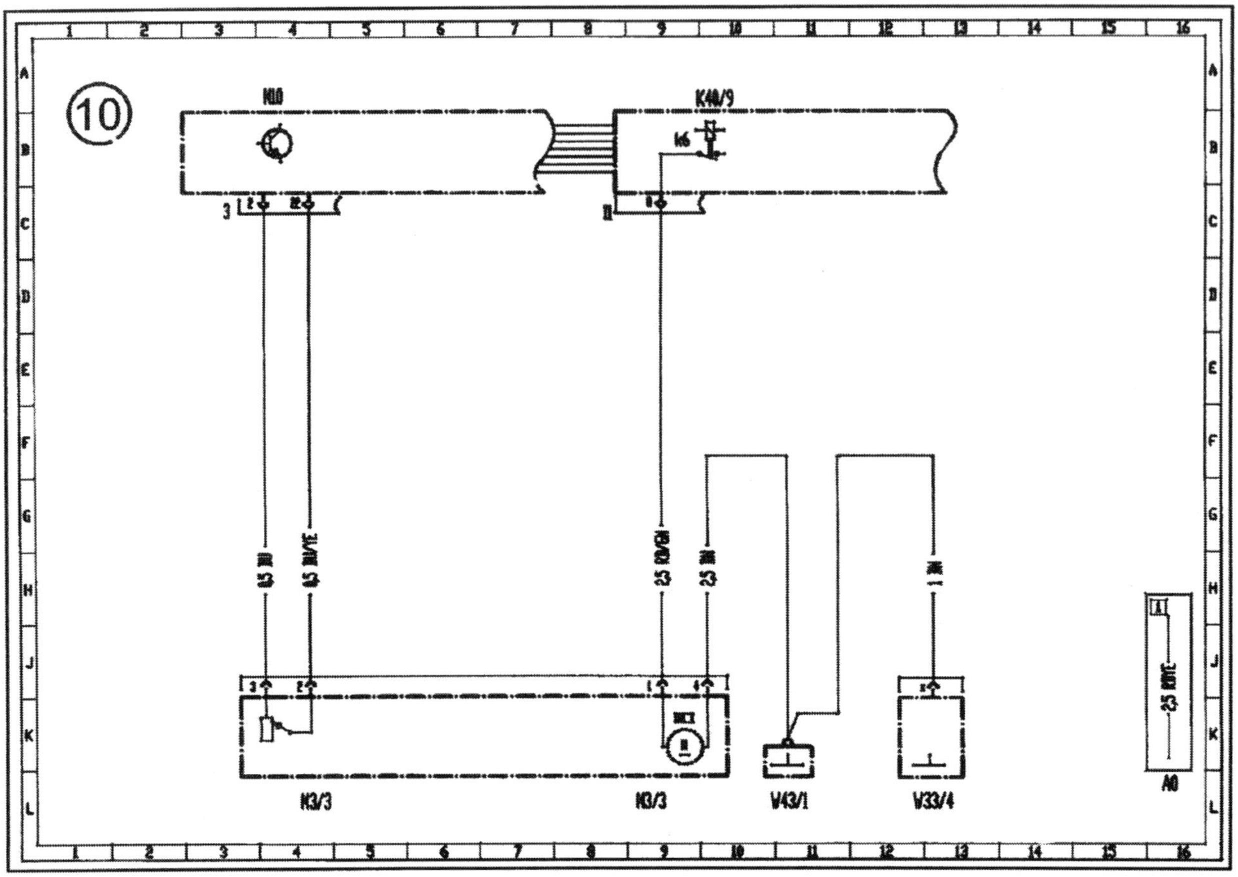
10
N10
K40/9
N3/3
N3/3
W43/1
W33/4
A0
0.5 BU
0.5 BU/YE
2.5 RD/GN
2.5 BN
1 BN
2.5 RD/YE

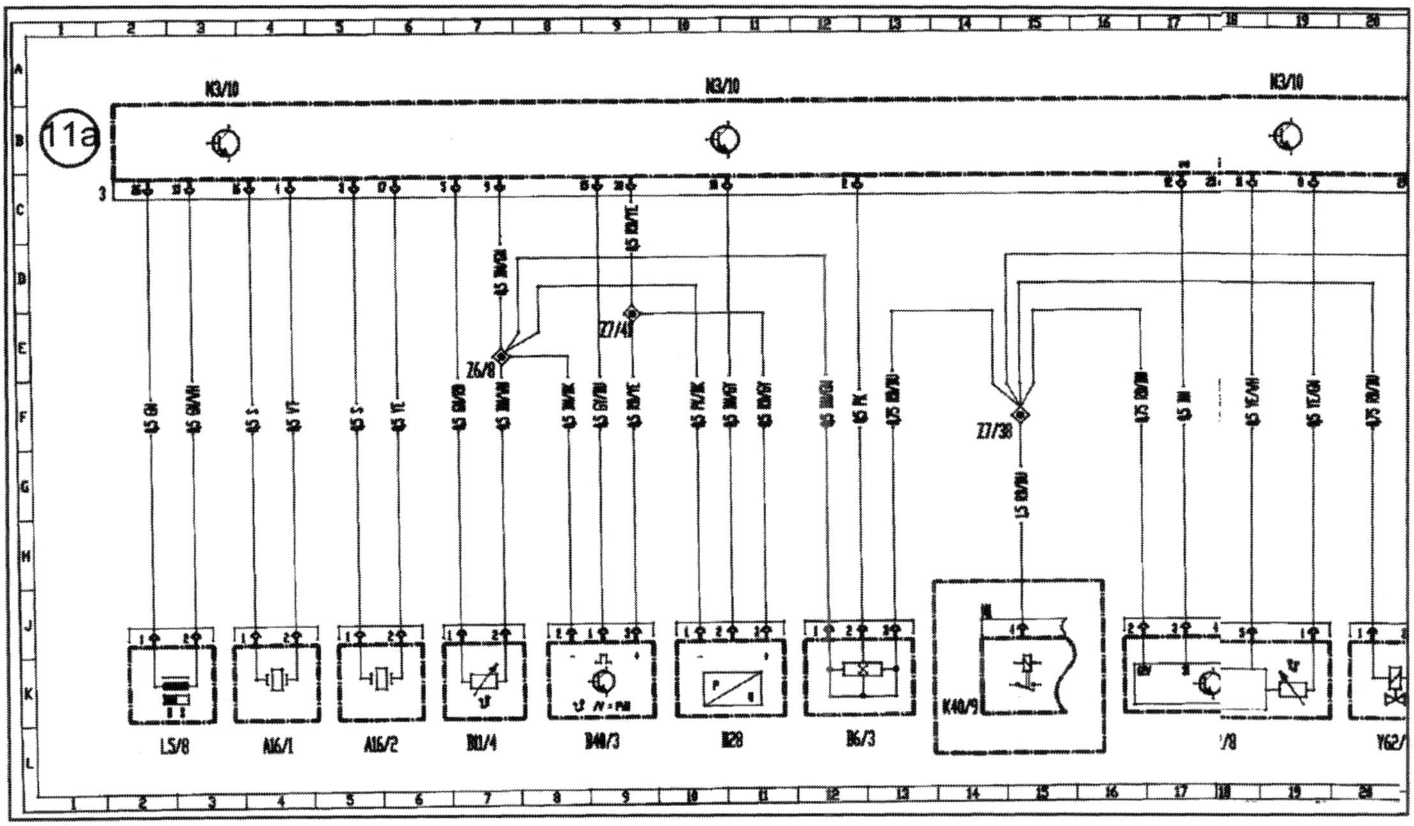
11a
N3/10
N3/10
N3/10
Z6/8
Z7/41
Z7/38
K40/9
L5/8
A16/1
A16/2
B11/4
B40/3
B28
B6/3

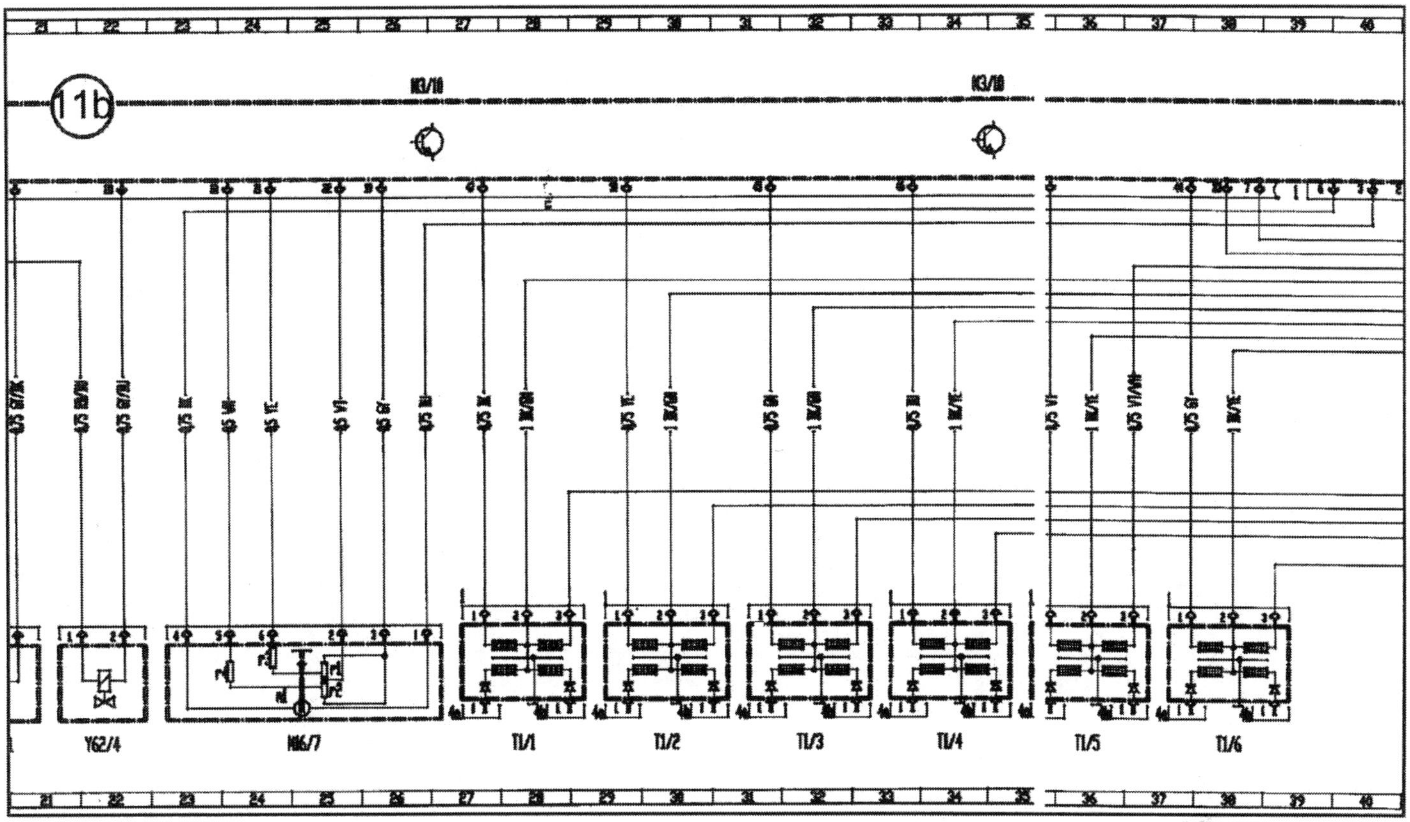
11b
N3/10
N3/10
Y62/4
N16/7
T1/1
T1/2
T1/3
T1/4
T1/5
T1/6

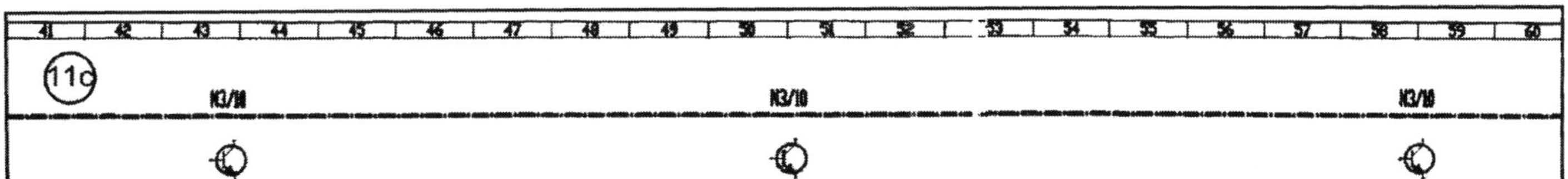
11c
N3/10
N3/10
N3/10
Z3/29
Z86
Z7/39
K40/9
G3/3
G3/4
Y59/2
Y87
Y22/5
K40/9
Y62/2

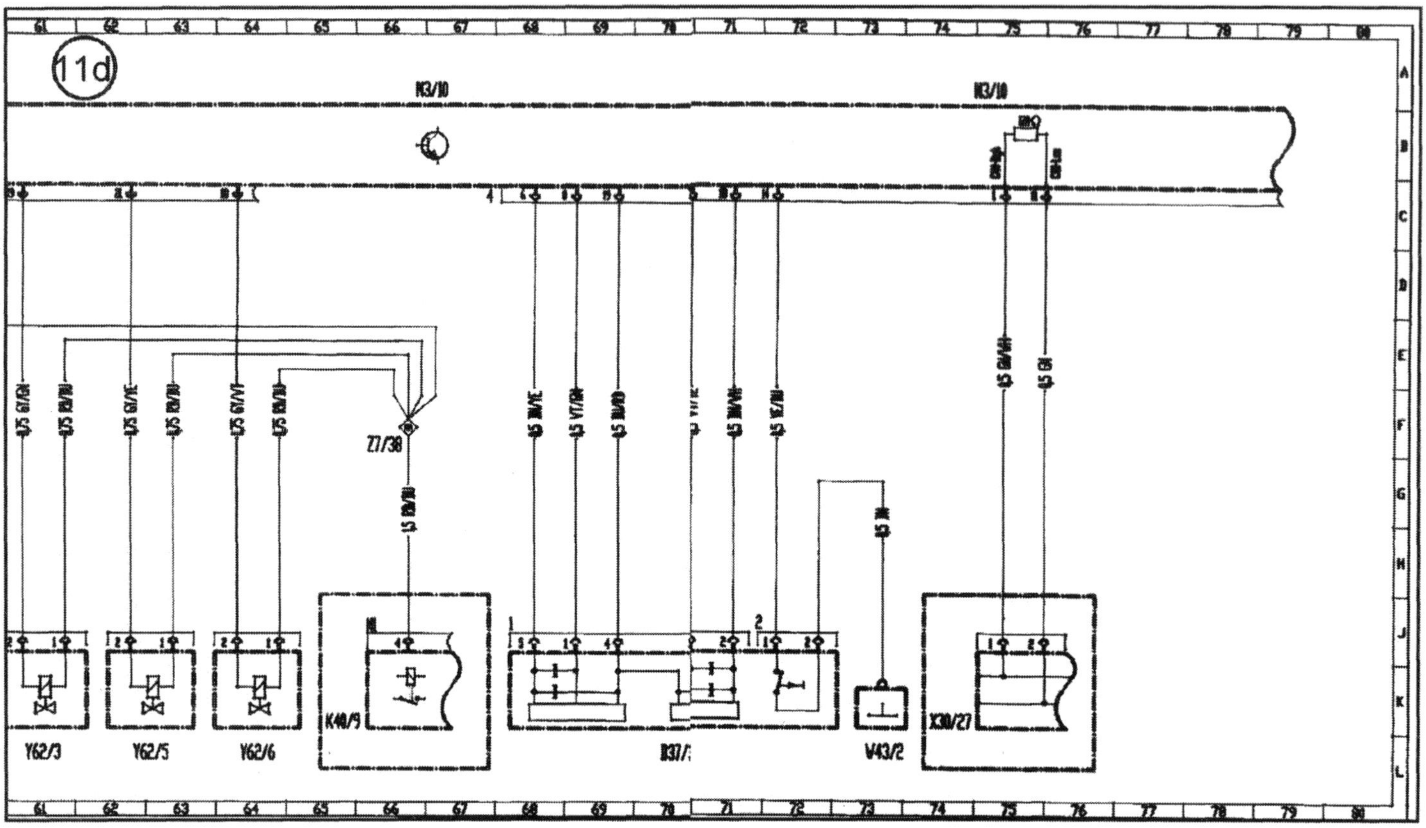

FURTHER READING ON MERCEDES-BENZ

9781855209381 Mercedes-Benz Automobiles 1926-1939 A Brooklands Portfolio
9781855209329 Mercedes-Benz 190SL 300SL 300SLR A Brooklands Portfolio
9781855206779 Mercedes-Benz SLs & SLCs Ultimate Port. 1971-1989
9781855202689 Mercedes-Benz SLs Performance Portfolio 1989-1994
9781855208865 Mercedes-Benz 230SL - 250SL - 280SL Ultimate Portfolio 1963-1971
9781855206991 Mercedes-Benz G-Wagen Gold Portfolio 1981-2005
9781855206939 Mercedes-Benz S-Class & 600 Lim. Ed. Ex. 1965-1972
9781855208896 Mercedes-Benz E-Class W124 1985-1995 Road Test Portfolio
9781855205819 Mercedes-Benz S-Class Limited Edition Extra 1980-1991
9781855207455 Mercedes-Benz AMG Gold Portfolio 1983-1999
9781855207486 Mercedes-Benz AMG Ultimate Portfolio 2000-2006
9781869826321 Mercedes-Benz Repair & Tune-Up Guide
9780620164436 Mercedes-Benz W123 Owners Workshop Manual 1976-1986
9780958402613 Mercedes-Benz W124 Owners Workshop Manual 1985-1995
9781855207684 Mercedes-Benz E-Class Petrol Workshop Manual W124 & W210 1993-2000
9781855207691 Mercedes-Benz E-Class Diesel Workshop Manual 1999-2006
9781855207295 Mercedes-Benz Sprinter Cdi Workshop Manual 2000-2006
9781855209558 Mercedes-Benz E-Class Diesel Workshop Manual W210 & W211 Series 2000-2006
9781855209619 Mercedes-Benz E-Class Petrol Workshop Manual W210 & W211 Series 2000-2006
9781783180097 Mercedes-Benz Vito & V-Class Workshop Manual 1996-1999
9781783180035 Mercedes-Benz Vito & V-Class Petrol & Diesel Models Workshop Manual 2000-2003
9781783180516 Mercedes Benz ML Diesel Models Series 163 & 164 Workshop Manual 1998-2006
9781783180523 Mercedes-Benz ML Petrol Models Series 163 & 164 Workshop Manual 1998-2006
9781783180547 Mercedes-Benz G-Wagen Workshop Manual 1979-1991
9781783182473 Mercedes-Benz 220B Owners Workshop Manual 1959-1965
9781783182480 Mercedes-Benz 230 Owners Workshop Manual 1963-1968
9781783182503 Mercedes-Benz 280 Owners Workshop Manual 1968-1972
9781783181209 Mercedes-Benz 250 Owners Workshop Manual 1965-1968

Made in United States
Orlando, FL
04 November 2026

90200819R00114